OREGON STATE UNIVERSITY
SCHOOL OF AGRICULTURE
SYMPOSIUM SERIES NO. 1

Controlled Atmospheres for Storage and Transport of Perishable Agricultural Commodities

EDITORS
D. G. Richardson
M. Meheriuk

TIMBER PRESS
Beaverton, Oregon
1982
in cooperation with
SCHOOL OF AGRICULTURE, OREGON STATE UNIVERSITY

TIMBER PRESS
P.O. Box 1631
Beaverton, Oregon 97075

Library of Congress Cataloging in Publication Data

National Controlled Atmosphere Research Conference on Controlled Atmospheres for Storage and Transport of Perishable Agricultural Commodities (3rd: 1981 : Oregon State University)
Proceedings of the Third National Controlled Atmosphere Research Conference on Controlled Atmospheres for Storage and Transport of Perishable Agricultural Commodities, July 22, 23, and 24, 1982, held at Oregon State University, Corvallis, Oregon.

1. Food — Storage — Congresses. 2. Food — Transportation — Congresses. 3. Protective atmospheres — Congresses. II. Richardson, D. G. (Daryl G.) II. Meheriuk, M. (Michael) III. Oregon State

University, Dept. of Horticulture. IV. Title.
TP373.3.N26 1981 664 82-723
ISBN 0-917304-26-8 AACR2

TABLE OF CONTENTS

Proceedings of the

THIRD NATIONAL CONTROLLED ATMOSPHERE RESEARCH CONFERENCE

on

Controlled Atmospheres for Storage and Transport of Perishable Agricultural Commodities

July 22, 23, and 24, 1981

held at

Oregon State University
Corvallis, Oregon

Edited by

D. G. Richardson
Associate Professor of Horticulture
Oregon State University

M. Meheriuk
Agriculture Canada, Research Station
Summerland, British Columbia, Canada

Sponsored by

Department of Horticulture
Oregon State University

ACKNOWLEDGEMENTS

It is with great pleasure that I take this opportunity to express my gratitude to the many people and organizations who contributed to the success of the III National Controlled Atmosphere Research Conference held July 22-24, 1981 at Oregon State University.

Dr. Michael Meheriuk served as co-chairman who, along with Dr. Ken Olsen, developed the program of speakers, handled the correspondence with the participants, and assisted in editing the proceedings. Dr. Don Dewey also provided valuable ideas from his experience as organizer of the previous conference. The participants themselves deserve credit for the quality of their presentations. Dean E. J. Briskey provided valuable advice and encouragement and assisted with arrangements for publishing the proceedings. Mr. Richard Abel, President of Timber Press - International Scholarly Book Services, publisher of the Proceedings of the conference, who provided the technical expertise and many excellent suggestions in editing.

Most of the local arrangements were capably and willingly accomplished by several of my own fine graduate students: Ms. Sylvia Meadows who handled transportation and much of the detail work in assembling the manuscripts and assisted substantially in the editing and proof-reading, Arazdordi Toumadje and Esmaeil Fallahi for handling the audio-visual equipment, Danai Boonyakiat for many miscellaneous duties, and Ms. Patty Bennett for typing of programs and manuscript revisions. Without the enthusiastic cooperation of these people, occasionally under time duress, things would not have run as smoothly.

Finally to Edward Hemway & Associates, to Food and Agro Systems, Inc., and to the Tillamook Cheese Co. our sincere thanks for their generous contributions to the Wine and Cheese reception and to Mr. Bill Critchfield, Corvallis Elks Club for catering the Banquet. Thanks also to the Bruce Eldridge Bluegrass Band and to Cartoonist James Cloutier for providing excellent entertainment.

Daryl G. Richardson
Dept. of Horticulture
Oregon State University
Corvallis, Oregon 97331

CONTROLLED ATMOSPHERE CONFERENCE PARTICIPANTS

Mary Lu Arpaia
c/o Department of Pomology
University of California
Davis, CA 95616

Daniel A. Baker
Sheild Brite Corporation
P.O. Box 558
Kirkland, WA 98033

Richard D. Bartram
Room 15-Fuller-Quigg Bldg.
103 Palouse Street
Wenatchee, WA 98801

James A. Bartsch
Department of Ag. Engineering
Cornell University
Ithaca, NY 14853

Luce S. Berard
Post-Harvest Physiologist
Agriculture Canada
Research Station
SL-Jean, Quebec, Canada
J3B 6Z8

G. D. Blanpied
Pomology Department
Cornell University
Ithaca, NY 14853

Danai Boonyakiat
Horticulture Department
Oregon State University
Corvallis, OR 97331

William J. Bramlage
Dept. of Plant & Soil Science
French Hall
University of Massachusetts
Amherst, MA 01003

Jeffrey K. Brecht
c/o Department of Pomology
University of California
Davis, CA 95616

Arthur Cameron
Dept. Environmental Horticulture
University of California
Davis, CA 95616

Paul M. Chen
Mid-Columbia Experiment Station
3005 Experiment Station Drive
Hood River, OR 97031

Chun-Lung (George) Chu
Ministry of Agriculture and Food
Box 587, Simcoe, Ontario, Canada
N3Y 4N5

Reine Dalton
Elgin Fruit-Packers Co-op
P.O. Box 14
Elgin
South Africa
0240 2620

Don H. Dewey
Department of Horticulture
Michigan State University
East Lansing, MI 48823

David R. Dilley
Michigan State University
East Lansing, MI 48823

David A. Dixon
Foodsource, Inc.
900 Larkspur Landing Circle
Larkspur, CA 94939

Robert Lee Dixon
Foodsource, Inc.
900 Larkspur Landing Circle
Larkspur, CA 94939

Esmaeil Fallahi
Horticulture Department
Oregon State University
Corvallis, OR 97331

Chaim Frenkel
Dept. of Horticulture & Forestry
Cook College
Rutgers University
P.O. Box 231
New Brunswick, NJ 08903

A. P. Fresno (Alex)
Graymead Farm
P.O. Elgin
South Africa
02252-260

R. C. Gilardi
Airco Industrial Gases
575 Mountain Ave.
Murray Hill, NJ 07974

Lazar Ginsburg
32 Rowan Street
Stellenbosch, South Africa 7600

Harold Gorfien
Food Engineering Lab
U.S. Army Natick Labs
Natick, MA 01760

Rich Harcombe
Box 1644
Summerland B.C.
Canada, VOH 120

Jane Harman
Glasshouse Crops Research Institute
Worthing Road
Rustington
Littlehampton,
Sussex, England

John M. Harvey
U.S. Dept. Agriculture
2021 S. Peach Avenue
Fresno, CA 93747

Frank Haun
Controlled Atmosphere Instrumentation, Inc.
Rt. 2 106 Tyee
Okanogan, WA 98840

James R. Hicks
Veg. Crops Dept.
Cornell University
Ithaca, NY 14853

D. L. Hunter
Food Plant Engineering, Inc.
P.O. Box 9906
Yakima, WA 98909

Fouad Abd El-Latif Ibrahiem
20 El-Hussin St. Dokky
Giza, Egypt - ARE.
707756

Yilmaz Ilker
Sea-Land Service, Inc.
P.O. Box 1050
Elizabeth, NJ 07207

Morris Ingle
Division of Plant & Soil Sciences
West Virginia University
Morgantown, WV 25606

Yeldez M. Ishak
Horticultural Institute
Agric. Res. Center
Ministry of Agriculture
University str, Giza,
Egypt
988169

John Jameson
Fruit Storage Division
East Malling Research Station
Maidstone
Kent
England
0732-843832

Joseph J. Jen
Campbell Inst. for Research & Tech.
Campbell Soup Company
Campbell Place
Camden, NJ 08101

Adel A. Kader
Department of Pomology
University of California
Davis, CA 95616

Stanley J. Kays
Department of Horticulture
Plant Science Bldg.
University of Georgia
Athens, GA 30605

Kazuhide Kawada
709-4 Kamifukuoka
Takamatsu, Kagawa
760 Japan
(0878) 31-4759

Mohamed Karmal El-Kazzaz
Department of Pomology
University of California
Davis, CA 95616

John Kelly
Department of Horticulture
2001 Fyffe Ct.
Ohio State University
Columbus, OH 43210

Ed Krueger
607 Brunken Avenue
P.O. Box 1906
Salinas, CA 93902

Donald Laatsch
11555 Elk Creek Road
Trail, OR 97541

O. L. (Sam) Lau
B.C. Tree Fruits LTD.
c/o Pomology Section
Agriculture Canada
Research Station
Summerland, B.C. Canada
VOH 120

P. D. Lidster
Post-Harvest Physiologist
Agriculture Canada
Research Station
Kentville, Nova Scotia
B4N 1J5

E. C. Lougheed
Horticulture Sci. Dept.
University of Guelph
Guelph, Ontario, Canada
N1G 2W1

Tom McClellan
P.O. Box 2921
Yakima, WA 98907

Barry McDonald
P.O. Box 2225
Auckland
New Zealand
34 116

Stella McLeod
New Zealand Apple & Pear Marketing
Board
Box 279
Hastings
New Zealand
68-149

L. M. Massey, Jr.
NYS Ag. Exp. Station
Cornell University
Gemeuz, NY 14456

George E. Mattus
Horticulture Dept.
VPI & PU
Blacksburg, VA 24061

Don Maxwell
General Mills, Inc.
9000 Plymouth Avenue N.
Minneapolis, MN 55427

Sylvia Meadows
Grad. Student - Horticulture Dept.
Oregon State University
Corvallis, OR 97331

Michael Meheriuk
RRZ Thornber Street
Summerland, B.C.
VOH 1Z0

Walt M. Mellenthin
Mid-Columbia Experiment Station
3005 Experiment Station Drive
Hood River, OR 97031

Diven Meredith
Pressure Cool Co.
P.O. Drawer Y.
Indio, CA 92202

Walt Mithaug
c/o Washington Fruit & Produce
Yakima, WA 98902

Leonard L. Morris
Dept. of Vegetable Crops
University of California
Davis, CA 95616

Laura Mrachek
Fruit Maturity Program
1104 N. Western Avenue
Wenatchee, WA 98801

Kenneth L. Olsen
1104 N. Western
Wenatchee, WA 98801

Thomas R. Parks
Food & Agro Systems Inc.
1289 Mandarin Drive
Sunnyvale, CA 94087

Max E. Patterson
Horticulture Department
Johnson Hall
Washington State University
Pullman, WA 99163

Clara Pelayo
Departamento de Fisiologia de Postcosecha
Subdireccion de Investigacion y Docencia
Comision Nacional de Fruticultura
Km 14½ Carretera Mexico - Tolmea
Mexico 18 D. F.
570-17-79

Peter W. Perrin
Research Branch
Agassiz Research Station
Box 1000
Agassiz, B.C.
VOM 1AO

Timothy A. Prince
Department of Horticulture
Michigan State University
East Lansing, MI 48864

M. Rangappa
P.O. Box 453
Virginia State University
Petersburg, VA 23803

Michael S. Reid
Dept. of Environmental Horticulture
U. C.
Davis, CA 95616

Daryl G. Richardson
Post-Harvest Physiologist
Department of Horticulture
Oregon State University
Corvallis, OR 97331

William R. Romig
General Foods Corp.
555 S. Broadway
Tarrytown, NY 10591

M. E. Saltveit
Dept. Horticultural Science
North Carolina State University
Raleigh, NC 27640

R. O. Sharple
Fruit Storage Division
East Malling Research Station
East Malling
Maidstone, Kent
England
0732-843833

Loretta Shum
Grumman-Dormavac
90 Crossways Park West
Woodbury, NY 11797

Jingtair Siriphanich
Pomology Department
University of California
Davis, CA 95616

Derek Snowball
Turners & Growers, LTD.
City Markets - Box 56
Auckland, New Zealand 34116

Theo Solomos
Department of Horticulture
University of Maryland
College Park, MD 20742

Noel Sommer
Pomology Department
University of California
Davis, CA 95616

Tom Spatz
327 E Fir Street
Medford, OR 97501

Mary Spencer
Department of Plant Science
University of Alberta
Edmonton, Alberta, Canada
T5R 0C1

Larry L. Strand
Department of Vegetable Crops
University of California
Davis, CA 95616

Ling-Yuan Su
Mann Lab
Department of Vegetable Crops
U.C. Davis
Davis, CA 95616

Stuart Thorne
Queen Elizabeth College
University of London
Campden Hill Road
London, W8 7AH
01-937 5411

Daniel Tompkins
Room 6440 South
CSRS
U.S. Dept. of Agriculture
Washington D.C. 20250

Arazdordi Toumadje
Horticulture Department
Oregon State University
Corvallis, OR 97331

Marita Cantwell de Trejo
Mann Lab
Department of Vegetable Crops
University of California
Davis, CA 95616

Rui Luiz Vaz
1850 Hanover Drive #125
Davis, CA 95616

Chien Yi Wang
USDA - ARS
Bldg. 002 BARC-West
Beltsville, MD 20705

Tsu-Tsuen Wang
Mann Lab
Department of Vegetable Crops
University of California
Davis, CA 95616

Alley E. Watada
USDA ARS
Bldg. 002 Room 113
Beltsville, MD 20705

Chris Watkins
c/o C. Frenkel
Department of Horticulture & Forestry
Rutgers University
New Brunswick, NJ 08903

Harold E. White
Life Science Department
Pasadena City College
1570 E. Colorado Street
Pasadena, CA 91106

Richard Wicks
C/A Services
315 No. 61st Avenue
Yakima, WA 98908

Henry Witte
Liquid Carbanic, Inc.
40 Gurholt Drive
Dartmouth, Nova Scotia
Canada

Richard E. Woodruff
P.O. Box 1788
Salinas, CA 93902

Elhadi M. Yahia
Viticulture Department
U.C. Davis
Davis, CA 95616

Shang Fa Yang
Mann Lab
Vegetable Crops Department
University of California
Davis, CA 95616

A. Yazdaniha
Department of Horticulture
Washington State University
Pullman, WA 99164

Li Yu
Department of Pomology
University of California
Davis, CA 95616

Highlights of The Third National Controlled Atmosphere Research Conference

M. Meheriuk and D. G. Richardson

Considerable emphasis was placed on the use of very low levels of oxygen for the storage of tree fruits and other commodities. The application of low oxygen (<2%) lengthens the storage period for most apple and pear cultivars, retains green stems in cherries and prevents superficial scald in d'Anjou pears. Senescent changes such as softening are retarded to a greater extent than with conventional controlled atmosphere (C.A.) Storage. One aspect associated with low oxygen storage is the delayed ability of apples to synthesize their characteristic aroma. Post-C.A. conditioning in air storage appears to overcome the aroma deficiency. Nevertheless, the significant improvement in texture and other quality parameters out-weighs the lack of a full aroma. Low oxygen storage is not without hazards and comtemplation of its use should be preceeded by thorough evaluation.

Properly constructed storages for low oxygen should limit a drop of 0.25 in of water (from 1.00 to 0.75 in) to 30 minutes. Mode of construction varies with country and region and no particular design is superior to others. With the advent of low oxygen storage it may be necessary to install automatic control systems such as those currently in use in England. The systems are accurate and substantially eliminate the 'operator' factor in maintenance of good storage environments. Removal of CO_2 may be difficult with carbon or molecular sieve scrubbers when CO_2 values must be less than 0.25%. Lime scrubbing is perhaps the best alternative and improved lime scrubbers are being developed.

Ethylene and relative humidities in storages were also discussed. Removal of ethylene in low oxygen storage is important in order to retain the high degree of quality in fruit during long storage periods. An exterior catalytic burner was described that may be more economical than the commercial ethylene scrubbers. It was pointed out that apples producing more than 1 ppm of internal ethylene are unlikely to retain good quality even if ethylene in the storage room is removed almost entirely. High relative humidities that leave a film of water on fruit promote disorders and units that provide high humidities with no free water on the fruit were recommended.

That fruit are insensitive to ethylene at low temperatures was amply disproved by the reports on kiwi fruit which soften rapidly in the presence of 1 ppm ethylene at 0^oC. The climacteric, however, occurs several weeks later. Levels below 20 ppb are, therefore, needed to prevent the premature softening of kiwi fruit in storage.

The use of rapid CA also offers considerable potential for long term storage of fruits. Loading, cooling and establishment of the atmosphere for apples should be completed within 5 to 7 days.

AVG (amino ethoxy vinyl glycine) is a useful tool to study the biochemical events in ripening and has been used experimentally as an orchard spray to delay the onset of the climacteric. A longer harvest period without loss of storage potential is, therefore, possible.

Respiration of fruit under CA and low O_2 conditions was covered in depth and of particular interest was the topic on the inhibitory biochemical effects of low oxygen. Apparently, oxidases other than the cytochromes are influenced by low oxygen because sufficient oxygen is present under low oxygen regimes to meet the apparent needs of the cytochrome chain. A practical means of measuring the resistance of gases across the skin of fruit may enable researchers to predict the lower limits for CA storage atmospheres for a given commodity, and may provide insights into the different season-to-season responses observed in commodity tolerances to CA conditions.

The use of CO in CA was dealt with at length. CO is converted by plant tissue into CO_2 and ultimately into the organic acid fraction. Presence of CO in the transit atmosphere prevents discoloration of several commodities. However, the control of some species of fungi by CO appears to be the most useful application of the gas in CA and transit atmospheres. Precaution must be exercised in its use because it can be somewhat stimulatory like ethylene.

CA storage has permitted the storage of tomatoes for several weeks. Peaches and nectarines are amenable to CA storage but require intermittent warming to retain quality. Vegetables differ in their response to CA, even with cultivars within a crop. Ethylene can cause physiological disorders with vegetables in air storage but the problem is diminished considerably when CA storage is involved.

Ornamentals and floral crops continue to receive CA attention and it was of interest to learn that low pressure storage is being used for preservation of cut and potted flowers.

Polymeric films are receiving more attention in regards to storage of several horticultural crops. Through the judicious choice of films relative to a given quantity of commodity one can obtain the correct atmosphere for the crop in question. Citrus fruit appear particularly amenable to individual wrapping with polymeric film. Quality is retained for longer periods and mould can be contained more readily.

Also mentioned was the role of mineral nutrition to storage potential and development of physiological disorders. Calcium continues to have a predominant role in this regard.

A brief report on the use of high CO_2 and low O_2 indicated control of some tree fruit insects, especially where export markets are involved.

From a personal point of view, the meeting was very satisfying and the favorable comments from many participants suggests that such gatherings be continued. It is unfortunate that travel restrictions prevented the attendance of notable researchers, extension and industry people but we look forward to their presence at the next conference, tentatively scheduled for 1985.

TOPIC I

CA STORAGE STRUCTURE

by

D L. HUNTER, P.E.

President, Food Plant Engineering, Inc.

Yakima, Washington

The increase in CA storage in the last four years especially in the Northwest part of the United States and in British Columbia has been dramatic. CA storage in other areas of the country has also increased, but not to the extent witnessed by Washington, D.C. and Oregon apple and pear shippers. The CA increase in the Northwest in the last two years appears to be in the order of 10,000,000 boxes or about 17%. Because of this large volume increase this paper and presentation will concern mainly observations in the Northwest where much of this design has been done as well as observations in several other fruit growing sections of the country like Michigan, New York, Pennsylvania-Virginia area where our firm also does design.

Building Materials

Even though common building materials used in CA storage include

Concrete Block

Wood Frame

Metal Buildings

Tilt-up Concrete

the vast majority of CA storages in the Northwest utilized tilt-up concrete buildings. In recent years techniques and developments have occurred which permit less manual work on the tilt-up buildings, such as eliminating most of the poured pilasters by bracing the walls through the floor and roof system. In addition, the surfaces of the walls provide an excellent base for vapor barriers, insulations and seals.

A further advantage of the tilt-up concrete is its relatively low insurance rate and the fact that dry rot does not occur within the walls. Maintenance costs on a tilt-up building are virtually non-existent except where they are damaged on the interior by lift trucks. An added factor is that movement in the walls is very minor compared to, say, metal buildings.

All the floors in CA's are concrete laid on grade and recently some double slab buildings have gone in with a built-up roofing membrane in between where low O_2 is desired. See Figures 1 and 2. Most of the CA's, however, still have only one concrete floor as an air seal as in Figure 3.

Roofs on top of CA's are either steel joists and steel deck or truss joists which appear similar to steel long span joists, but have wood members for the top and bottom cord with pipe sections for the interiors. Consequently, the deck on the roof is steel deck where steel joists are used and plywood when truss joists are used. Right at the present time truss joists and plywood deck are far the most common in the Northwest and the lowest cost, even though they usually require an extra layer of sprinkler in the attic area compared to steel joists. In the East and South steel joists and deck are still most commonly used although pre-stressed concrete roofs are being used.

New systems of roofing are being tried especially by our firm because we are generally dissatisfied with normal built-up roofing with asphalt and felts.

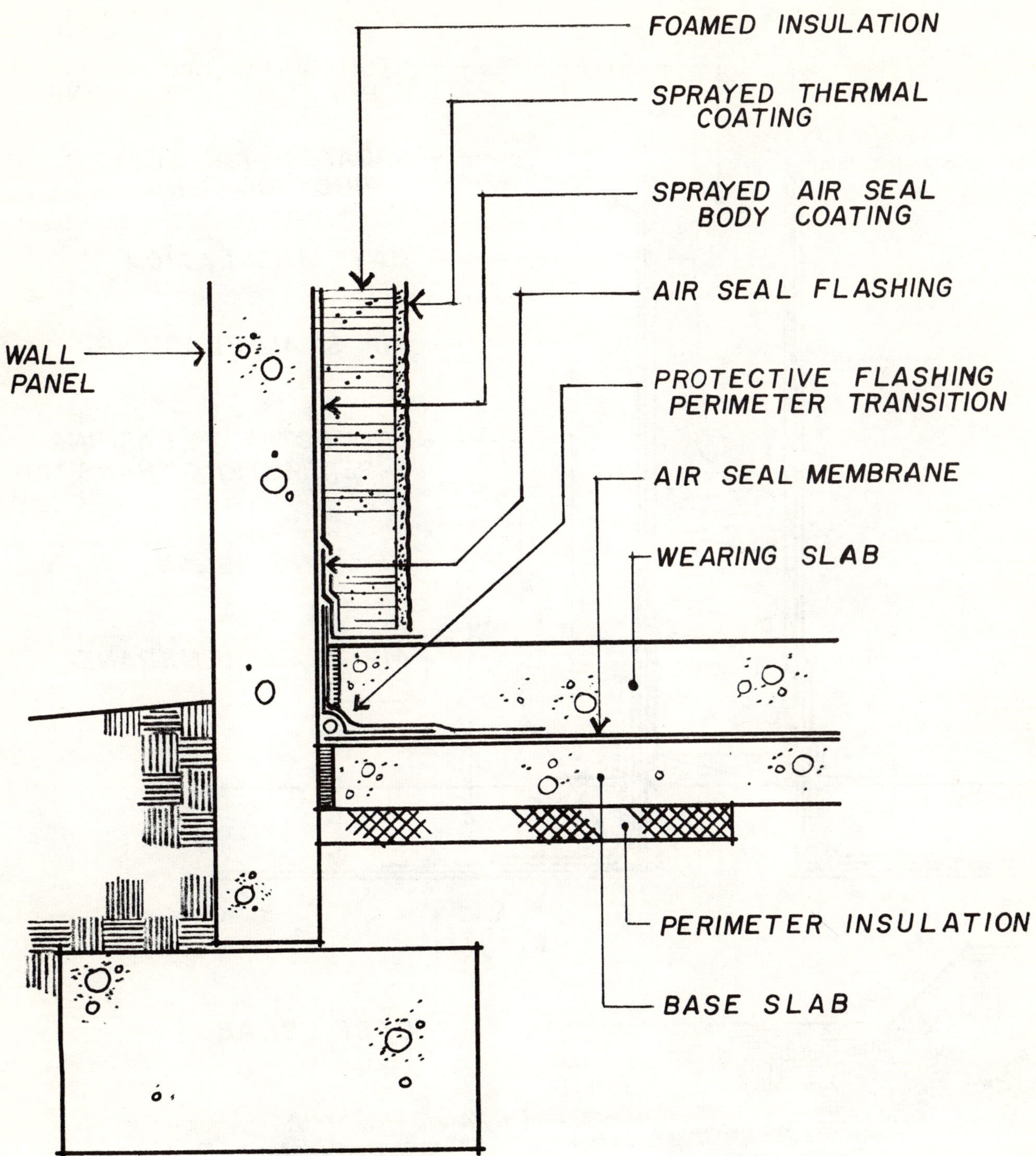

WALL - FLOOR JUNCTURE
AT DOUBLE FLOOR SLAB

Figure 1.

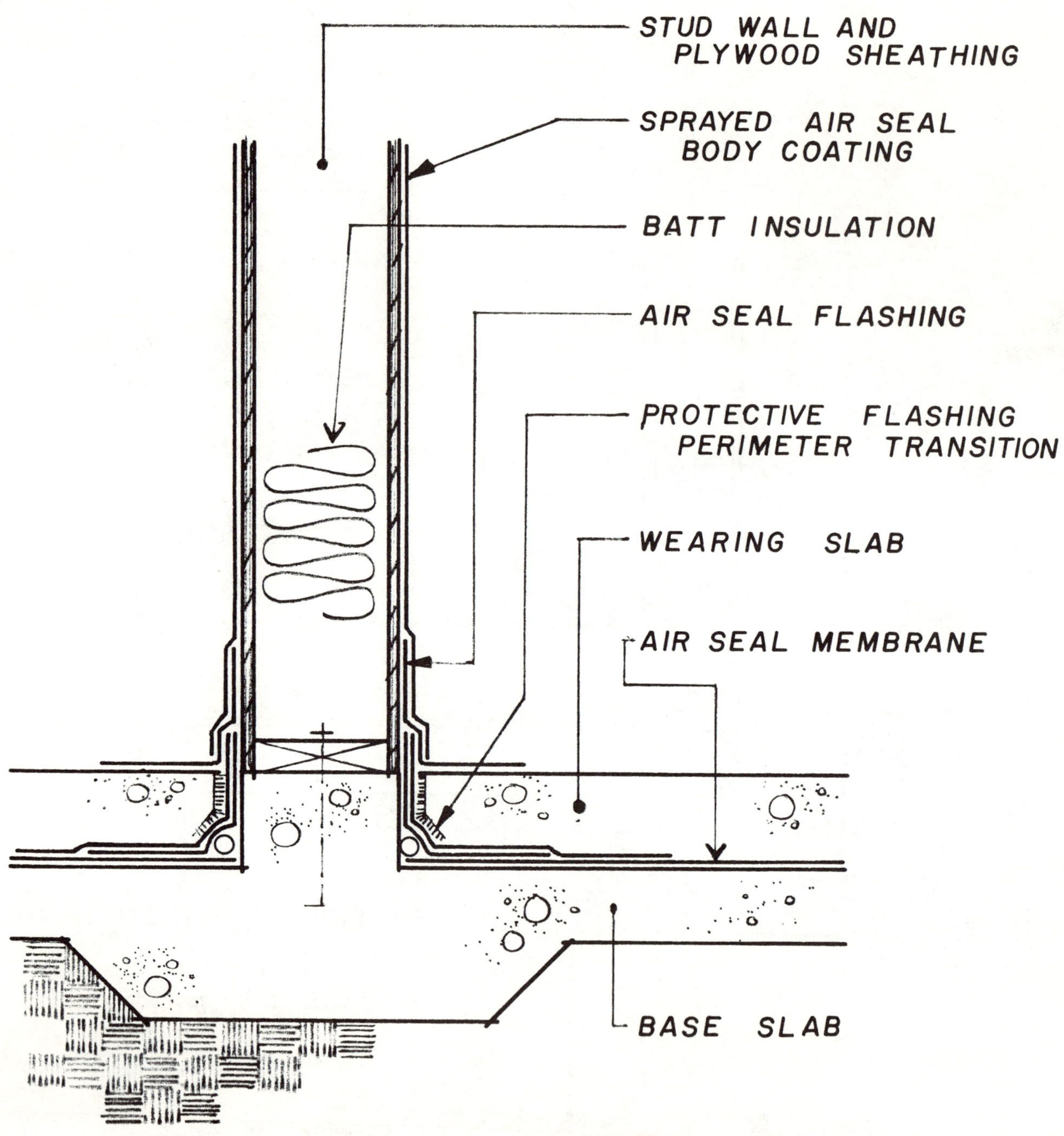

WALL - FLOOR JUNCTURE
AT DOUBLE FLOOR SLAB

Figure 2.

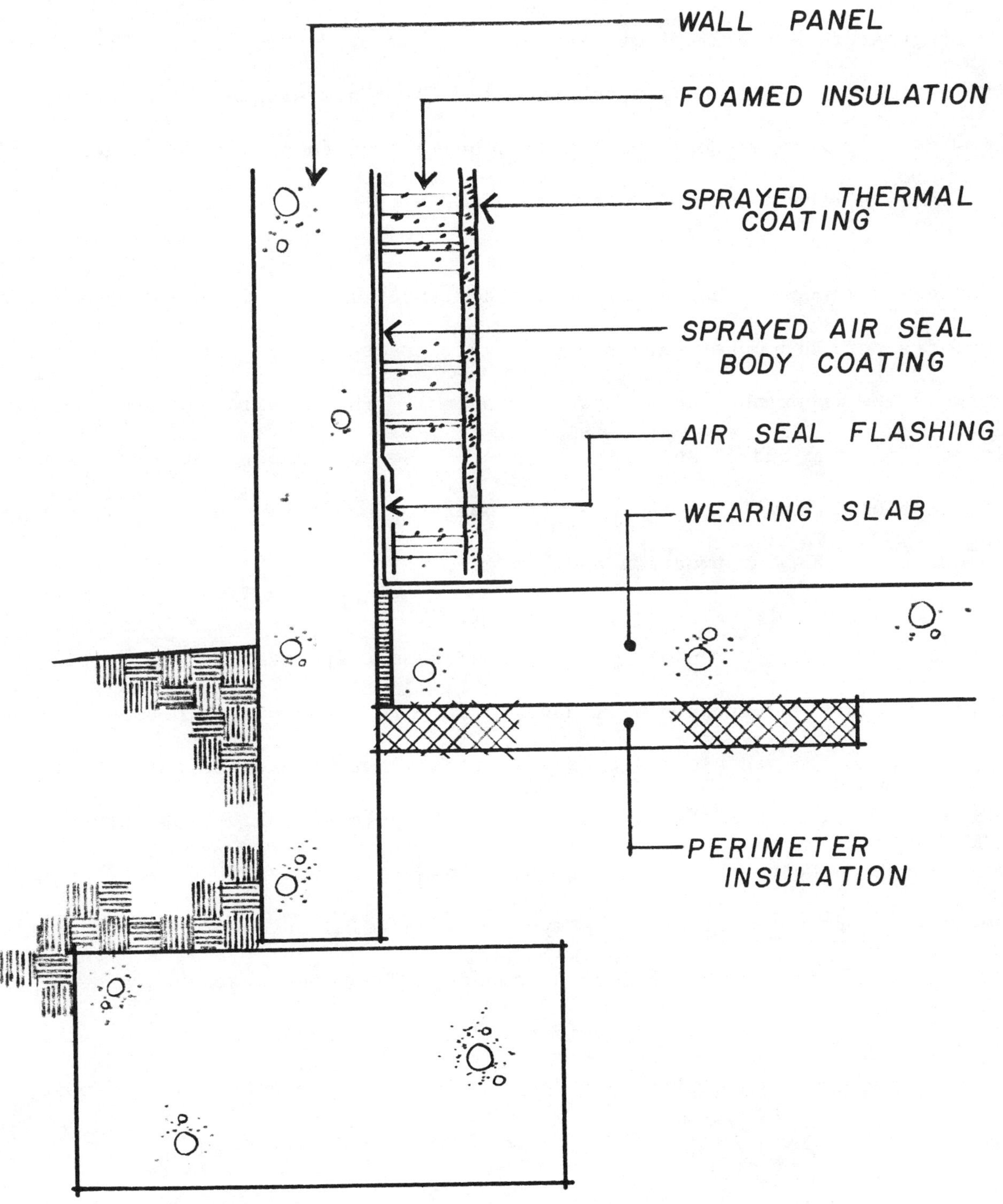
WALL PANEL
FOAMED INSULATION
SPRAYED THERMAL COATING
SPRAYED AIR SEAL BODY COATING
AIR SEAL FLASHING
WEARING SLAB
PERIMETER INSULATION
WALL - FLOOR JUNCTURE

Figure 3.

Food Plant Engineering is advising more single membrane roofing materials as a means of greater reliability for leak-proof roofs over structures. These are shown in slides given in this presentation. These new single layer roofing systems usually cost a little more money, but appear to provide a longer lasting roof with less repair and maintenance needed.

CA storage construction has been given a new facet in that it would appear that some contractors and owners are speculatively erecting CA's for lease at the time of harvest for fruit they do not own or control. These buildings are usually built on a minimum budget and there is some cause for concern inasmuch as most are minimum cost facilities and what one operator can't figure to leave out or cut down in good CA construction another can.

In general, however, it would appear that different storages may be suited to different product and growers will have to be more conscious of possible different storage. In most cases a storage user is really not able to compare the real weight loss in different stages even though it could vary 1 or 2 percent. Red Delicious can be kept until March or April in a relatively low quality CA and the effect is not noticeable to the eye. Golden Delicious and McIntosh on the other hand, may have shrinkage show up sooner. The point that should be of some concern is that fruit is sold by weight and 1 or 2 percent more weight loss difference between good quality and low quality CA represents a lot of money even though the difference isn't noticeable to normal observation.

Layout Systems

Floor plans for CA's sometimes develop from the amount of space a company has to erect a facility, but more and more the layout is chosen as a matter of economy. Most of these are still based on one of those shown in Figure 4, attached.

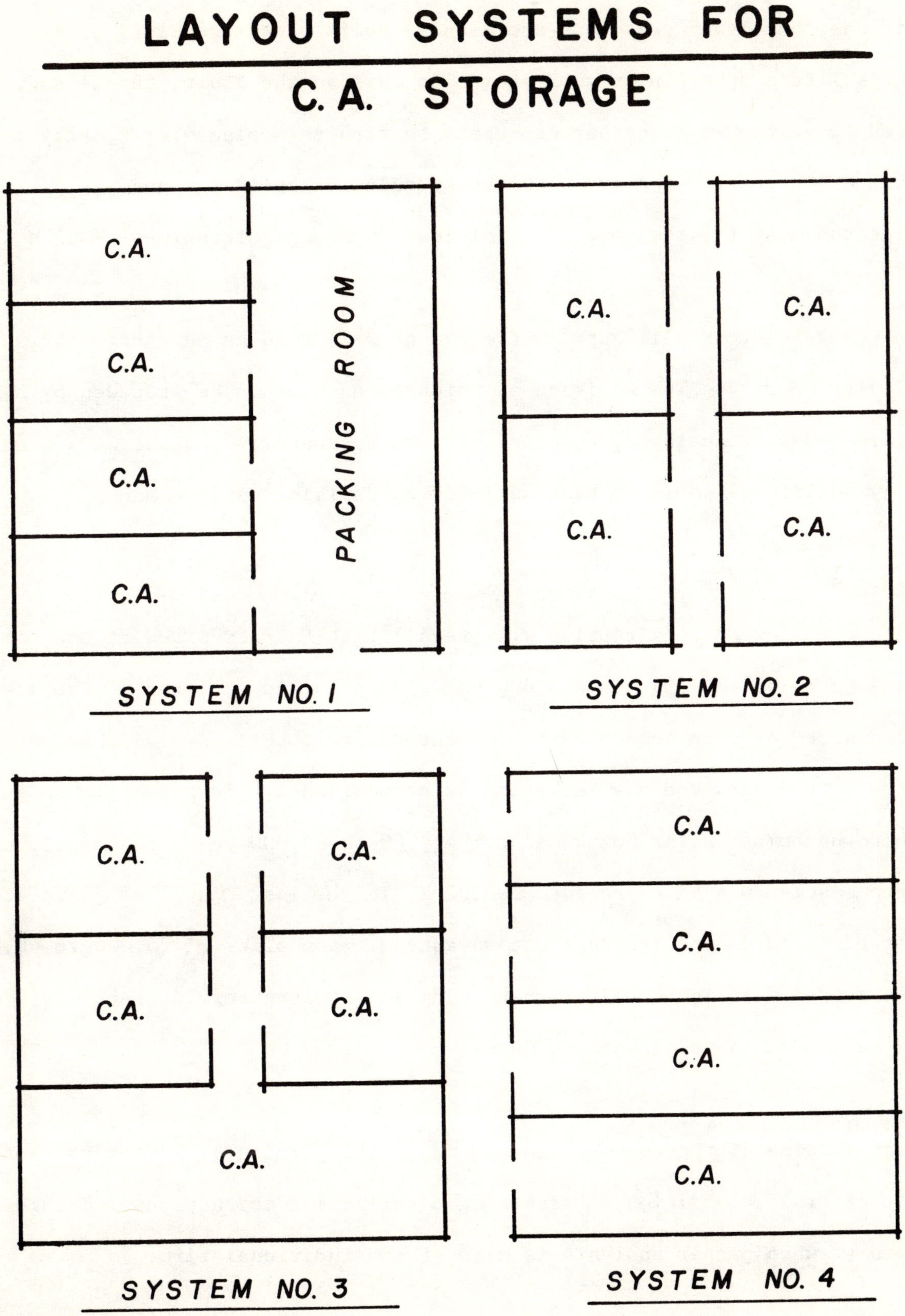

LAYOUT SYSTEMS FOR C. A. STORAGE

C.A.
C.A.
C.A.
C.A.
PACKING ROOM

SYSTEM NO. 1

C.A.
C.A.
C.A.
C.A.

SYSTEM NO. 2

C.A.
C.A.
C.A.
C.A.
C.A.

SYSTEM NO. 3

C.A.
C.A.
C.A.
C.A.

SYSTEM NO. 4

Figure 4.

Systems 2 and 3 shown appear to offer the greatest control and best operating conditions, but lower cost CA's are usually built along the lines of System 4. There is little opportunity in System 4 to observe the fruit, the CA doors are exposed to weather and further CA expansion with this plan will usually result in some extra cost. However, with the speculative buildings and the pressure on getting the most for the least initial cost, System 4 oftentimes results.

Any of these systems will work, but each one will find an advantage for a particular company. The decision of which to use should be preceded by a cost analysis and systems layout to consider future expansion and the work should be done by qualified people so that an informed decision can be made.

Room Sizes

Those who follow CA construction will recall that CA room sizes at one time were thought to be quite large when 5,000 boxes were put in a storage. The most favored size seems to depend on the amount of fruit that is available, how fast the fruit is loaded and how marketing is accomplished. In recent years it has not been uncommon in the Northwest for 40,000 to 80,000 boxes to be put in CA rooms and rooms sometimes have exceeded 100,000. In the last couple of years with the advent of rapid CA and its apparent advantage, room sizes of 25,000 to 40,000 appear to be favored for high quality fruit going in for long term storage where it could be loaded in perhaps three days.

The actual size of the room has to depend on the particular varieties, total amount of fruit, condition of marketing a particular company has and this can be arrived at when proper analysis is made of the individual firm.

It is usually the feeling at Food Plant Engineering that rooms should be as large as practical to minimize cost that comes from partitions, doors, and controls. Most companies given an informed choice will want to maintain the option of keeping fruit for the maximum period of time even though in some years they may sell early. To do this, "Rapid CA" practices would appear to be in order and therefore, reasonable limits need to be placed on the size of the rooms.

Stacking Heights

In the last four years almost any CA storage of reasonable size is filled with the bins 10 high in the storage room. These are not experimental nor considered dangerous, but do permit an owner to maximize his use of cash. Rooms of small capacity or whose bin heights are greater than 30" should probably be 6 to 8 bins high.

Room Loading Rates

The rate at which rooms are loaded has customarily been in terms of about 7-8 days. The daily rate is obtained by dividing the room capacity by the number of days loading. Recently, however, with the advent of rapid CA this practice is being changed considerably by plants who are trying to load in three days and bring the atmosphere down to less than 5% oxygen in no more than another three days. Where this practice has been done there are generally favorable results and it would appear that this practice will increase.

Those plants that have skimped on refrigeration capacity in the past will not find this program to their advantage and will be forced to slower loading rates in order to attain proper temperatures.

Air Seal on Walls and Ceiling

Practices for air seals in the Northwest have become more restricted in recent years because of economics as well as developing satisfactory procedures. Air seals on concrete walls have generally gone to urethane with 1/2" of fire retardant lightweight concrete over them. Plywood walls are commonly covered with a rubberlike material which seals the walls quite adequately. Our firm has used "Gaco" rubber for several years and have not used any metal walls because of their tendency to develop leaks and require maintenance each year. The rubber compound treated walls go along as long as five years or more without any after-season repairs or tightening except for damaged areas.

Metal lined walls are simply not used. High density plywood will not last as long as other alternatives because it is a wood product and high moisture exposure to it is harmful in the long run. Some do use it, however.

The rubber-like films that have been developed are usually sprayed or rolled on the walls. There is a concept used of placing these films underneath insulation to act as both a vapor barrier and an air seal and this was developed by Food Plant Engineering and used widely throughout the Northwest. No dry rot can occur, it is less subject to damage and the air seal integrity has been very high.

Polyurethane has done a very good job, but has required fireproofing in order to be acceptable. The cementitious protecting coating can be used in conjunction with Gaco rubber on plywood walls with no problems. Slides shown in connection with this presentation will illustrate some recent jobs where it has been done.

Because of the length of seasons being nearly year around Food Plant has made a change in recommendations for the seal on ceilings of CA storages in the Northwest. We prefer urethane on a plywood ceiling or Gaco with an underhung insulated stud system of fiberglass rather than the common system of blowing loose fill insulation over the top of a plywood ceiling. Too much moisture has been observed to collect, the plywood will tend to warp and the seal will not last as long nor will the room be as tight if this practice continues.

The advent of low oxygen in CA's has caused our company to raise its room test requirement because of the exceptionally tight room that appears to be necessary for fine control of oxygen for apples and pears. In the past .25 inches of water has been satisfactory after a one hour pressure test and has been our standard. For low oxygen CA we now call for .75 inches after an hour and are able to achieve it.

To assure the extra tightness in CA's, it may be necessary to two layer floors with a built-up roofing material in between similar to Figures 1 and 2. This may depend on the type of soil in the area and previous experience. Some operators have been able to go to lower oxygen requirements with normal CA rooms having some tightening and special care.

Refrigeration

Nearly all refrigeration in Northwest CA rooms is ammonia and the interior of the room is served by either multiple smaller air units or sometimes by one large single unit. There is some exposure in the large single units because if problems develop with the one unit, it is the only means of refrigeration available within the room itself. On smaller units, one unit could fail and the other units could still maintain the refrigeration in the atmosphere even at slightly low capacity.

Other disadvantages of the large units is the increased horsepower required for cubic foot per minute of air circulated and the usual higher temperature difference across the cores causing slightly lower humidities. In addition, on comparable basis electric costs are usually higher for the type of fans utilized compared to the propeller fans on the smaller units.

Large low cost storages, however, are commonly using large single units. As pointed out earlier, it is infrequent that an owner can check the weight of his fruit coming out of one storage versus the other. However, it would appear that, in most cases of design, greater shrinkage and weight loss will occur in single unit rooms compared with multiple units, but this, of course will vary depending on the specific case.

Scrubbing Material

Scrubbing equipment commonly used is made by two manufacturers for the Northwest. Both of these have done a very good job quickly lowering the atmospheres in rooms and permitting the owner to establish CA conditions sooner than under circumstances where the fruit itself brings the room atmosphere down.

Many storages use desiccant scrubbing and may still use lime which appears to be the best means of obtaining extra low CO_2 in storage rooms itself. During this year Food Plant Engineering is installing a device which is intended to more fully utilize lime than the placing of 50 lb. lime bags on pallets in the CA room or separate lime room as done in the past. This is because it has been illustrated that when bags of lime are used oftentimes the lime itself is crusted with an inch or two on the outside surface and sometimes the interior is not fully utilized. In an effort to decrease costs and more fully utilize the lime, a new concept is being explored at this time. This concept utilizes a device to accept bulk lime from trucks and gradually move the lime through an enclosure with the air from the CA rooms passed over it permitting the lime to be handled in bulk and exposed to the air more fully and readily than has been the case in the past. This is illustrated in a slide during the presentation.

CONSUMPTION AND LOSS OF ENERGY IN COMMERCIAL STORAGE UNITS

James A. Bartsch
Department of Agricultural Engineering
Cornell University
Ithaca, New York

Acknowledgements

The financial support of the Cohn Foundation through the Cornell University Agricultural Experiment Station is acknowledged. The assistance and cooperation of the growers and management of Lake Country Storage Cooperative, Inc. Wolcott, N.Y. is appreciated.

Introduction

Approximately ten million bushels of New York State apples were placed in cold storage this past year (State of New York, Department of Ag. and Markets, 1980). Based on this volume and the average holding patterns for the year, 18.5 million kilowatt hours (kWh) of electricity was consumed by the refrigeration compressors to cool and maintain the crop at the required storage temperature. Additional energy was also consumed by the refrigeration support systems, air handlers, condensors, defrost equipment and related storage equipment. Based on studies by Price (1969), the 18.5 kWh estimate for refrigeration would indicate an annual electrical consumption on the order of 38.4 millon kWh for the ten million bushel industry. Costs of this energy range from approximately 4.5 cents per kWh in Western New York to upwards of 7 cents per kWh in the lower Hudson River Valley. If a statewide average of 5.5 cents per kWh is used, the total energy cost for storage runs around 2.11 million dollars annually.

Growers are continually seeking ways to improve efficiency and reduce energy costs through conservation, heat recovery and load leveling techniques. For these measures to be effective it is necessary to know something about the cold storage energy budget and the factors which influence it.

Objectives

The purpose of this research is to determine the present day use of electrical energy in an operating cold storage facility. Energy budgets are being prepared for the current year and will be developed for the next three years. The information will be analyzed to identify the factors and factor interactions which affect energy consumption. The results will be integrated into cooperative extension and other educational programs.

Procedure

A cold storage facility located west of Wolcott, NY was selected for the energy audit. The structure was built in 1979, is of concrete block construction and has a total CA storage capacity of 112,000 bushels in five rooms.

Figure 1 is a sketch of the storage floor plan indicating the capacity, layout and orientation of the structure. The major refrigeration and refrigeration support equipment is listed in Table 1. This equipment is physically located in the machine room on the southeast corner of the facility. The condenser unit is mounted on the roof of the machine room. The oxygen and CO_2 scrubber units are located on the mezzanine over the central corridor of the building. The gas analysis hardware, room temperature controls, and data acquisition system are also located in the mezzanine area.

Table 1. Equipment Monitored During 1980-81 Storage Period

40	Hp Compressor with Full-Half Capacity Control
75	Hp Compressor with Full-Half Capacity Control
3	Hp Defrost Pump
1.5	Hp Condenser Pump
10	Hp Condenser Fan
1	Hp Oxygen Burner Coolant Pump
.5	Hp Compressor Coolant Pump

The data acquisition system used at the cold storage is described in Figure 2. The system was centered around a Kaye Instruments Incorporated Digistrip II digital multipoint recorder. This recorder has a capacity to scan, acquire and store 48 channels of data input every 6 seconds. It is user programmable to enable conversion of analog input voltages to printed digital output in the units of the calibrated sensor from which the signal originated. The computational capabilities of this recorder make it possible to perform arithmetic operations including averages and accumulation of data as it is being received. Output from the 48 channels is printed in a digital array on an 8 1/2" by 11" bifold page at a user selected print rate. The date, time and measurement units are also included in the recorder printout.

The data recorder was interfaced to a MFE model 2500 buffered data terminal. The MFE tape recorder stored the printed information from the Digistrip II during each print cycle. The data acquistion system and tape unit was operated by storage manager and completed tapes and printed output were returned to campus for analysis. Data from the tapes were transcribed to a floppy disk via a Digital Equipment Corporation Minc 11 computer interfaced to an identical MFE 2500 terminal on campus. Data reduction and analysis was preformed with the Minc computer system maintained in the Department of Agricultural Engineering.

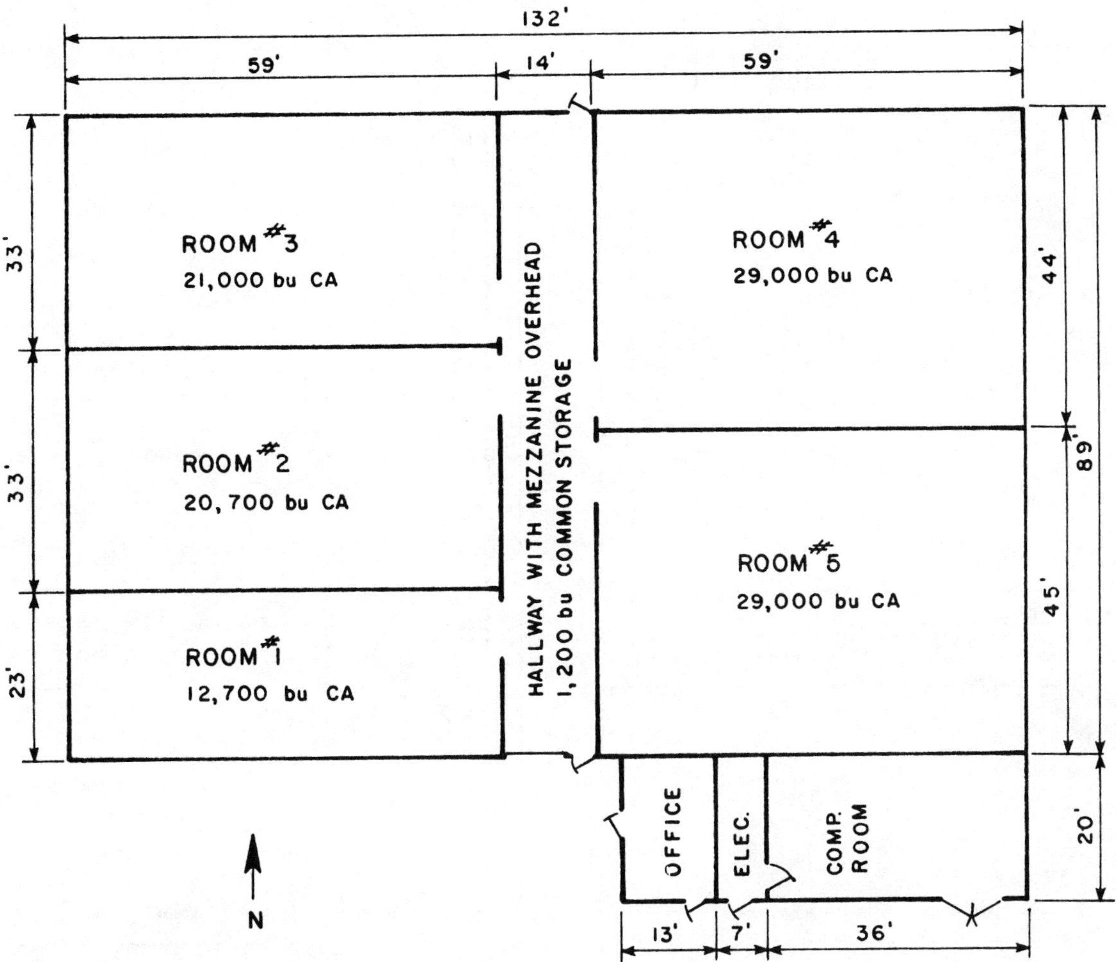

Figure 1. Plan of Cold Storage Facility.

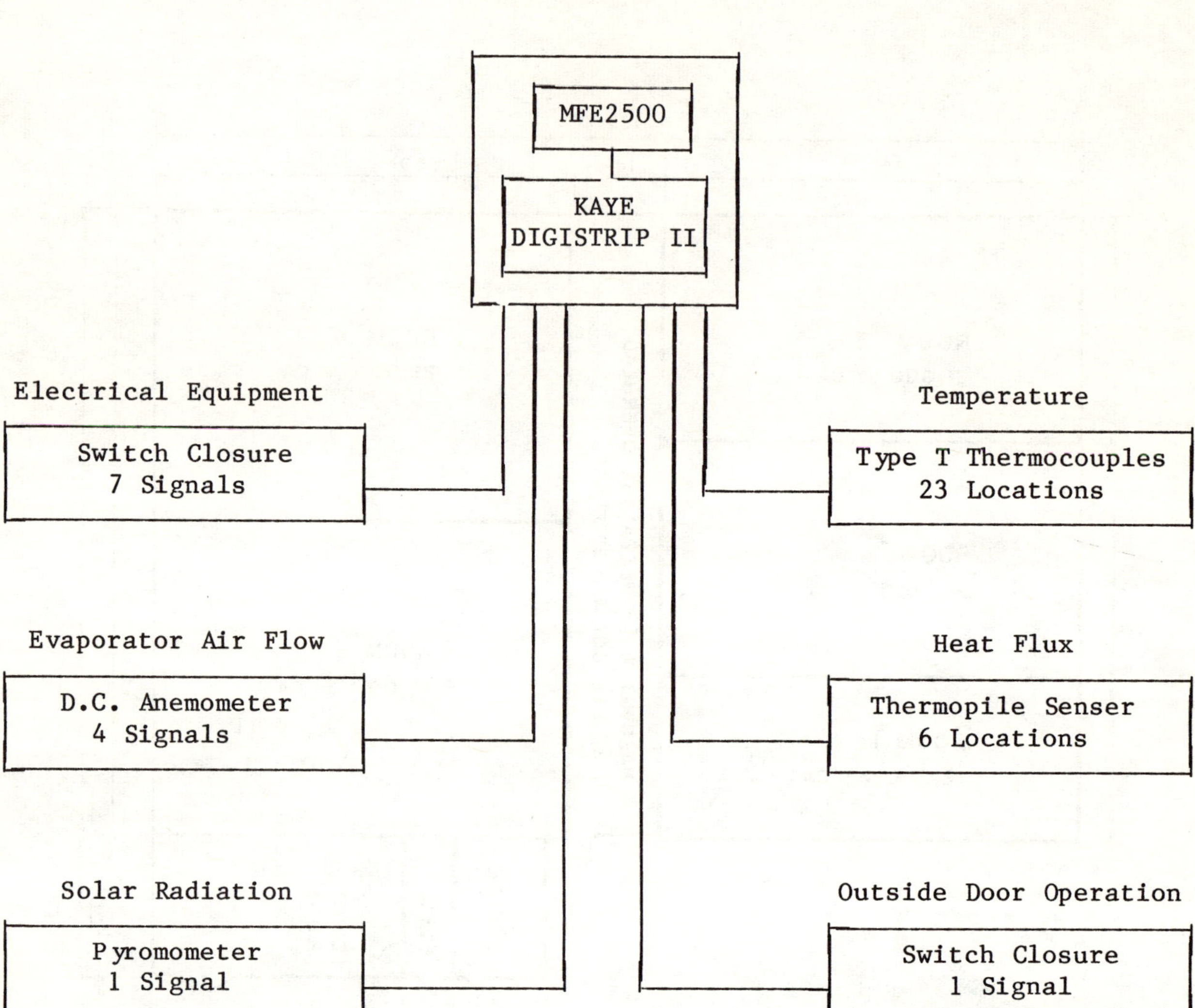

Figure 2. Sensor and Data Acquisition System Used in the Energy Audit

The operating characteristics of the major refrigeration equipment was monitored to determine the run time and cycle frequency of each piece of equipment. Since the project was initiated after the 1980 loading had begun the first year data was not obtained for the catalatic oxygen burner and CO_2 scrubber which operated during room pull down. In addition to the equipment listed in Table 1 continuous observations were also made of the heat flux through the floor and temperatures in rooms 3 and 4 (Figure 1).

Inside wall temperatures were recorded on the east, north and west sides of the building. Outside wall temperatures and roof surface temperatures were also recorded for rooms 3 and 4. Remote soil temperature at the 12 inch depth and ambient air temperatures were recorded on the north side of the structure. Solar radiation was detected by a sensor placed on a mast above the roof of the building.

Data on air flow through the evaporator units in rooms 3 & 4 was also recorded. Anemometers were constructed and calibrated by the Department of Agricultural Engineering and placed on the discharge side of the evaporators. A bank of averaging thermocouples (three pair on each evaporator) was used to obtain the mean dry bulb temperature of the atmosphere entering and leaving the evaporator. This information was used to determine the temperature change of atmosphere passing through the evaporator unit. The identity, type, and number of installed sensors is summarized in Figure 2.

The 42 analog signals were continuously monitored by the data acquistion system. Instantaneous readings and accumulated run time data were printed out and recorded on magnetic tape once each hour. Daily summaries of run time data and averages of flux values and solar radiation were printed out daily and also recorded on magnetic tape. In the time since the instrumentation was installed in October of 1980, data has been collected for over 200 days of continuous operation. In the first 200 days over 1.5 million observations have been made by the data acquistion system and over 300,000 data points have been logged and placed on magnetic tape.

Data Analysis

Raw data was analyzed using the Minc 11 computer systems in the Department of Agricultural Engineering. Daily operational sequences for the monitored equipment (compressors, pumps and fans) were determined from the data. The total hours of daily operation for each device was multiplied by the appropriate electrical consumption factor to determine the daily kWh consumption of the respective equipment.

The power consumption factors listed in Table 2 were obtained from three sources. Operating current and voltage levels were recorded for the 40Hp compressor, the condenser fan and pump, and the defrost pump.

Table 2. Power Consumption Factors for Electrical Equipment

Unit	Horsepower	Consumption, kW	Comments
Blower Fan	1/2 each	.86 kW/Hp[d]	50 Units Total
Compressor	40	27.4[m]	@ 1/2 capacity
Compressor	40	35.1[m]	@ Full capacity
Compressor	75	53.7[c]	@ 1/2 capacity
Compressor	75	68.8[c]	@ Full capacity
Defrost Pump	3	2.84[m]	
Condenser Pump	1.5	2.49[m]	
Condenser Fan	10	11.55[m]	Centrifugal
Burner Pump	1.0	.56[d]	Cools C.O.B.
Jacket Pump	.5	1.0[m]	Cools compressors

d - manufacturers data
m - measured on site
c - calculated

Manufacturers, data for the blower units (Krack Corporation) and burner coolant pump (Stahlman Engineering Incorporated) were used to determine the power consumption factors for those pieces of equipment. Manufacturers data for the 75 Hp compressor (Frick Corporation) and the known characteristics of the 40 Hp compressor (also Frick) were used to calculate the consumption factors for the 75 Hp compressor.

The power consumed by these units was then computed by the following relationship for three phase circuits:

$$P = \sqrt{3}\ EI \cos \phi$$

where P = power in watts

E = line voltage

I = line current

ϕ = current-voltage phase shift angle

(assumped to be equal to 0°)

A list of equipment and respective consumption factors used in the energy budget analysis are summarized in Table 2.

Preliminary evaluation of heat flux gain through the floor was also completed. This information was of interest because of the potential energy demand the floor heat gain places on the refrigeration system during pull down. This phenomena is of particular interest in this study since only perimeter insulation was installed in the building and the floors themselves were not insulated. The relationship between foundation temperature, mean air temperature, and heat flux were evaluated from the data by graphical techniques. Final analysis and evaluation will be reported in future writings.

Results

In Figure 3, the daily energy consumption of each monitored electrical component is presented for the 1980-81 storage season. Since the Data Acquisition System was installed during the loading and pull down period (days 265-315 of 1980) data for that portion of the year is not available. This was the only period of operation of the 75 Hp compressor. The energy consumption of this device will be monitored during future seasons.

The contribution of each operating component to the energy budget is evident from Figure 3. The consistent demand of the 50 air handling motors is also apparent. These motors comprise the bulk of the energy budget and consume approximately 10,300 kWh daily.

The cyclic nature of electrical consumption by the 40 Hp compressor during the winter period is also noted in Figure 3. Daily electrical demand by this compressor varied from less than 100 kWh per day during very cold periods to 840 kWh daily when operating at full capacity for 24 hours. Both the condenser pump and fan operated on similar temperature related cycles although the total magnitude of demand for both of these components represented less than 10% of the total energy budget.

The total energy budget is presented in Figure 4 along with the instantaneous fruit inventory, ambient temperature, and power company billing data for the storage period. The effect of low ambient temperature upon total consumption is evident. Reductions in fruit inventory during mid-summer have little influence upon consumption. Reducing inventories to 35% of storage capacity did not reduce energy consumption appreciably because all 5 rooms were still operational and ambient temperatures were increasing.

Monthly electrical consumption as reported by the electric utility is also present in Figure 4. This value which includes the overall power factor (a value always less than unity) for the facility averages 20% less than the value determined from the audit data which is computed from Equation 1. What this really means is that the facility is using the quantity of energy shown in the Energy Data Curve of Figure 4 and being changed for the consumption indicated by the energy meter curve of the same figure. The energy meter data reflects this influence of average temperature and fruit inventory levels discussed above.

Future Studies

Based upon the contribution of the blower units to the total energy budget air handling efficiency deserves futher detailed study. The cyclic energy demand of the compressor during periods of low refrigeration demand also requires

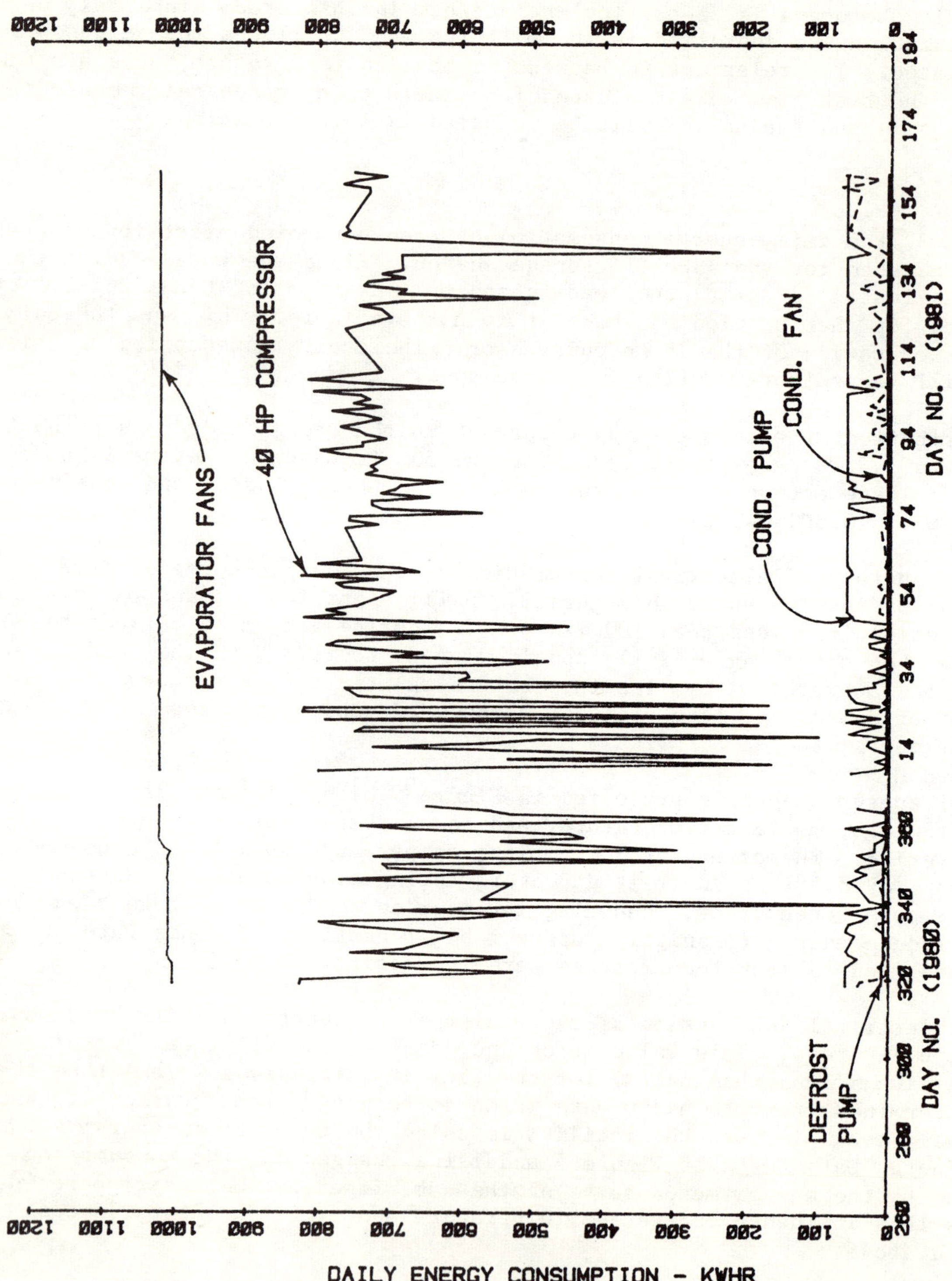

Figure 3. Energy Consumption for 1980-81.

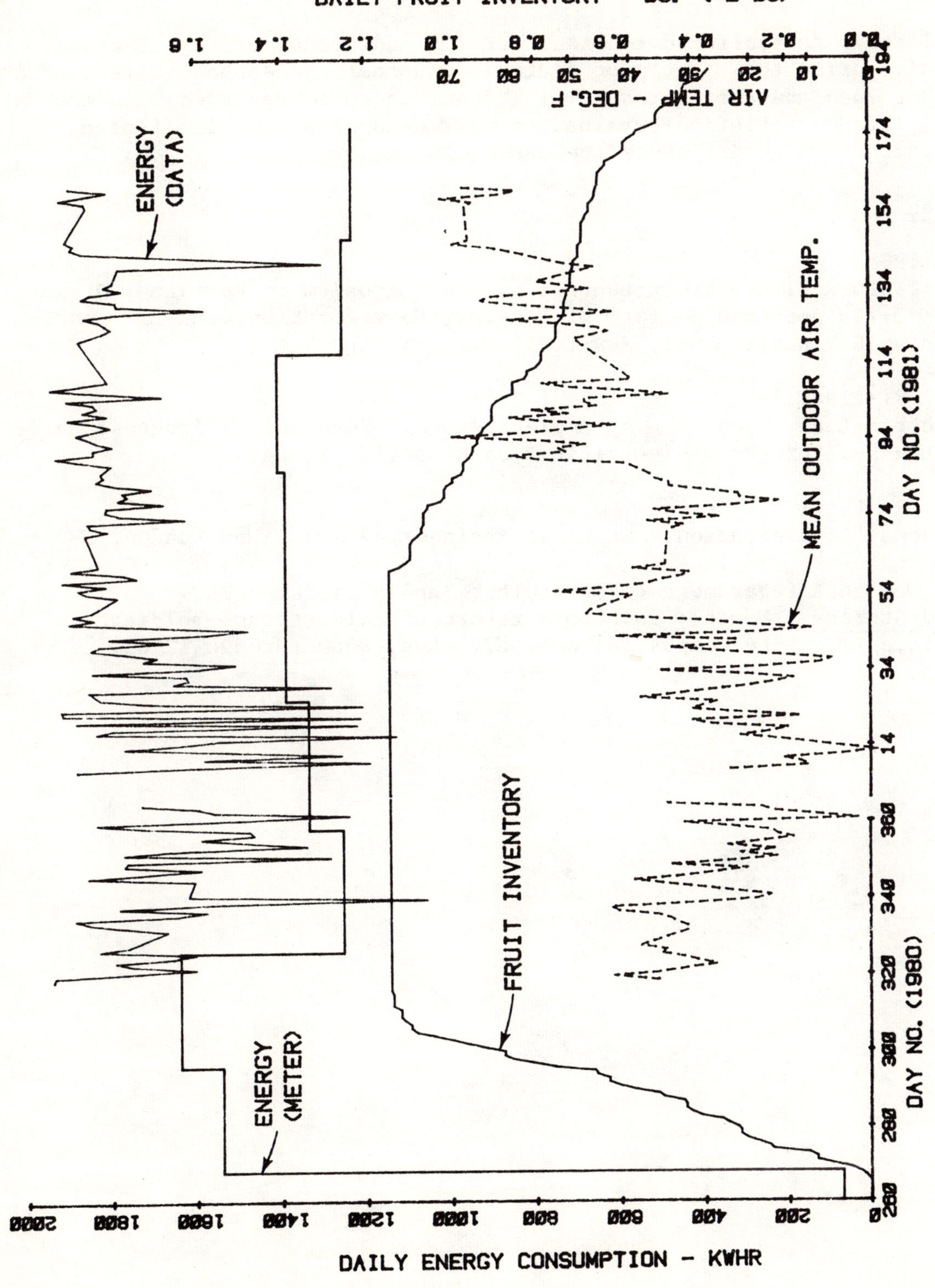

Figure 4. Total Energy Use and Fruit Inventory for 1980-81.

additional attention. In this particular case electrical demand charge reductions may be possible since this demand is based upon 20 minute duration spikes in electrical use.

This project is scheduled to continue for two additional storage seasons. During this period the effects of diurnal, seasonal and annual variations in the weather upon the energy budget will continue to be recorded. As the data base is enlarged statistical evaluation will determine the significance of various factors which influence the energy budget.

References

ASHRAE. 1965.
Fruit Storage. Papers presented at the Symposium in Portland, Oregon, July 5-7. American Society of Heating, Refrigeration, and Air-Conditioning Engineers, Inc., New York, NY. 25 pages.

Price, D. R. 1969.
Electric Load Study of C.A. Apple Storage. 26th Annual Progress Report to the New York Farm Electrification Council. 2 pages.

Stahlman, R. L. 1981.
Personal Communication. Stahlman Engineering Inc. New London, NH.

State of New York Department of Agriculture and Markets. 1980.
Cold Storage. Monthly inventory report of cold storage holdings. Building 8, State Campus, Albany, NY. Nos. 9-80 through 12-80. 2 pages each.

THE AUTOMATED CONTROL OF CONTROLLED ATMOSPHERE STORAGE CONDITIONS

J. Jameson
Fruit Storage Division
East Malling Research Station
Maidstone, England

It is generally accepted that restriction of O_2 supply to apples only starts to become effective in delaying senescence changes when the concentration is reduced below 5% and it is necessary to lower O_2 concentrations below 3% to obtain significant commercial benefits (Knee, 1980). A study of many store log books shows that, in practice, operators often fail to maintain conditions sufficiently close to those recommended. For example, where a storage regime requires 2% $\pm$ 0.25% O_2, log books often show readings ranging from 1.5-3.5% with a mean value which may be as much as 0.75% too high. As a consequence, a loss of quality is incurred which may severely restrict marketing flexibility. Moreover, as storage periods are extended and wage costs become higher there is increasing economic pressure to minimise the amount of labour required to operate CA stores.

Cox's Orange Pippin, the most important UK dessert variety, is normally stored for up to 6 months at an O_2 concentration of 2%. Recent research in England has shown that the storage life can be extended for at least a further four weeks if stored in an O_2 concentration of 1.25% (Sharples, 1981). In both regimes CO_2 was maintained below 1%. However, below 1.25% O_2, the variety may fail to develop its full aromatic qualities and, since in unfavourable circumstances the threshold for anaerobic respiration may be as high as 0.8% O_2, precise control of O_2 is essential.

Most of the problems of O_2 control can be resolved by automation which enables concentration to be maintained to better than $\pm$ 0.2% of the target figure with virtually no demands on the operator. However, when working within such precise specifications, regular and frequent measurement of store gas concentrations is essential and a sampling frequency of at least one reading per store per hour has been adopted for all commercial installations in the UK.

This paper describes the fundamental design of automatic control systems and gives details of the present extent and probable expansion of commercial CA fruit stores in the UK. It is based on experience obtained in designing and using laboratory scale control systems as well as from working closely with manufacturers in the design of systems for commercial fruit stores.

Which variables are controlled?

The three most important parameters involved in the CA storage of apples, temperature, O_2 and CO_2 are amenable to automatic control. Temperature is normally controlled by a thermostat and, other than the progressive replacement of the capillary type of thermostat by solid state electronic devices, there seems no reason to alter the mechanism of temperature control. At present, both O_2 and CO_2 are measured using complex electronic analysers which provide signals suitable for control purposes. The recommendation for late-stored Cox is to maintain CO_2 levels below 1% and for this no form of control is necessary. Whilst many operators maintain this low level of CO_2 by placing sacks of hydrated lime directly in the store, others remove CO_2 continuously by mechanical scrubbers. Although these latter devices are not immediately amenable to direct control they could be modified for this purpose if necessary. Hence, to date only automation of O_2 control has been installed on fruit stores in the UK and it is this aspect which will now be considered in detail.

Design criteria

The working principle on which all store control systems are based is that the O_2 depletion rate by the fruit is sufficiently predictable to enable a control system to be produced which will accommodate the variations in respiration rate and store leakage. In practice, the depletion rate may be found by observing the rate of O_2 utilisation in a sealed contained of fruit after allowing conditions to normalise at the intended storage condition. Alternatively, the rate can be calculated using published laboratory measurements of respiration rate (see Appendix 1). Both methods indicate that, for Cox stored at 3.4-4.0^{o}C in 1% O_2, the depletion rate varies between 0.05% and 0.07% per hour and observation has shown depletion rates of 0.1% per hour at 1.25% O_2. From this, a 100 tonne commercial Cox fruit store can be shown (Appendix 2) to have a theoretical air input requirement of approximately 600 to 800 $\ell\ h^{-1}$(17 to 23 $ft^3\ h^{-1}$). Depending on the type of control system, the time available for admitting air to the store can range from 10 minutes to 1 hour and thus the required pump capacity will range from 5000-600 $\ell\,h^{-1}$. In practice, pumps have been oversized by a factor of 2 or 3 to permit other varieties and storage regimes to be accommodated.

If stores are remote from the analyser, or a small pump is fitted, sampling times could take as long as 10 minutes thereby restricting the number of stores on any one control unit to only six. By reducing the sampling period more stores can be sampled in the same time with a consequent saving in investment on expensive instruments and control systems. Since, in a perfectly sealed store, O_2 depletion rates can be up to 0.1% per hour and the maximum tolerance for Cox in 1.25% O_2 is only $\pm$ 0.15%, it follows that readings should be obtained from each store once every hour.

When gas is drawn from a store to an analyser and then discharged to waste, it is replaced by an equal volume of air leaking into the store. In unfavourable conditions this volume of air could add more O_2 to the store than is required by the fruit. This situation can occur in a small 80 kg experimental cabinet but, of course, in a 100 tonne store, the O_2 required for respiration is such that this is not a problem. It should also be noted that, even in a small experimental cabinet, there are significant O_2 gradients for 10 to 15 minutes following the input of air, even with an air circulating fan running at about 20 store volumes per hour. Hence it is important to ensure that adequate mixing of the admitted air has taken place before re-sampling from the same store.

Types of system

Certain features are common to all types of automatic control system. In all cases, gas is taken from the selected store to the gas analysers and, after sufficient time has elapsed, the output from the analyser is compared with the reference level for that store. If the O_2 concentration is too low, air is admitted. Data recording is used with all systems. The general outline of such systems is shown in Figure 1.

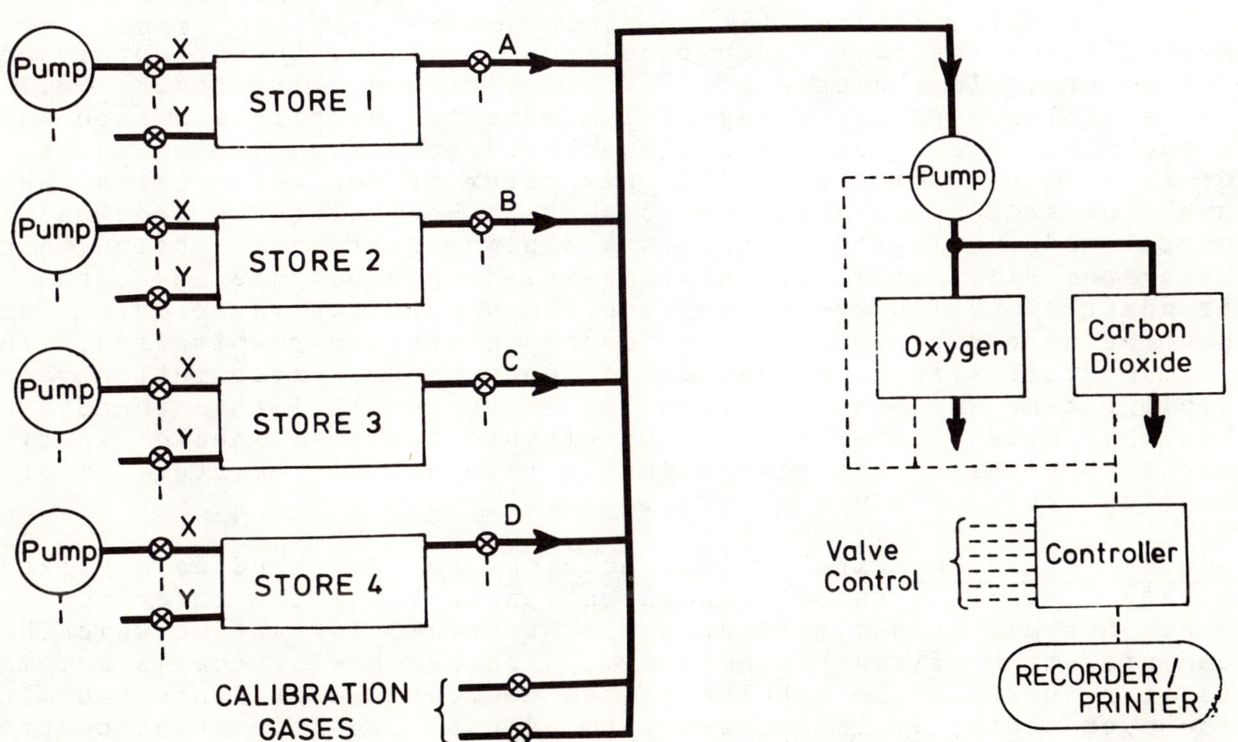

Fig. 1. Valves A, B, C or D are opened according to the store to be sampled (1, 2, 3 or 4) and a sample of store gas is pumped to the analysers. Once a representative sample has been obtained, the analyser output is compared to a reference value for that store.

Several modes of operation are possible, comparison of the measured and reference values may or may not be quantified and the consequences of that comparison may vary. In practice, four modes of operation have been tried in commercial stores and all have been made to work satisfactorily.

Mode 1. "Feed-back control": The feed-back control system is the most simple but least effective method. The controller selects the store to be sampled and gas is drawn from that store until the time runs out and the controller selects another store. Air is admitted to the store whenever the analyser output is less than the reference level. This was the type of control system used in 1966/67 on the semi-commercial scale ultra-low O_2 trial at East Malling (Sharples, 1981). There are three disadvantages of this type of system. Firstly, once air is admitted to the store, an imbalance is set up between the O_2 concentration in the gas within the bins and that surrounding them. This imbalance persists for 10-15 minutes after ventilation ceases and thus analyser readings during ventilation are optimistically high and consequently the ventilation process is often terminated too soon. Secondly, there is a finite period (known as the distance-velocity lag) involved in pumping a sample from the store to the analyser. Thus the O_2 analyser is always out of date in the information it receives so that, during ventilation, the analyser always reads low and, in waiting for the reference level to be matched at the analyser, the O_2 level in the store over-shoots the required concentration. In process control systems it is generally accepted that for stable performances the distance-velocity lag should not exceed 10% of the time to correct from the lower to the upper control points. In the case of a fruit store operating at 2% $\pm$ 0.1% with a distance-velocity lag of 1 minute the ventilation rate must be such that the O_2 concentration should move from 1.9% to 2.1% in not less than 10 minutes. This parameter effectively limits the number of stores per analyser to 4 or 5 when using the feed-back control mode. Finally, there is a minimum store size, below which the system will not work. Since sampling periods are long, the air admitted by leakage to replace the sample gas may provide more than the necessary input of O_2 to compensate for respiration. The minimum store size (see Appendix 3) depends upon respiration rate, sampling time and other factors but is typically 1 to 3 tonne and thus this mode of operation in unsuitable for small-scale experimental work unless the sample gas is returned to the store of origin (and other precautions are taken).

Mode 2. "Fixed package": This mode may be described as a "fixed package", since if the O_2 concentration is below the reference level then air is admitted at a flow rate and for a time which has been pre-set or fixed by the user. Although the system is automatic it requires a clear thinking and dedicated user since, to minimise over-shoot, it is necessary to adjust each store vent control specifically for each store and to modify the adjustment as the storage season progresses. However, the fixed package mode permits a large number of stores to be committed to one system, the only restriction is the number of valid samples which can be presented to the analyser within 1 hour.

Mode 3. Vent-Time Control: In this mode the ventilation time is made proportional to the "error"* (Bishop, 1981). Thus, when the O_2 concentration is too low, the error signal is used to determine the time for which air will be admitted. The advantage of this mode is its proportioning capability whereby small deviations from the reference value result in only a small quantity of air being admitted to the store, thereby avoiding over-shoot. The limitation on the number of stores is the same as for mode 2.

Mode 4. Vent-Rate Control: In this mode the ventilation rate is made proportional to the error. Thus, air is constantly pumped into the store and the deviation either upward or downward from the reference concentration is used to vary proportionally the rate of air admittance by slightly opening or closing a gate valve in the store-air supply line. This system, although expensive, provides exceptionally uniform store conditions as air reaches the store at approximately the correct rate, even when the store is not being sampled.

All four systems have been made to work successfully although in the future it is likely that most new installations will be based on proportional controllers (modes 3 and 4). The precise control obtainable by rate proportional control (mode 4) is at present unequalled but, by using a technique known as "pulse width modulation", equipment using time-proportional control (mode 3) could be enhanced to give a similar performance.

Types of controller

Whichever sampling and ventilation arrangements are used, it is necessary to have some form of controller to suppress the analyser output until the information is valid for the store being monitored. The analyser output is then compared with a reference relating specifically to that store and, if there is a significant error signal, corrective action is initiated. Controllers differ in the way in which the comparison is made. The simplest system comprises a chart recorded with one pen dedicated to each store and calibrated to read directly in percentage O_2. The reference is simply a movable buffer which can be locked in any position so that when the pen contacts the buffer a switch operates to initiate air admission. As this system makes no measurement of O_2 deficiency, it is suitable for use only in modes 1 and 2.

Alternatively a purpose built dedicated electronic controller can be used in which the reference is a variable voltage, set by turning a pointer across a scale calibrated in percentage O_2. By means of suitable electronics, this type of equipment may conveniently be used to correct O_2 deficiency by modes 1, 2 and 4, and permits corrective action in proportion to the O_2 deficiency.

---

*The term error is used to describe the difference between measured O_2 concentration and the reference level.

The third type is based on a micro-computer which controls the sampling of the gas from the store, the reading of the analyser and compares the value obtained with a pre-set reference entered via a key pad. The controller also initiates the corrective action and recording of the data. The micro-computer is able to work in any of the previously outlined modes and both of the systems installed in the UK work in mode 3 where correction is proportional in time to the O_2 deficiency.

There is virtually no limit to the degree of automation which is possible using a micro-computer and for small-scale laboratory systems it is probably the only economic way to provide precise control over 2 or more variables.

Instrumentation

The satisfactory functioning of any control system is entirely dependent on the accuracy and reliability of the information which is fed to it from the store thermometers, CO_2 and O_2 analysers. Although platinum resistance thermometers are undoubtedly the most accurate for measuring temperature, their use is declining because of expense, adverse effects of moisture, and errors due to cable resistance. More recently, thermistors have been installed in fruit stores in the UK since they can be readily encapsulated and standardised to provide accurate, tough, reliable and comparatively cheap devices ideally suited to the agricultural environment.

The traditional CO_2 analyser which measures changes in thermal conductivity is unsuitable for use with atmospheres containing variable concentrations of O_2. A practical alternative is the single-beam infra-red absorptiometer which is now available providing accuracies of better than 2% of full scale deflection. However, none of the control systems currently in use provide automatic control of CO_2.

With few exceptions the O_2 analysers found in fruit stores throughout the UK use a magno-dynamic cell to measure the paramagnetic susceptance of O_2. The accuracy, reliability and general ruggedness of this type of instrument is unequalled by other types. Even so, it is advisable to use an industrial quality temperature-regulated analyser for automatic control equipment to ensure adequate long term stability. Various types of O_2 analyser have become available in the last few years which utilise an oxy-voltaic cell and, although they are adequate for periodically confirming the readings obtained from the regular sampling system, those that have been tested in our laboratory lack the stability required for use as the "primary" measuring instrument.

Several options are available for presentation of the data. A micro-computer based system offers the greatest range of possibilities. It is possible to record all conditions, or to record only when outside the required limits. Arrangements can be made to trigger alarm signals when the limits are exceeded by a wider margin. Most micro-computer systems print data, but in other types the output is generally plotted on a chart: a micro-computer could equally well plot a chart if that was preferred.

Commercial CA control systems

The first automatic control system in the UK was installed on a fruit farm in Kent in 1978. Since then the number of installations increased until, by 1980, over 4000 tonnes of storage capacity was being controlled automatically (Table 1).

TABLE 1

Commercial systems (UK)

Year	Tonnes	Controller
1978	4 x 80	Dedicated electronics
1979	4 x 80	Electro-mechanical
1980	4 x 120	Micro-computer
1980	12 x 100	Micro-computer
1980	10 x 180	Dedicated electronics

Current indications suggest that an increasing number of stores will be equipped with automatic control systems over the next two years (Fig. 2).

Estimates for 1982 are based on serious enquiries to both East Malling and the manufacturers for large systems requiring more than one year's financial planning. In addition, there are several growers who are likely to install units of up to 600 tonnes so that the 1982 total will probably approach 16000 tonnes. By 1984 sufficient stores are likely to be equipped to operate ultra-low O_2 conditions to meet the market demand for late-stored Cox. Thereafter the installation of automatic control systems will tend to increase more slowly on stores operating at higher O_2 concentrations to improve the precision of control and to release staff for other farm and packhouse duties.

Current UK costs for a typical automatic O_2 control system complete with O_2 and CO_2 recording is in the range $24-28,000 (£12-14,000) depending on type. Such an installation will control 12 x 100 tonne stores which, if all the fruit realises a premium of just 1 cent per lb (½p sterling), the capital investment will be repaid in the first year.

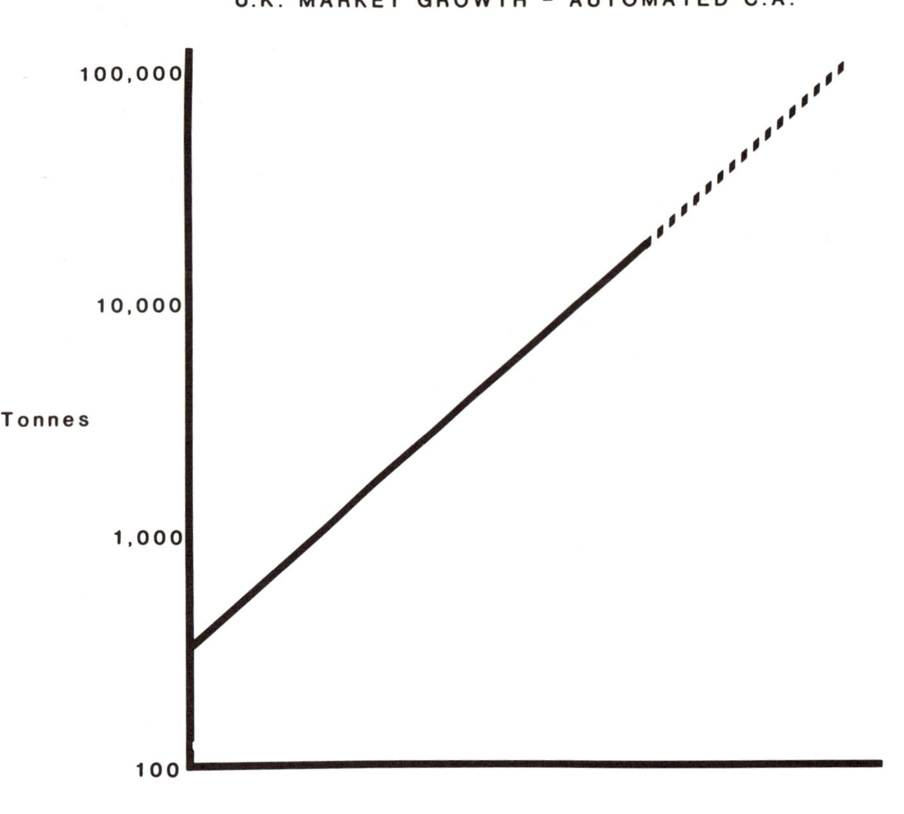

Fig. 2. The total installed capacity of automated fruit stores in the UK has increased logarithmically since 1978

Special considerations for commercial systems

Sampling lines often have to be outdoors and so every time that ambient temperatures drop below store temperature (eg 4°C) condensation may occur in the sampling line. It is therefore essential that lines run gently downhill with a suitable catch-pot inserted at the lowest point in the system so that condensate may be collected and drained off; if water is allowed to accumulate in the pipe it may eventually be pumped around the system until it reaches and seriously damages the O_2 analyser. The risk of the pipe becoming blocked by water is greatly reduced if pipe of 9 mm (3/8") o.d. is used, instead of the more usual 6 mm(¼" pipe).

Outside lines may also be subject to sub-zero temperatures and the sampling line may be sealed off by internal frosting. The problem can be circumvented by insulating the sampling pipes with closed-cell foam rubber. Furthermore, where low temperatures occur regularly pipes should be equipped with low rating heating tape which provides sufficient heat to keep the pipes above freezing point; the heaters are switched on by a thermostat as the ambient temperature approaches zero.

Store operators are strongly advised to cross check O_2 concentrations by drawing samples directly out of the store into a portable O_2 analyser. This provides an entirely independent check of the conditions being recorded by the "primary" instrument. They are also advised to remove samples of fruit at intervals through the hatch of the store to check its condition and, if necessary, send it to the Extension Service laboratory for a check on back-ground alcohol levels.

A great deal of work goes into ensuring that low O_2 stores are reasonably gas-tight and it is therefore essential to fit a purpose-made safety valve to protect the store from collapse or structural damage in the event of any malfunctioning of the valves in the sampling and control systems.

Future developments

The next development is likely to be the inclusion of automatic control of scrubbing for those stores with mechanical scrubbers, thus extending the usefulness of the equipment to cover varieties requiring higher CO_2 concentrations. It is also possible that control of humidity and ethylene levels may be added at some future date as may the on-line measurement of background alcohol vapour concentrations in the store atmosphere. However, it is likely that these developments will first take place at a Research Institute. The techniques involved, along with those problems associated with automatic control systems for small-scale experimental work are a subject in their own right.

Acknowledgements

I would like to thank Mr D J Chappell and Mr P Gull, both of East Malling Research Station, for their interest in this project, and also to the pioneering growers who first installed automatic CA control systems on their fruit stores.

References

Bishop, D.J. The application of microprocessors to controlled atmosphere fruit storage. Proc. Agric. Eng. (1981), 36, 51-54.

Knee, M. Physiological responses of apple fruits to oxygen concentrations. Ann. appl. Biol. (1980), 96, 243-253.

Sharples, R.O. Effects of ultra-low oxygen conditions on the storage quality of English Cox's Orange Pippin apples. Proceedings of 1981 National C.A. Research Conference, Oregon State Univ. Corvallis, Oregon, USA,

Thorne, S. and La Laguna, F. (1981). Private communication.

Appendix 1

Oxygen depletion rate

Using as a basis Cox apples stored at 3.5-4.0°C in 1% O_2:

Respiration rate = 15 - 20 mℓ 10 kg^{-1} h^{-1} RQ = 1.25 (Thorne and La Laguna, 1981)

∴ O_2 utilisation = 12 - 16 mℓ 10 kg^{-1} h^{-1}

Taking density as 0.9 g mℓ$^{-1}$ vol. of 10 kg = 11.1ℓ

In practice, apples occupy approx. 33% of enclosed store volume

∴ 10 kg fruit require store vol. of 33.3 ℓ and free space = 22.2ℓ

$$O_2 \text{ depletion rate} = \frac{O_2 \text{ utilisation rate}}{\text{store free space}} = 0.054 - 0.072 \text{ per h.}$$

Appendix 2

Air admittance rate

Assume fruit has O_2 utilisation rate = 12 - 16 mℓ 10 kg^{-1} h^{-1}

& for 100 tonne store, enclosed vol. = 340 m^3 (12,000 ft^3)

Fruit displacement = 33% ∴ free volume = 227 m^3 (8000 ft^3)

O_2 requirement = free vol. x O_2 depletion rate = 123 - 163 ℓ h^{-1}

∴ requirement = 5 x oxygen requirement = 615 - 815 ℓ h^{-1}

Appendix 3

Minimum store size - Mode 1 controller

Assume air requirement as 6.15 to 8.2 ℓ tonne^{-1} h^{-1}

& assume gas sampling pipe is 4.5 mm bore x 50 m long.

Vol. = 0.795 ℓ^3

If sampling continues for 10 min h^{-1}, vol. displaced = 7.95 ℓ h^{-1}

$$\text{Minimum store size} = \frac{\text{displaced volume per hour}}{\text{air requirement}} = \underline{\underline{1.3 \text{ tonnes}}}$$

Allowing for store leakage, suggested min. store size = 2.5 tonnes.

Relative Humidity in CA Apple Storage Rooms

G. D. Blanpied
Pomology Department
Cornell University
Ithaca, New York

For many years I unquestioningly accepted the recommendation of 90-95 percent relative humidity for the storage of apples. This recommendation was based on the observations that very high relative humidity resulted in free water collecting on the surfaces of apples. The free water increased decay and caused apples to split open, particularly Golden Delicious. Furthermore, it was observed that humidity levels below 85 percent could have no beneficial effects and caused skin shrivel to develop during storage. Subsequently I have learned the recommendation and corroborative observations may not be valid under all conditions. For example, our dew-point measurements in a modern storage facility equipped with modulating EPR valves indicated a constant relative humidity of 98 percent in the sealed CA rooms. When the CA rooms were opened we saw no condensed water on the apples, decay was not abnormally high, and there was no evidence of Golden Delicious skin splitting in the rooms. Further, our storage studies confirm observations in other countries that lowering the humidity to about 80 percent may result in significant reductions of storage breakdown in the apples.

Interest in the responses to relative humidity were heightened in 1978-79 when we observed significant reductions of senescent (McIntosh breakdown) and low temperature breakdown (brown core) in McIntosh stored in air at 80 percent relative humidity (Table 1). These observations were confirmed in two subsequent seasons.

Table 1. Relationship between McIntosh weight loss and breakdown in air storage at 32-34°F.

	Crop year					
	1978		1979		1980	
Weight loss (%)	0.0a	3.9b	0.3a	2.7b	0.1a	3.8b
Senescent breakdown (%)	23b	2a	27b	8a	27b	16a
Low temp. breakdown (%)	71b	45a	0a	0a	61b	34a

In several other tests prior to 1980 we reduced incidences of senescent breakdown in McIntosh with prestorage calcium chloride dip treatments. The two types of treatment were combined in 1980 (Table 2). When apples were not treated with three percent calcium chloride significant reductions in senescent and low temperature breakdown were associated with increased weight loss caused by low relative humidity during storage. When apples were treated with calcium chloride before storage, the weight loss had no significant effect on breakdown. The data indicated the calcium chloride dip treatment could be substituted for low relative humidity in storage. That experiment diminished our interest in reduced levels of relative humidity in storage.

Table 2. **Weight loss and storage breakdown of McIntosh apples.** 1980-81.

Prestorage treatment	Weight loss during storage (%)	Senescent breakdown (%)	Low temp. breakdown (%)
None	0.1a	27b	61b
None	3.8b	16a	34a
3% $CaCl_2$ dip	0.1a	6a	47a
3% $CaCl_2$ dip	4.6c	7a	41a

New York and New England CA operators have historically used brine-spray defrost evaporators because most of the early refrigeration equipment was installed by an engineer who developed a carbon dioxide scrubbing system that used the brine-spray evaporators for carbon dioxide absorption. Many of these storage operators associated continuous pumping of water over the coil with high relative humidities and were therefore reluctant to replace rusted-out evaporators with dry coil units. They wanted to eliminate the salt and its corrosion of equipment, but feared their apples would shrivel. One of my extension activities during the past two winters was a survey study to determine if their fears were justified.

The survey study included 40 sealed CA rooms located at 24 storage establishments in New York and Vermont. Twenty were 37^oF McIntosh rooms and twenty were 32^oF rooms for other varieties. There were ten wet and ten dry coil rooms in each of the two temperature groups. A General Eastern dew-point meter and a calibrated reference thermometer were used to determine the relative humidity inside the CA rooms. An electric heating tape prevented condensation inside the plastic tube that carried the CA atmosphere from the CA room to the dew-point meter. The plastic tube and reference thermometer were inserted through holes in a Styrofoam plug that was placed into the porthole of the CA doors. Dew-point and temperature readings were recorded when the refrigerator turned "on" and when it turned "off". The relative humidity for the room was the average for these two sets of readings.

Table 3. Temperature, dew point, and relative humidity in three selected sealed CA rooms.

CA Room	End of refrig. cycle	Temp. (^oF) Air	Dew Point	Relative humidity (%)
A	off	33.0	30.5	91
	on	31.0	28.4	94
B	off	32.8	30.8	93
	on	31.4	29.4	93
C	off	33.7	31.2	91
	on	33.0	30.0	89

In a few rooms the relative humidity was highest at the end of the "on" refrigerations cycle (room A in Table 3). In a few other rooms the relative humidity held constant (room B in Table 3). Data for most rooms were similar to those shown for room C in Table 3: the humidity was highest when the refrigeration turned "on" and lowest when it turned "off".

Table 4. Swing of air temperature, dew point, and relative humidity. Survey study - 1979 to 1981.

	Temperature (^oF) Air	Dew point	Relative humidity (%)
Average swing	0.8	0.9	2
Range	0.1-3.3	0.0-4.5	0 - 5

The average and range of the swings in air temperature, dew point, and relative humidity are shown in Table 4. The average swing of air temperature was only 0.8^o ($\pm$ 0.4^o), and of relative humidity was only 2% ($\pm$ 1%).

One limitation of the survey study was that all data were collected at a location about 30 cm inside the sealed CA door. It should be noted in defense of this procedure that there was usually good air circulation at this location because the evaporators were almost always located above or adjacent to the CA doors. However, data gathered in an opened, 80 percent filled CA room indicated there are small variations in relative humidity within a storage room (Table 5).

Table 5. Within room variations of relative humidity in an opened, 80% filled CA room.

Location of measurement	Temperature (°F) Air	Temperature (°F) Dew point	Relative humidity (%)
Blower intake (refrig. on)	33.3	29.1	85
Blower exhaust (refrig. on)	30.5	27.8	90
Pallet opening at aisle			
Slow moving air	35.0	30.1	82
Fast moving air	33.8	29.3	83
Center of room	33.4	29.7	86
Center of filled bin	33.3	30.3	89

The results of the survey are presented graphically in Fig. 1. There was a considerable spread in relative humidity within each of the four groups of rooms. The high relative humidity readings in each group usually occurred in rooms equipped with modulating EPR valves. Statistical analyses of the data (Table 6) indicated the following significant differences in relative humidity: it was higher in the 37° dry coil rooms than in the 37° and 32° wet coil rooms; it was higher in the 32° dry coil rooms than in the 37° wet coil rooms; the average relative humidity in all dry coil rooms was higher than the average relative humidity in all wet coil rooms.

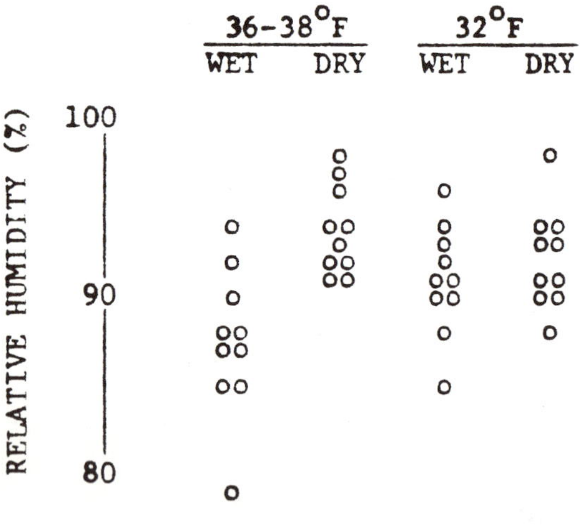

36-38°F WET DRY
32°F WET DRY
RELATIVE HUMIDITY (%)
100
90
80

Figure 1. Relative humidity in 40 sealed CA rooms at 24 storage locations.

Table 6. Average relative humidity in 40 sealed CA rooms. 1979-81.

Type of Evap. coil	CA room temperature 37°	32°	Average
Wet	88a	91b	89y
Dry	94c	92bc	93z
Average	91q	92q	

Carbon dioxide concentrations were controlled by water scrubbing in wet coil rooms. Either lime or carbon scrubbers were used to control carbon dioxide concentrations in the dry coil rooms. Average relative humidities in lime and carbon scrubbed rooms were not significantly different (Table 7). The system for defrosting dry coils did not appear to influence the relative humidity in the dry coil rooms, either.

Table 7. Relative humidity in study survey, dry coil CA rooms.

	Number of rooms	Average relative humidity (%)
Rooms with lime scrubbers	14	92a
Rooms with carbon scrubbers	6	95a
Rooms with electric or hot gas defrost	13	93a
Rooms with water defrost	7	93a

Relative humidity measurements were made in an air apple storage room equipped with a two-speed motor on the evaporator fan. At the slow fan speed the split was greater, but the average relative humidity was not affected (Table 8, upper). Relative humidity was monitored during three 8-hour test runs at slow and at fast fan speeds. Each test run was preceeded by a 48-hour equilibration period at the fan speed to be used in the subsequent test run. The average relative humidity at the slow and fast fan speeds were not significantly different during the 8-hour test runs (Table 8, lower).

Table 8. Relationship between evaporator fan CFM and relative humidity.

Measurement	Evaporator fan CFM	
	18,700	10,800
Air temperature at evaporator intake (°F)	31.3	31.5
Relative humidity at evaporator intake (%)	90	89
Temperature at evaporator discharge (°F)	29.7	27.0
Relative humidity at evaporator discharge (%)	94	95
Average relative humidity	92	92
(average conditions during 8-hour test runs)		
Evaporator coil temperatures (°F)	16a	13a
Hours refrigeration was "on"	3.7a	2.9a
Average relative humidity in center of room (%)	90a	91a

While collecting data for the study survey at one storage establishment, it was observed that adjacent rooms with different operating back pressures had similar levels of relative humidity. The refrigeration "on" periods in the low back pressure room were noticably shorter than the refrigeration "on" periods in the high back pressure room. The differences in the lengths of the refrigeration "on" periods could have been caused by differences in thermostat sensitivity. However, if the thermostats were equally sensitive, one should then ask if a cold coil, which is "on" for fewer hours each day, will remove more water or the same amount of water as a warm coil, which will be "on" for more hours each day? We tried to collect real data to answer this question but found that two evaporators and controlling thermostats that were probably similar when they were installed in adjacent CA rooms in 1954, did not perform comparably in 1981. The question was therefore answered by modeling. Under the conditions set into the model, significantly more water was removed by the colder coil, even though its "on" period per 24 hours was considerably less than the "on" period of the warmer coil (Table 9).

Table 9. Calculated water removal from a 15,000 bushel capacity CA room equipped with a 25,000 CFM - 11,200 BTU/°TD evaporator. Outside temperature 50°F.

Coil temp.	Evaporator				Refrig. "on" (24 hr. period)	Water removed from air (24 hr. period)
	Intake		Outlet			
	Air	Dew-point	Air	Dew-Point		
°F	°F	°F	°F[1/]	°F[1/]	hours[1,2/]	pounds[3/]
20	32.0	31.1	29.5	28.4	4.30	205
12	32.0	31.1	27.0	25.7	2.65	232

1/Information provided by Krack Corp., Addison, Ill.
2/Hours required to remove sensible and latent heat load, assuming continuous delivery of 25,000 CFM during 16/24 hours.
3/Calculated by J. A. Bartsch, Agric. Eng., Cornell Univ.

The calculated water removal (24.6 gallons per day at 20° coil temperature) appeared to be rather high. This amount of water loss from the room atmosphere would result in a 6.6 percent weight loss by the apples during a 200 day storage period. However, during the winter months, when outside temperatures are below 50°, the refrigeration "on" period would be shorter, resulting in less water loss. The model, therefore, appears to be valid.

CONTROLLED ATMOSPHERE STORAGE DEVELOPMENT IN SOUTH AFRICA

by (a) L Ginsburg, Stellenbosch, S A;
(b) P Worthington Smith, Cape Town, S A;
(c) A B Truter, Fruit and Fruit Technology Research Institute, Stellenbosch, S A.

Controlled Atmosphere (CA) Storage was introduced to South Africa in 1934. Two such stores were built during that year. They are still in use today but as Regular Atmosphere Stores. Their use as CA stores was short lived. Retention of gas tightness of the storage spaces and the use of concentrated sodium hydroxide solution scrubbers were the two major reasons for the failure of these stores. Regular Atmosphere remained the only system used for the cold storage of Apples until the late 1970's.

Why did South Africa not again build CA stores when gas tightness problems were overcome and simplistic gas scrubbing methods were available? The answer is that after the 1939-45 war, South Africa realised the potential market for fresh Southern Hemisphere apples in England and Europe during their winters.

The apple exports given in 1 000's of cartons (18, 2 kg per carton) for 1950, 1960, 1970 and 1980 are - 204; 2 499; 6 482; 9 882 cartons respectively. A tenfold increase took place from 1950 to 1960 and a further triple increase from 1960 to 1980.

Prices were lucrative and this led to greatly increased plantings of apples. Well designed RA stores capable of rapid precooling with effective air movement for holding of fruit at a RH of 90% permitted storage of from 4 to 8 months depending on cultivars. These stores were also used for inland retention of fruit destined for export. This catered well for apple marketing in South Africa until the early 1970's.

Three important factors which changed the outlook on storage of apples in South Africa were as follow -

(i) the advent of the European Common Market with it's high import levies and the threat that import quotas may be introduced.

(ii) The rapid rise in fuel costs and the consequent big rise in freight rates has lowered the profitability of fruit sold on the export markets.

(iii) The development of an all-year-round market for high quality apples and pears in South Africa.

The need for CA storage is not only confined to some fruits. If adapted to sea transport in one form or another, it could play an important role in the export and local marketing of avocados, mangoes, capsicums and other fruit and vegetables. The more perishable the crop, the greater the need for modified gas holding systems during transport.

CA STORAGE RESEARCH IN SOUTH AFRICA:

During the 1960's:

Research at the Fruit and Fruit Technology Research Institute showed that CA-stored Golden Delicious and Starking Apples retained their quality much better than when they were held in RA cold stores[(1)]. The results are given in Table 1 (Addendum 1). The most striking difference was the better acid content retention for Golden Delicious Apples viz 0,24% for CA and 0, 12% for RA. Both an expert and a mass Taste Panel found a highly significant preference for CA-stored Goldens compared to the R A fruit but they were unable to detect a dessert quality difference in Starking Apples irrespective of storage technique.

Decay control was much better for CA than RA stored fruit, especially after extended storage life at ambient temperature. The differences on removal from 7½ months storage were small viz 0,9 and 0,3% for RA and CA Golden Delicious apples and 4,1% (RA) and 0,9% (CA) for Starking apples. The decay after 22 days at ambient temperature was 15,2 (RA) and 4,2 (CA) for Golden Delicious and 11,4 (RA) and 2,9 (CA) for Starking[2].

The results for Granny Smith apples were also in favour of CA. A very good retention of firmness and juiciness after 10 months' storage was the most striking benefit gained by this cultivar when stored under CA conditions.

The gas compositions recommended were 3,0 (O_2) and 5,0 (CO_2) for Golden Delicious, 3,0 (O_2) and 3,0 (CO_2) for Starking, and 3,0 (O_2) and 0,0 (CO_2) for Granny Smith apples.

C A Storage Research 1970-1981:

The importance of CA storage for the continued health of the South African apple industry was realised during the early 1970's. This brought with it the need for further CA research under South African conditions. Research is being conducted both by the FFTRI and technical personnel attached to private and cooperative apple organisations. Maximum cooperation exists between the different research groups.

Picking stage of Maturity:

The protracted blossoming season experienced in South Africa makes the determination of an optimum apple-picking stage a very complex problem. A certain variation in maturity stage of apples harvested for storage must be lived with.

Use of %TSS, iodine test, acidity, firmness, and colour remain the most important indices. The relative importance of these indices change according to cultivar. One large apple-packing and storage cooperative has found e g that the I_2 test works well for Starking apples while colour is the more important factor for Golden Delicious apples. Goldens at the silver stage are ideal for long-term CA storage.

Use of initial ethylene development has under South African conditions to date proved disappointing. High values and big variations are found in specific samples.

Days from full bloom is used as a dominant maturity index determinant by some growers while others regard it as a good check against accepted indices. Granny Smith is an exception. Here 178 days from full bloom is found to be a reliable index.

CA Gas Consistencies:

Work at the FFTRI has shown that the following O_2 and CO_2 concentrations are optimum for long term storage:

3:3 for Golden Delicious and Starking apples and Packhams Triumph pears.

3:0 for Granny Smith apples.

The use of ultra low i e 0,5% O_2 for the first 30 days storage and zero initial O_2, 1,0 to 1.5% for the first 3 months storage have not yet been looked at. Commercial experience tends to show that O_2 values below 2,5% for Granny Smith apples may result in alcoholic flavours. This shows that a high risk factor may be faced when using O_2 from 1,0 to 1,5% for certain cultivars.

Core flush:

Core flush development varies from season to season and is regarded as a serious disorder during the storage of Granny Smith apples.

The FFTRI have shown that an increased incidence of core flush develops as the CO_2 concentration in CA stores holding Granny Smith apples rises from 1,0% to 4,0%.

Investigation at the Elgin Fruit Growers Cooperative [2] have established the following:

Core flush is more prevalent in RA than in CA- stored apples.
The incidence and severity increases rapidly once the cold stored apples are held at ambient temperature for a few days.
Post harvest delays from 48 to 96 hours before precooling reduces the core flush noticeably. The longer the delay the less the core flush.
Intermittent warming after 1, 3 and 5 months CA storage is the most effective core flush control. It also tends to suggest a link in the nature of core flush and low temperature break down. Both respond positively to intermittent warming.
Delays and intermittent warming are both done at the expense of a lowering of textural firmness which means compromise values must be determined.
Commercial experience has shown that the RA accepted optimum storage temperature of -0,5^{o}C encourages core flush development. Raising the temperature by 1,0^{o}C for CA results in lowering of core flush.

Ethylene in CA Stores:

Ethylene build-up in CA stores during the 1981 season reached approximately 1 000 ppm within 3 to 4 months storage. Scrubbing by means of Ethysorb filters has been applied in a commercial store to study the effects of lowering the ethylene concentration on the fruit quality.

BON CHRETIEN CA STORAGE:

A weakening market for canners' fruit may call for extended storage life for Bon Chretien pears.

Bon Chretien (BC) pears were stored for 68 days at -0,5^{o}C in 3,0% O_2 and 0,0% CO_2 and compared to similar fruit stored at -0,5^{o}C and in regular atmosphere storage. Senescent scald exceeded 5,0% for the RA stored BC pears while only a trace was present on the CA pears. Core breakdown was above 10,0% for RA stored pears compared to less than 2,0% for CA. This result is particularly promising for 1981 which was an abnormal year with heavy unseasonal rainfall and cool weather during the fruit development stage.

Utilisation of CA stores for BC pears until mid-April and then for Granny Smith apples will result in all the year utilisation for such CA stores. It will also provide the market with high quality out of season BC pears.

AVOCADO STORAGE TRIALS:

Avocados stored at 5,5^{o}C for 23 days with the O_2 maintenance at 2,0%, 5,0% and 8,0% and the CO_2 at 10,0% for all the samples were compared with RA stored avocadoes held at 5,5% (FFTRI tests).

The CA-stored samples were all superior to those held in CA stores. The best gas consistency was 2,0% of O_2, 10% CO_2 and 88,0% N_2. This treatment provided a 10 days shelf life at ambient temperature compared with 4 days for the RA stored samples.

The CA stored Avocados when ripened were free from pulp spotting and lead discoloration compared to a high incidence of these disorders found in the RA samples. This tends to suggest that CA storage retards cell wall deterioration as there are indications that pulp spot development is linked with thinning of the cell wall(3).

MODIFIED ATMOSPHERE SEA SHIPMENT AVOCADO TEST:

A Transfresh test was decided on in 1980 with a view to improving the avocado outturn condition on the European Markets.

The Transfresh system was applied to a container carried on deck and cooled by means of a clip-on unit. The average fruit temperature throughout the voyage was 3,0^{o}C above the recommended carrying temperature of 5,5°C. The gas consistency at the start of the voyage was 5,0% O_2, 7,0% CO_2 and 8.0% CO. Ventilation had to be applied frequently to maintain the O_2 level above 1,5%. The CO_2 concentration remained relatively constant for the first 5 days at 5,0% and then rose steadily to a value of 14,0%. The CO_2 content could be rectified easily by increasing the dry lime charge. The CO concentration dropped to 4,0% after 10 days and finished at 2,0%.

The outturn was promising despite the relatively high and fluctuating shipping temperature and varying gas concentrations used. An average 5 to 6 days extra shelf life was found for the Transfresh shipped samples on arrival in Paris compared with the avocados carried at 5,5°C under RA contitions in the refrigeration bay of the container ship.

CONSTRUCTION OF CONTROLLED ATMOSPHERE STORES IN SOUTH AFRICA:

Construction:
Controlled Atmosphere stores in South Africa are either new constructions or conversions of large RA to CA stores.

New cold stores are generally constructed with a steel frame building to which are bolted sandwich panels consisting of a core of expanded polystyrene insulation between flat galvanising metal sheets. These panels have tongue and groove joints along the long length of panels. These are sealed with a sealant and the end joints which butt are secured to the steel frame building. Once the insulation has been thus effected, all joints are taped using a bandage with a stretch, only in the width and a taint-free paint with elastic properties when dry.

Portions of existing large RA stores are being converted to CA stores. They generally end up as a type of jacketed store. The internal jacket is made of wood with the inside surface coated with paint as described earlier.

Both systems are easy to seal and operate very satisfactorily.

Refrigeration and Air circulation:
The coolers and air circulation are so arranged (see Fig 1 Addendum 2) that the best possible use is made of the internal volume of the room as well as ensuring very even temperature distribution throughout the stack. The arrangement is also aimed at producing very high relative humidities.

The air is taken from the top of the fruit stack immediately under the roof of the cold room and is drawn to the coolers by the fan on the air-entering side. The air, having passed through the cooler, is delivered into a smaller plenum from where it is forced to enter the fruit stack by a curtain hanging from below the cooler part way down the front of the stack. Most of the sensible heat gains such as through the roof insulation and from fan motors, enters the air stream before the air is refrigerated in the coolers thus ensuring a high relative humidity of the air leaving the coolers. The air is then immediately fed to the fruit stack before picking up heat from insulation etc and returning to the coolers.

In general Ammonia is used as the refrigerant and the coolers themselves are flooded by means of pumped recircualtion. Temperature control is achieved by sensing the air leaving the cooler and controlling a modulating back pressure regulator to provide the precise amount of refrigeration on a continuous basis to keep temperatures absolutely even.

This system optimises relative humidity. The air temperatures leaving the coolers are recorded continuously.

The refrigeration system applied to the cold room is designed to do a small amount of cooling but mainly to maintain temperatures. In most South African installations the apples and pears are precooled in other stores prior to them being transferred to the CA storage room. This system results in storage complex in South Africa is designed for precooling to be done in the CA store.

Scrubbing:

The three types of scrubbers used in South Africa at present are as follows:

(i) Lime Scrubbers:

Lime scrubbing is used in stores where the CO_2 must be kept as near to 0,0% as possible as is required for Granny Smith apples. The general practice in South Africa is to introduce the bags of lime in bins into the cold room. A ratio of approximately one 20 kg bag of lime is used per bulk bin of fruit. This charge of lime has been found to keep the CO_2 level at a trace for at least 6 months storage. Thought is presently being given to using lime in an external scrubber outift. The spent lime is used on orchards which lowers the overall cost of scrubbing and also overcomes the disposal problem.

(ii) Activated carbon Scrubbers:

Activated carbon scrubbers imported from Australia are currently used in most South African CA cold stores. They are readily able to reduce the CO_2 concentration to just below 2,0%. Lime must be introduced when lower CO_2 levels are required.

(iii) Water Scrubbers:

Two CA storage plants have water scrubbers in addition to activated carbon scrubbers. This type of scrubber is a very simple design and reliable. With further development it is possible that they may come into more general use. There are, however, no proprietary makes of water scrubbers available and are tailor made for given installations.

CA Atmosphere Generators:

Most of the South African CA atmosphere generators are imported from Australia and are proving to be reliable.

Gas Analysis:

Gas analysis was initially done by means of the Orsat analyser. Infra red analyses are now generally used for CO_2 determination and para-magnetic analyses for O_2. They are either manually operated or automatically recorded.

Growth rate of CA storage Space:
All South African apples were held in RA cold stores until 1977. A growth rate from nil to 5,0% took place from 1978 to 1981. A further growth to 6,5% for CA stores is projected for 1982. A growth rate of 15% was projected for the 1980's[4]. The present trends indicate that 20,0% of all apples stored is South Africa may be held in CA stores by 1990.

LITERATURE CITED:

1. Ginsburg L, Visagie T R and Truter A B, 1969. Controlled Atmosphere storage in the Apple Industry in the Republic of South Africa. Deciduous Fruit Grower 19: 132-139.
2. Dalton R and Nel P J, 1981. Personal Communication. Elgin Fruit Packers Cooperative.
3. Kotze J M, 1981. Personal Communication. Pretoria University, South Africa.
4. Ginsburg L and Eksteen G J, 1980. Trends in Apple and Pear Storage in South Africa. Proceedings FRIGAIR 1980.

ADDENDUM 1

TABLE 1:

STORAGE OF GOLDEN DELICIOUS AND STARKING APPLES DURING 1969 IN RA AND CA STORES FOR 7½ MONTHS AT -0,5°C PLUS 2 DAYS AT 4°C:

DIFFERENT INDICES	CULTIVARS AND TYPE OF STORAGE			
	GOLDEN DELIC.		STARKING	
	RA	CA	RA	CA
Pressure (16)	8,7	11,0	10,8	12,0
Total Soluble Solids %	11,0	11,5	13,0	13,5
Acidity %	0,12	0,24	0,15	0,20
ph Index	4,5	4,0	4,3	4,1
*Colour Score Index	4,4	5,5	5,0	6,5

* A score of 10 for colour represents maximum colour score.

ADDENDUM 2

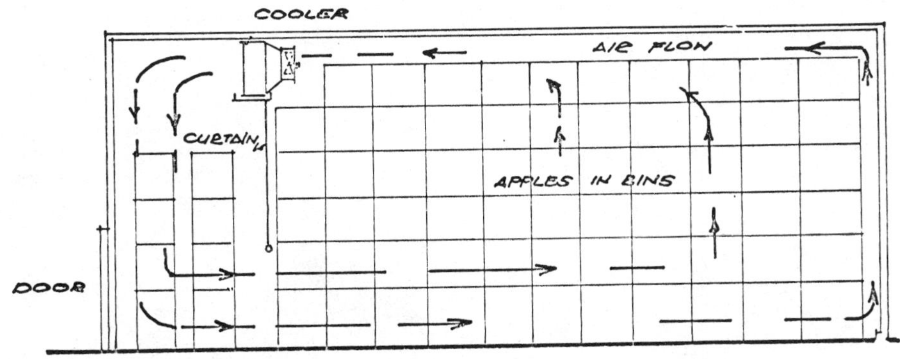

SECTION THRO C.A. CHAMBER

CA STORAGE STRUCTURES

Donald H. Dewey
Department of Horticulture
Michigan State University
East Lansing, MI 48824

Horticulturists with responsibilities for the commercial application of controlled atmospheres frequently are confronted with requests for information on the design and construction of suitable facilites. Although many of us at one time or another have had cooperators with suitable training, interest, and experience, changing times, other responsibilities and new assignments constantly divert engineers elsewhere so that we find ourselves again and again dealing with the same old problem of how to build an air-tight structure. We remain envolved with the problem because we deal continually with the commodity requirement of a specified low and regulated level of oxygen that is essential for the successful operation of a CA storage. Often the air tightness problem is the major concern in planning a storage since the options are neither unlimited nor infalible. Most storage organizations cannot afford and have no desire to develop new methods, therefore commercial firms or publicly supported organizations are called upon to devise, test and implement new ideas and concepts for air seals.

Structures

Many of the CA structures in the United States and Canada utilize load bearing walls, whereas abroad curtain or non-load bearing walls are common. The latter type of structure seemingly offers more options and greater freedoms for selecting the type of insulation and air seals that can be utilized. The post-frame structure, which is now very popular in Michigan, is of this type even though less obviously so since the framework of the building is hidden within the cladding, whereas often the structural framing is exposed in overseas buildings. Ventilated attics are typical for such structures, whereas buildings with load-bearing walls frequently have flat roofs with the insulation and airseals integrated with the roof-ceiling structure. In both types, the insulation and airseal are integral parts of the walls and ceiling.

Jacketed and modified jacketed storages, the latter of which have a limited amount of internal refrigeration, are utilized to a limited extent. Several in use in South Africa are constructed somewhat similar to the plywood jacketed storage described by Phillips et al. (4) for Canada. The so-called tents of plastic film, erected within refrigerated storages, for enclosing the fruit with or without air circulation within the enclosure (2) fall into the category of jacketed rooms. They are used in South Africa and Australia to a limited extent. Major shortcomings of jacketed storages include low cooling capacity, relatively poor air circulation and moisture condensation on the inside surface of the ceiling. For tents, there is the additional nuisance factor of handling and closing the tent at a time when operators are extremely busy with other storage requirements.

Air Seals

Sheet metal room linings, as notes by Dalyrymple (1) in his report on the development of CA storage, was the early traditional means for providing an air seal. Steel was employed as early as 1865 for a commercial structure in the United States and as an unique exterior shell for the first commercial CA storage in 1935 at Elgin, South Africa. In addition to sheet metal linings, Smock and Blanpied (6), Sive (5), and Hunter (3) reported at previous CA research conferences that commonly used materials for air seals included mastics, laminated synthetic films, plywood, prefabricated panels and sprayed-on polyurethane. There seems to be no other system to add to the list even though the process of trial and error with new ideas and materials has continued. At home and abroad, the methods employed are determined by factors that range from economics through engineering design and application to personal desires. There is no single best method; each has its own weaknesses and faults with the key to success being proper engineering and construction. Probably the lack of adequate attention to the meticulous detail required for the construction and application of the air tight material is the greatest limitation. General contractors often lack insight as to the quality and degree of tightness needed. Perhaps we should convey to them the need for a "hermetic" rather than an "airtight" structure. Unfortunately, there are no scientifically developed standards of airtightness that serve as a basis for the design and construction and application of the air tight material is the greatest limitation. General contractors often lack insight as to the quality and degree of tightness needed. Perhaps we should convey to them the need for a "hermetic" rather than an "airtight" structure. Unfortunately, there are no scientifically developed stanards of airtightness that serve as a basis for the design and construction of CA buildings.

"Inside-Outside" Method

A construction system widely used in Ontario and Michigan is the "inside-outside" method of construction in which sprayed-on polyurethane and plywood sheeting are utilized in a post-frame wooden building to provide an airseal in combination with good insulation adequately protected by a suitable interior thermobarrier (Stone, 7). As shown in the partial wall section, Fig. 1, the polyurethane is applied to the outside of the interior plywood lining which has been previously nailed to the girt. The exterior sheeting is applies last. The ceiling is completed in a similar manner. The joints and nail marks of the interior plywood lining are covered with self-adhering tape to provide additional air tightness if needed. Simplicity of construction, economical cost and dependability for tightness account for the wide acceptance of the "inside-outside" method. The durability factor is still unknown, but installations in use for at least 5 years have shown little deterioration and have required a minimum of maintenance.

Literature Cited

1. Dalyrymple, D. G. 1967. The development of controlled atmosphere storage of fruit. U.S. Dept. Agr., Div. of Mktg. and Util. Sciences, Federal Ext. Service. 56 pp.

2. Holligan, P. J. and K. J. Scott. 1971. A prefabricated plastic room and open flame generator for the controlled atmosphere storage of apples. Food Tech. in Australia 7:336-339.

3. Hunter, D. L. 1977. CA storage structure. Proc. Second National Controlled Atmosphere Research Conference. Michigan State Univ. Horticultural Report No. 28: 9-15.

4. Phillips, W. R., C. P. Lentz, E. A. Rooke and W. M. Rutherford. 1961. The use of the jacketed room system for the storage of apples. Can. Regrig. & Air Cond. 27(12):20-23.

5. Sive, Arye. 1969. CA systems in Israel and the possible application of some new European developments. Proc. National Controlled Atmosphere Research Conference. Michigan State Univ. Horticultural Report 9:26-31.

6. Smock, R. M. and G. D. Blanpied. 1969. Methods of gas sealing commercial and experimental CA rooms. Proc. National Controlled Atmosphere Research Conference. Michigan State Univ. Horticultural Report No. 9:20-23.

7. Stone, R. P. 1979. New methods in storage construction. Proc. of the Ontario Hort. Conf., Ontario Ministry of Agric. and Food:18-20.

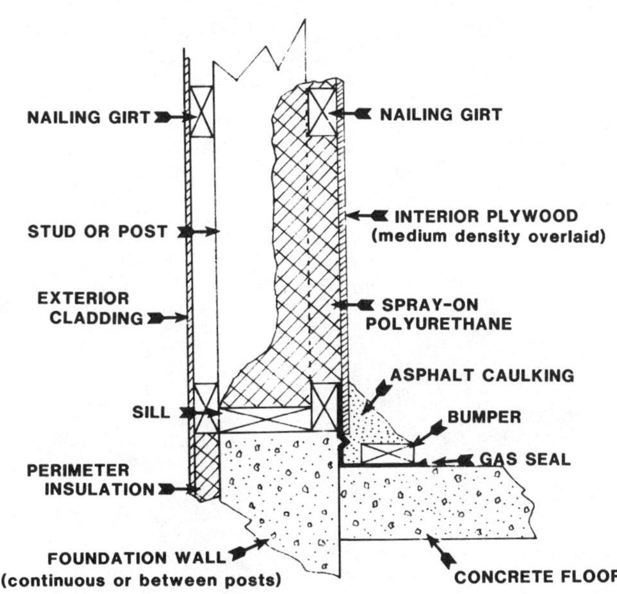

Fig. 1. Detail of partial wall-floor section of "inside-outside" method of construction of CA storage room [adapted from Stone (7)].

STORAGE OF SPECIAL PRECOOLED TULIP BULBS IN CONTROLLED ATMOSPHERE PACKAGES

T.A. Prince and R.C. Herner
Department of Horticulture

S.W. Gyeszly
Department of Packaging

Michigan State University
East Lansing, Michigan
(Presented by T. Prince)

Special precooling is defined as storage of tulip bulbs in a dry unplanted state at 5°C for a minimum of 12 weeks which satisfies the cooling requirement for proper elongation and flowering (2). The marketing of these bulbs for indoor forcing by the consumer has had limited study. Current recommendations suggest that special precooled bulbs be held at 15°C or lower, maximum ventilation be provided, and the shipping period not exceed 8 days (5). We have previously demonstrated that atmospheres of 3 to 5% O_2 in a flow-through system can enhance the flowering ability of special precooled tulip bulbsheld up to 4 weeks at room temperatures (6). We are now attempting to use polymeric films to produce a modified atmosphere in market packages of precooled bulbs. We outline here an analytical method for predicting the permeability requirements of a polymeric film and our results of packaging of precooled tulip bulbs.

Predicting Permeability Requirements

In a sealed market package of tulip bulbs or other commodities CO_2 accumulation and O_2 reduction is expected. Methods have been previously published for predicting these levels (4). We employed a simplified method to aid in choosing the proper film for our package. The method involved a preliminary test of small containers of bulbs sealed with film and using the data obtained for estimating film requirements.

Four films with varied permeabilities to CO_2 and O_2 were chosen for this preliminary study. The films and their permeabilities are listed in Table 1.

Table 1. Permeabilities of 4 films used for prediction experiments.[z]

Film	Thickness (mils)	Permeability (cc/m^2 atm day) O_2	CO_2
Mylar	0.5	80	218
Polypropylene	1.0	2,800	3,300
Polyethylene	1.0	6,500	31,000
Pliofilm	1.0	28,000	101,700

[z]Values of O_2 and CO_2 obtained with Oxtran-100 and Permatran respectively (Modern Controls, U.S.A.) at $20^{\circ}C$ and 0% R.H.

Bulbs of 'Kees Nelis' were special precooled for 13 weeks at $5^{\circ}C$. One or two bulbs were sealed in 1 pint Mason jars with 40 cm^2 of film. Sampling ports were applied to the films as small droplets of silicone rubber caulking (Dow) and the jars were placed randomly in rooms at $20^{\circ}C$. The CO_2 and O_2 levels in the jars were monitored over a 15 day period (Table 2). The trends in the data reflect the CO_2 and O_2 permeabilities of the films. Lower CO_2 permeability led to higher CO_2 accumulations while lower O_2 permeability led to greater O_2 depletion. The data from jars containing 2 bulbs sealed with pliofilm were not used for subsequent calculations since a great deal of stretching of the film occurred which obviously changed its permeability during the experiment. The remaining data was used to calculate the desired CO_2 and O_2 permeabilities of a film for a modified atmosphere package. The parameters required for the calculation were the CO_2 and O_2 levels at measured time intervals, the headspace in the jars, and the permeabilities and surface area of the films. The following relationships were utilized:

(1) Δ cc CO_2 = RR(Δ t) - P $[CO_2]$ (Δ t), where Δ cc CO_2 = $\Delta [CO_2] \cdot$(headspace), RR = respiration rate in cc CO_2/bulb·day, P = permeability for a known area and thickness of film in cc CO_2/atm $CO_2 \cdot$day and $[CO_2]$ = $\%CO_2$ in the jar.

Solving the equation for RR yielded:

(2) $RR = \Delta \text{ cc } CO_2 / \Delta t + P\,[CO_2]$

Table 2. Levels of CO_2 and O_2 obtained within 1 pint Mason jars containing 'Kees Nelis' bulbs sealed with 40 cm^2 of film at 20°C.[z]

Days	1 bulb/jar		2 bulbs/jar	
	$\%CO_2$	$\%O_2$	$\%CO_2$	$\%O_2$
		Mylar		
1	4.2 ± 0.7	17.3 ± 0.5	5.6 ± 1.6	15.5 ± 2.0
4	11.4 ± 1.1	8.2 ± 1.9	14.9 ± 1.9	2.2 ± 1.2
7	13.6 ± 0.4	4.3 ± 1.6	19.4 ± 2.3	1.5 ± 0.5
15	18.2 ± 3.0	2.7 ± 2.4	27.3 ± 2.7	0.9 ± 0.3
		Polypropylene		
1	3.8 ± 0.2	17.6 ± 0.9	5.6 ± 2.4	15.9 ± 3.2
4	10.8 ± 0.3	8.6 ± 0.9	13.3 ± 2.0	3.5 ± 2.8
7	12.8 ± 0.2	3.4 ± 0.5	17.4 ± 1.1	1.8 ± 0.7
15	16.2 ± 0.2	1.8 ± 0.1	24.2 ± 1.3	1.2 ± 0.1
		Polyethylene		
1	3.3 ± 0.4	17.9 ± 0.3	5.9 ± 0.8	14.6 ± 0.9
4	7.4 ± 1.1	10.6 ± 0.7	10.3 ± 1.3	4.2 ± 1.5
7	8.5 ± 0.7	5.8 ± 0.4	9.9 ± 2.5	2.5 ± 0.6
15	6.4 ± 1.6	2.2 ± 1.1	10.7 ± 3.7	2.2 ± 1.1
		Pliofilm		
1	2.3 ± 0.0	18.1 ± 0.2	4.1 ± 0.2	15.4 ± 0.4
4	3.0 ± 0.1	11.2 ± 0.8	2.5 ± 0.1	8.4 ± 0.2
7	2.2 ± 0.4	7.7 ± 1.4	1.8 ± 0.1	7.8 ± 0.2
15	2.6 ± 0.9	6.4 ± 1.9	2.0 ± 0.1	10.4 ± 2.0

[z]Values are means of no more than 3 reps ± 1 standard deviation.

Multiple regression (3) was performed with the data (Tab. 2) to express RR as a function of the CO_2 and O_2 levels such that:

(3) $RR = f\,[CO_2]\,[O_2]$

Similiar equations were developed for O_2 consumption (O_2 cons) of the bulbs.

(4) $\Delta\,cc\,O_2 = P(0.21 - [O_2])\,(\Delta t) - O_2\ cons\,(\Delta t)$,

where $O_2\ cons = cc\ O_2/bulb \cdot day$

Solving this equation for O_2 cons yielded

(5) $O_2\ cons = P(0.21 - [O_2]) - cc\ O_2/\Delta t$

Multiple regression yielded

(6) $O_2\ cons = f'\,[CO_2]\,[O_2]$.

At equilibrium of a package system, the respiration rate and O_2 consumption rate are equal to their corresponding permeation rates through the film (7) such that

(7) $RR = P\,[CO_2]$, and

(8) $O_2\ cons = P(0.21 - [O_2])$

These relationships allowed us to set up the following 2 equations:

(9) $f\,[CO_2]\,[O_2] = P\,[CO_2]$ and

(10) $f'\,[CO_2]\,[O_2] = P(0.21 - [O_2])$

Finally by solving the prediction equations and setting acceptable limits on CO_2 and O_2 levels within a package the desired permeability ranges of a film were predicted. For our data the solutions to equations (3) and (7) yielded

(11) $RR = 9.1 - 0.18\,[CO_2] + 0.059\,[CO_2]\,[O_2]^2 - 0.00798\,[CO_2]^2\,[O_2]$ ($R^2 = .81$)

(12) $O_2\ cons = 15.1 - 2.89\,[CO_2] + 1.01\,[O_2 + 0.176]\,[CO_2]^2 - 0.00349\,[CO_2]^3 - 0.0026\,[O_2]^3$ ($R^2 = .88$)

The above 2 expressions were not an attempt to explain a biological relationship between variables; they were only for the purpose of prediction.

We selected a range of 2-5% for both CO_2 and O_2 and an average package size of 20 x 20 cm to contain 5 precooled bulbs. Thus our calculations yielded permeabilities of 10,000 - 30,000 cc CO_2/day·atm·m^2 and 2,000 - 6,000 cc O_2/day·atm·m^2.

Packaging of Precooled Bulbs

Dow Chemical Corporation supplied us with 3 films they believed would be within our permeability requirements. All were types of low density polyethylene and had good heat sealable properties. The permeabilities were measured with a custom-made permeability cell. The cell was divided into two sections by the film; through one section either pure CO_2 or O_2 was passed. Pure N_2 or He of a known flow rate was passed through the other section which was then sampled for CO_2 or O_2 analysis until a constant value was reached. Three separate determinations were made for each film (Table 3). Only the 2 mil film had acceptable permeabilities based on our predictions.

Five 'Kees Nelis' bulbs were heat sealed in each of 4, 20 x 20 cm bags of each film and held at 20 or 25°C. This was to determine if our selected films could be adaptable to higher temperatures than used in our prediction experiment. Bulbs were to be stored for 2, 3, and 4 weeks but a malfunctioning temperature control ended the experiment at 25 days. The CO_2 and O_2 levels in the bags were monitored during storage (Fig. 1). At 20°, the O_2 level fell to around 12 - 14 % after 1 day for all films and declined further in the 2 and 3 mil film bags through 24 days. The thicker the film, the faster was the decline in O_2 level and the greater was the increase in CO_2 level. Only the 6 mil film bags appeared to reach an equilibrium level of O_2 by 24 days. The 2 and 3 mil films were observed to shrink around the bulbs during storage. This may have slowly decreased the effective surface area for permeation causing a delay in attaining equilibrium. At 25° the O_2 level was below 7% in all bags by day 2. The 2 mil bags showed a slight rise and then a slow decline in O_2 during the remainder of storage while the 3 mil bags showed a slow increase and the 6 mil bags exhibited a steady value. The 3 and 6 mil bags at 25° yielded higher CO_2 levels than at 20°. However, at 25° the CO_2 level within the 2 mil bags was similar to the level at 20° while the O_2 level declined less than at 20°. This could have been attributed to increased permeabilities of the 2 mil film at the higher temperature which may have acted to compensate for an increased respiration rate of the bulbs at 25°. This factor may prove to be important in the further design of modified atmosphere packages in that the films may help the system adjust to fluctuations in temperature.

The ethylene levels remained below 1 ppm (data not shown) in most of the packages during storage. We have already determined that at low O_2 atmospheres, precooled bulbs can tolerate 5 ppm ethylene without significant detriment to flowering (6).

Table 3. Permeabilities of 3 films from Dow Chemical used for CA packages.[z]

Film	Thickness (mils)	Permeability (cc/m^2 atm day) O_2	CO_2
LDF 301	2.0	3700	13,000
LDF 550	3.0	2800	6,500
PSD 599	6.0	2200	3,400

[z]Values are means of 3 determinations at 20°C and 0% R.H.

Table 4. Flowering of special precooled tulip bulbs after storage in CA packages.[z]

Film	Thickness (mils)	Temperature (°C)	% normal flowers wks of storage 2	3
LDF 301	2.0	20	76	70
LDF 550	3.0	20	80	6
PSD 599	6.0	20	56	0
	open	20	70	0
	initial	--	96	
LDF 301	2.0	25	84	46
LDF 550	3.0	25	20	0
PSD 599	6.0	25	0	0
	open	25	36	0
	initial	--	96	

[z]Values are means of 4 reps of 5 bulbs planted after storage in a 20 x 20 cm bag.

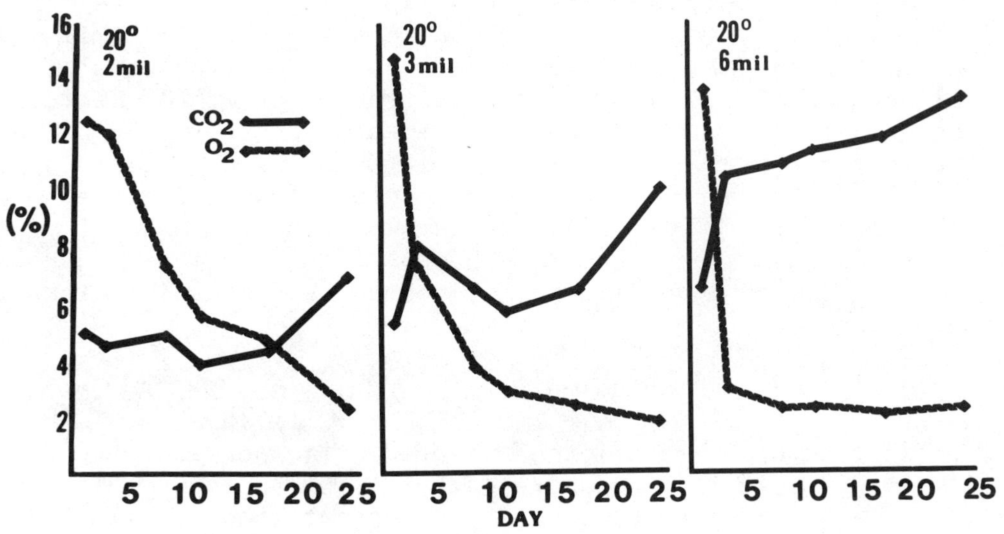

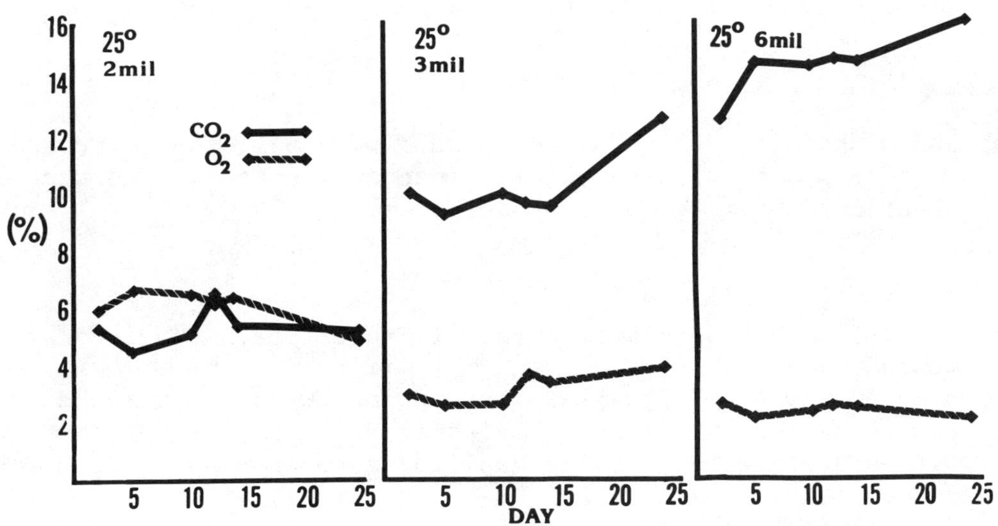

Fig. 1. CO_2 and O_2 levels within 20 x 20 cm packages of LDF 301 (2 mil), LDF 550 (3 mil), and PSD 599 (6 mil) films during 25 days of storage at 20 and 25°C. Values are means of 4 replications.

The bulbs were planted after 2 and 3 weeks of storage. Controls of bulbs planted immediately after special precooling and those stored in open air were also planted. The 2 mil film maintained 70 % flowering after 3 weeks of storage at 20° (Table 4). While the flowering was less at 25°, the 2 mil film again gave the best results. However, no film gave flowering equal to the initial control levels. This may have been due to the growth of penicillium on the bulbs during storage. It was observed to grow under the bulb tunic and invade both the root plate and shoot of some of the bulbs during storage. In the future we will investigate the use of appropriate fungicidal treatments to control this fungal growth. We will also examine the response of other cultivars to this packaging technique in hopes of finding a more vigorous cultivar that will yield a higher flowering percent. The use of other films also will be investigated.

Conclusions

We have demonstrated a simple method of predicting permeability requirements of a film for a modified atmosphere package. While the results were less than perfect, we are now closer to package development than if films were chosen at random and tested. With the availability of laminates of films with great variation in permeabilities and with the development of custom-made films to meet permeability requirements (1) it is becoming necessary and advantageous to use prediction methods for film selection. This will not only save development time, but may also allow the full potential of polymeric film packaging to be realized.

Acknowledgments

The authors wish to thank the Netherlands Flower Bulb Institute and the Dutch Bulb Exporters Association for their financial support. The technical assistance of Denise Cerny is greatly appreciated.

Literature Cited

(1) Anonymous. 1980. Visqueen Films Typical Properties. Ethyl Corporation, Richmond, Va.

(2) DeHertogh, A.A. 1974. Principles for forcing tulips, hyacinths, daffodils, Easter lilies and irises. Scientia Hort 2:313-355.

(3) Draper, N.R. and H. Smith. 1966. Applied Regression Analysis. John Wiley and Sons, Inc., N.Y. 407p.

(4) Hayakawa, K., Y.S. Henig, and S.G. Gilbert. 1975. Formulae for predicting gas exchange of fresh produce in polymeric film package. J. Food Sci 40:186-191.

(5) Moe, R., A.A. DeHertogh, and D.R. Dilley. 1978. Influence on growth and flowering of special precooled tulips of growth regulators and sucrose injections, and of temperature and ethylene exposures under ventilated and non-ventilated conditions. Sci. Rep. Agr. Univ. Nor 57:1-19.

(6) Prince, T.A., R.C. Herner, and A.A. DeHertogh. 1981. Low oxygen storage of special precooled tulip bulbs, cvs. Kees Nelis and Prominence. J. Amer. Soc. Hort. Sci. In Press.

(7) Tomkins, R.G. 1962. The conditions produced in film packages by fresh fruits and vegetables and the effect of these conditions on storage life. J. Appl. Bacteriol. 25:290-293.

USE OF POLYMERIC FILMS TO EXTEND POSTHARVEST LIFE AND IMPROVE MARKETABILITY OF FRUITS AND VEGETABLES --- UNIPACK: INDIVIDUALLY WRAPPED STORAGE OF TOMATOES, ORIENTAL PERSIMMONS AND GRAPEFRUIT

Kazuhide Kawada[1,2]
Postharvest Biology Group
Mann Laboratory
University of California
Davis, CA

The use of polymeric film in produce packaging increased greatly during the last two decades. This is due mainly to the rapid development of new films and packaging technology, together with changes in produce marketing systems. Semi-permeable films are used in several ways to modify in-package environment such as: 1) pallet load shrouds (9), 2) box liners (8), 3) shipping bags (6), 4) consumer packaging (5,7), and 5) individual wrapping (19,22). The author has termed the last system "UNIPACK" (12). The objective of this paper is to show existence of further potential in developing polymeric film packaging to maintain keeping quality and to improve marketability of fresh fruits and vegetables. Some data on UNIPACK of tomatoes, oriental persimmons and grapefruit are reviewed herein.

Some Factors Affecting In-Package Atmospheres and Ripening of Tomatoes

Retardation of senescence through modification of in-package atmospheres is one of the potential benefits of film packaging (7,10). Factors in modification of in-package atmospheres are summarized in Table 1 (23). Some of these factors have been studied in relation to ripening, coloring, of tomatoes (16).

Thickness of film: Medium size, breaker stage 'Kyoryoku goko' tomatoes were sealed individually with 500 ml air in 0.02, 0.04, 0.06, or 0.08 mm thick, low-density polyethylene (LD-PE) bags. The film surface area was 800 cm^2. Changes in in-package O_2 and CO_2 concentrations during storage at 20°C for 10 days are shown in Fig. 1. The thicker the film, the less the O_2 and the more the color changes (Fig. 2). Tomatoes in 0.08 mm thick PE turned to soft and dull green peel might due to too high the CO_2.

---

[1]The author gratefully acknowledges the advice of Drs. T. Tarutani and H. Kitagawa, Kagawa University, Japan and Dr. W. Grierson, University of Florida.

[2]Permanent address: 709-4 Kamifukuoka, Takamatsu, Kagawa, 760 JAPAN.

Number of fruit in a package: One to 6 breaker-stage tomatoes were sealed in a 0.03 mm thick, 20 x 16, 8 cm LD-PE bag, then stored at 20°C. The O_2 concentration in the bag decreased and the CO_2 increased as the fruit number increased (Fig. 3). Thus, the more the fruit in a bag, the more retardation of ripening (Fig. 4). Fruit number in a package affects 2 variables: film surface area per mass of respiring tissue, and in-package free air volume.

Fruit size and filn surface area: Three sizes of breaker-stage tomatoes, L (about 160g), M (120 g), S (85 g), were sealed individually with 500 ml air each in a 0.04 mm LD-PE bag having 400 or 800 cm^2 surface area, then stored at 25°C. The O_2 concentration in the bag decreased and the CO_2 increased as the fruit size increased or the film surface area decreased (Fig. 5).

Initial in-package free air volume: Breaker-stage tomatoes were sealed individually in a 0.04 mm thick, 800 cm^2 LD-PE bag with 0, 250, 500 or 1,000 ml air, then stored at 25°C. The steady state in-package O_2 and CO_2 concentration were achieved more rapidly as the initial in-package free air volume decreased (Fig. 6). Accordingly the CO_2 production from the fruit decreased (Fig. 7) and the degree of ripening retardation increased as the initial in-package free air volume decreased (Fig. 8). When the bags were opened after 5 days, every tomato, including those sealed without air, ripened normally at the same rate as the non-sealed control (Fig. 7,8).

These and other published (20, 23, 24) data should be useful to formulate and test mathematical models to predict in-package atmospheres and to find the optimum film specifications for particular products (11). The basic physical nature of fruits and vegetables with regard to gas exchange needs to be studied more. These and other data indicate that tight UNIPACK with 0.03 mm LD-PE or 0.04 mm 11% ethylene vinyl acetate (EVA) seems to be preferred film for tomatoes.

UNIPACK of Oriental Persimmons (*Diopyros kaki*)

Storage life of persimmons can be prolonged by controlled atmospheres (CA) containing 3 to 5% O_2 and 5 to 10% CO_2 with 95 to 100% R.H. at -1° to 1°C (21). Tarutani (22) developed a simple system he called "polyethylene cold storage which can establish such CA conditions easily and inexpensively by sealing individual fruit in a 0.06 mm thick LD-PE bag and storing at 0°C. 'Fuyu' non-astringent type persimmons, which has the largest production in Japan, has been stored commercially from November to the next spring by this method. During that study, he recognized that some packages occasionally lost their in-package atmosphere and those kept fruit better. On the other hand, the studies on UNIPACK of tomatoes revealed that steady state in-package atmospheres can be established quicker by reducing the initial in-package air volume (Fig. 6). Thus, tightly collapsed packaging (CP) and vacuum packaging (VP) were compared to conventional air enclosed packaging (AEP) and unpackaged control (UPC) (14).

Individual 'Fuyu' persimmons were sealed in 0.06 mm thick LD-PE bags after evacuation until collapsed (CP), under high vacuum (VP), or with 500 ml air (AEP). In summary, VP modified in-package atmospheres the most rapidly and best maintained fruit quality, firmness and ascorbic acid, followed by CP, AEP, and UPC (Fig. 9, Table 2). These effects of VP and CP are synonymous to that of BANAVAC (25), prestorage high-CO_2 treatments (4) and rapid CA (17).

UNIPACK of Grapefruit

As early as 1936 Stahl and Fifield (18) reported beneficial effects of film wrapping on preservation of cold stored Florida citrus fruits. Kitagawa (15) adapted Tarutani's "PE cold storage" concept (22) to citrus fruits. Japanese late-season citrus Hassaku and Amanatsu are commercially stored in film-lined field boxes or by non-sealed UNIPACK with LD-PE bags.

The author tested such storage methods with Florida grapefruit (12) resulting in a successful application of UNIPACK (1,13). Ben-Yehoshua in Israel independently reported that deterioration of citrus fruits is delayed by individual "seal-packaging" in a high density PE film (HD-PE), and equipment has been developed to wrap fruit mechanically (2).

Florida grapefruit are tightly wrapped individually in a 0.6 mil thick LD-PE bag which then can be heat shrunk. In contrast to tomatoes and persimmons, grapefruit does not respond well to CA. Thus, the principal benefit of UNIPACK on grapefruit is reduction in weight loss. Which, in turn, minimizes not only softening, deformation and chilling injury (Table 3, Fig. 10), but also keeps fiberboard containers drier thus maintaining box strength (13). UNIPACK could eliminate waxing which becomes an increasing problem due to air pollution and increasing oil price for the "solvent waxes," and the increasing fuel costs to dry "water-emulsion waxes." When adequate fungicides were used decay was often decreased rather than increased by UNIPACK because there was no water condensation between the tightly wrapped film and the peel. Another factor might be because UNIPACK maintained the grapefruit's own inherent resistance to stem-end rot due to *Diplodia natalensis* or *Phomopsis citri* by retarding senescence as indicated by the green calyces. UNIPACK also prevent decay problems such as cross infection, so called "soilage" (blemishing of sound fruit by mold spores) and the wetting of fiberboard containers, and makes it easy to discard decayed fruit (12, 13).

Discussion and Conclusion

With certain fruits such as tomatoes, oriental persimmons and grapefruit, UNIPACK appeared to be very effective in extending postharvest life. So called "hot house" cucumbers are already very commonly UNIPACKed. Chaplin and Hawson (3) reported that UNIPACK extended the postharvest life of avocados.

UNIPACK not only extends postharvest life, but also can improve marketability by using very attractive brand-named, pre-printed film for consumer packaging. The additional cost of UNIPACK may be easily recovered on selected high-quality produce particularly if a wrapping machine is employed. *Careful harvesting and handling plus proper decay control as well as choosing the right film is critical.*

If necerrary UNIPACKed fruit can be shipped or stored for short periods without refrigeration or humidity control. Proper refrigeration, however, is the prerequisite for prolonged storage. UNIPACK and other ways of using polymeric films should be considered as a *supplement* to refrigeration.

In conclusion, there is significant potential in developing UNIPACK and other polymeric film packaging to extend postharvest life and to improve marketability of fresh fruits and vegetables.

Literature Cited

1. Albrigo, L. G., K. Kawada, P. W. Hale, J. J. Smoot and T. T. Hatton, Jr. 1980. Effect of harvest date and preharvest and postharvest treatments on Florida grapefruit condition in export to Japan. Proc. Fla. State Hort. Soc. 93:323-327.

2. Ben-Yehoshua, S., I. Kobiler and B. Shapiro. 1979. Some physiological effects of delaying deterioration of citrus fruits by individual seal packaging in high density polyethylene film. J. Amer. Soc. Hort. Sci. 104:868-872.

3. Chaplin, G. R. and M. G. Hawson. 1981. Extending the postharvest life of unrefrigerated avocado (Persea americana Mill.) fruit by storage in polyethylene bags. Scientia Hort. 14:219-226.

4. Couey, H. M. and K. L. Olsen. 1975. Storage response of 'Golden Delicious' apples after high-carbon dioxide treatment. J. Amer. Soc. Hort. Sci. 100(2):148-150.

5. Grierson, W. 1968. Consumer packaging of citrus fruits. Proc. 1st Int. Citrus Sym. III:1389-1401.

6. Hardenburg, R. E. 1966. Packaging and protection. U.S. Dept. Agr. Yearbk. 1966:102-117.

7. Hardenburg, R. E. 1971. Effect of in-package environment on keeping quality of fruits and vegetables. HortScience 6(3):198-201.

8. Hardenburg, R. E. and N. W. Siegelman. 1957. Effects of polyethylene box liners on scald, firmness, weight loss, and decay of stored eastern apples. Proc. Amer. Soc. Hort. Sci. 69:75-80.

9. Harvey, J. M. 1977. In-transit atmosphere modification -- Effects on quality of fruits and vegetables. Proc. 2nd Nat. CA Conf. (Michigan State Univ. Hort. Rept. 28) 71-78.

10. Henig, Y. S. 1975. Storage stability and quality of produce packaged in polymeric films. p. 144-152. In N. F. Haard and D. K. Salunkhe (eds.) Postharvest biology and handling of fruits and vegetables. AVI Publishing, Westport, CT.

11. Henig, Y. S. and S. G. Gilbert. 1975. Computer analysis of the variables affecting respiration and quality of produce packaged in polymeric films. J. Food Sci. 40:1033-1035.

12. Kawada, K. and L. G. Albrigo. 1979. Effects of film packaging, in-carton air filters, and storage temperatures on the keeping quality of Florida grapefruit. Proc. Fla. State Hort. Soc. 92:209-212.

13. Kawada, K. and P. W. Hale. 1980. Effect of individual wrapping and relative humidity on quality of Florida grapefruit and condition of fiberboard boxes in simulated export tests. Proc. Fla. State Hort. Soc. 93:319-323.

14. Kawada, K., T. Tarutani and H. Kitagawa. 1978. Effect of vacuum packaging on keeping quality of Japanese persimmons. HortScience 13(3):390. (Abstr.)

15. Kitagawa, H. 1970. Storage and quality of Amanatsu. Kajitsu Nippon 25(4):22-25. (in Jap.)

16. Kitagawa, H., K. Kawada, T. Tani and T. Tarutani. 1978. Effects of polyethylene film package on the ripening of tomatoes. Tech. Bull. Faculty Agr. Kagawa Univ. 29(2):269-275.

17. Lau, O. L. 1981. The use of rapid CA storage for apples. Proc. 3rd Nat. CA Conf. (in press).

18. Stahl, A. L. and W. M. Fifield. 1936. Cold storage studies of Florida citrus fruits. II. Effect of various wrappings and temperatures on the preservation of citrus fruits in storage. Fla. Agr. Expt. Sta. Bull. 304.

19. Stahl, A. L. and P. J. Vaughan. 1942. Pliofilm in the preservation of Florida fruits and vegetables. Fla. Agr. Expt. Sta. Bull. 369.

20. Scott, L. E. and S. Tewfik. 1947. Atmospheric changes occurring in film-wrapped packages of fruits and vegetables. Proc. Amer. Soc. Hort. Sci. 49:130-136.

21. Tanaka, Y., N. Takase and J. Sato. 1971. Studies on the CA-storage of fruits and vegetables. III. Effect of CA-storage on the quality of persimmons (Diospyrus kaki L. f.) Res. Bull. Aichi-ken Agr. Res. Center Series B, 3:100-106.

22. Tarutani, T. 1965. Studies on the storage of persimmon fruits. Kagawa Univ. Faculty Agr. Mem. 19 (in Jap. with Eng. abstract).

23. Tolle, W. E. 1071. Variables affecting film permeability requirements for modified-atomosphere storage of apples. U.S. Dept. Agr. Tech. Bull. 1422.

24. Tomkins, R. G. 1962. The conditions produced in film packages by fresh fruits and vegetables and the effects of these conditions on storage life. J. Appl. Bact. 25(2):290-307.

25. Woodruff, R. E. 1969. Modified atmosphere storage of bananas. Proc. 1st Nat. CA Conf. (Michigan State Univ. Hort. Rep. 9) 80-94.

Table 1. Factors involved in modification of in-package atmospheres in polymeric film packaging of fruits and vegetables.

Permeability of film

- Type of film
- Thickness of film
- Surface area of film

Respiration and gas exchange system of the commodity

- Type of commodity
- Maturity of commodity
- Size and amount of commodity in a package

Others

- Initial free air volume in a package
- Initial in-package atmospheres
- External conditions
  - Temperature
  - Humidity
  - Partial pressure of gas
  - etc.

Table 2. Effect of packaging style on the keeping quality of 'Fuyu' persimmons after 5 months at 0°C. From (14).

Packaging[z] style	Firm-ness (kg)	Ascorbic acid (mg%)	Weight loss (%)	Marketable fruit (%)
Initial	2.9	53.2	-	-
Control	0.3 a[y]	19 6 a	10.7 a	0 a
With air	1.4 b	39.4 b	0.7 b	22.7 b
Collapsed	2.0 c	45.7 bc	0.6 b	57.4 c
Vacuum	2.5 d	55.3 c	0.7 b	72.8 d

[z]Packaged with 0.06 mm low-density polyethylene.

[y]Mean separation, within columns, by Duncan's multiple range test, 5% level.

Table 3. Effects of UNIPACK and storage temperatures on the keeping quality of Florida late-season 'Ruby Red' grapefruit[z]. From (12).

Treatments	Delay	Wt. loss Arrival (%)	Wt. loss Final	Decay (%)	Peel[y] color	Peel[x] gloss	CI[w] score	Firmness[v] (%)	In-fruit gas concn O_2 (%)	In-fruit gas concn CO_2 (%)	In-fruit gas concn C_2H_4 (ppm)
Expt. 1	No delay + 10°C, 5 wk + 21°C, 3 wk										
Control	—	3.6a[u]	5.5a	23.3a	3.8a	3.4d	1.3abc	100d	16.2a	4.2a	.15a
PE uni-pack	—	0.5d	0.9d	3.3b	3.4b	4.1bc	1.7abc	172bc	17.5a	3.5a	.03b
PVC uni-pack	—	1.2c	2.2c	10.0b	3.8a	3.9c	1.0bc	140c	15.4a	4.0a	.17a
Expt. 2	No delay or 3-day delay at 29°C, 80% RH + 0°C, 3 wk + 5°C, 1 wk + 10°C, 1 wk + 21°C, 3 wk										
Control											
Immediate	—	2.1b	4.0b	18.3a	2.9c	4.5a	3.7a	165bc	—	—	—
Delayed	1.8	3.2a	4.6b	8.3b	2.6cd	4.3ab	1.0bc	152c	—	—	—
PE uni-pack											
Immediate	—	0.4d	0.8d	6.7b	2.6d	4.6a	3.3ab	233a	—	—	—
Delayed	1.7	1.9b	2.1c	3.3b	2.6cd	4.5a	0.7c	202ab	—	—	—

[z]At the final examination except weight loss. Data based on 3 cartons per treatment.
[y]Peel color: 1—greenish yellow, 2—pale yellow, 3—deep yellow, 4—orange yellow, 5—light orange
[x]Peel gloss: 1—dull, 3—moderate, 5—bright.
[w]Chilling injury scored on a 1-100 scale, CI = 10 is onset of CI (5).

$$^{v}\text{Fruit firmness} = \frac{\text{Permanent deformation of the Expt. 1 control}}{\text{Permanent deformation of the treatment}} \times 100\ (\%).$$

Permanent deformation measured by the Grierson Creep Tester (10).
[u]Mean separation within columns by Duncan's multiple range test at 5% level.

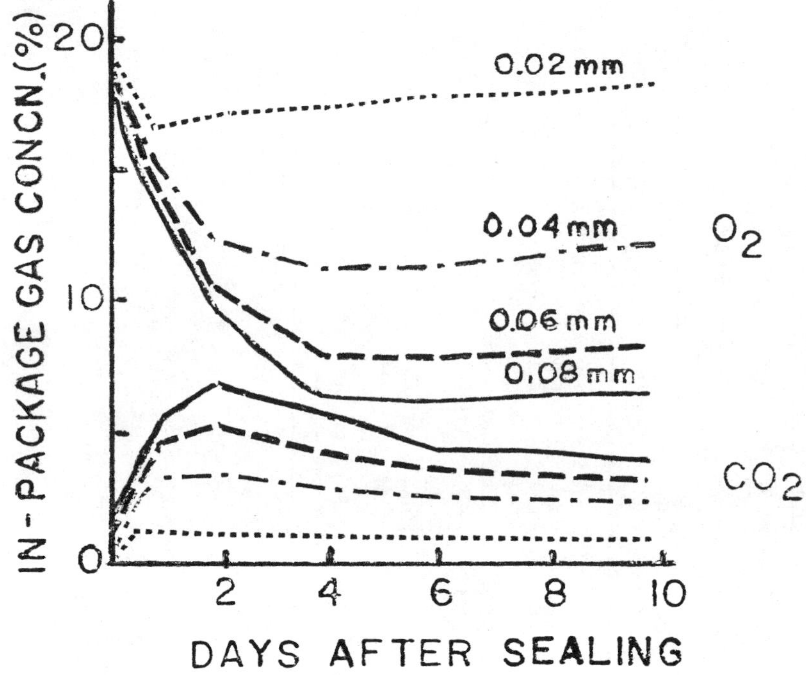

Fig. 1. Effects of PE film thickness on in-package O_2 and CO_2 concentrations of tomato UNIPACK. From (16).

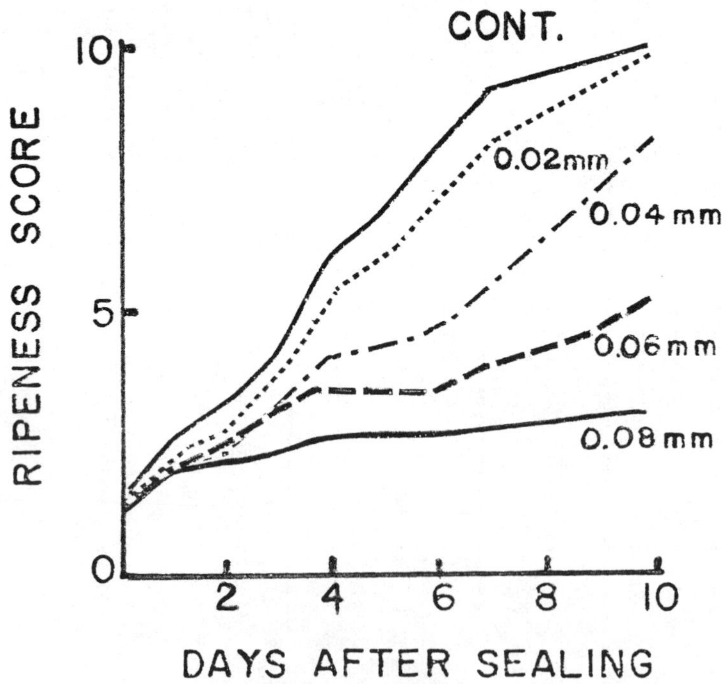

Fig. 2. Effects of film thickness on the ripening of tomatoes packaged in a PE bag. Score 0: mature green, 5: table ripe, 10: over ripe. From (16).

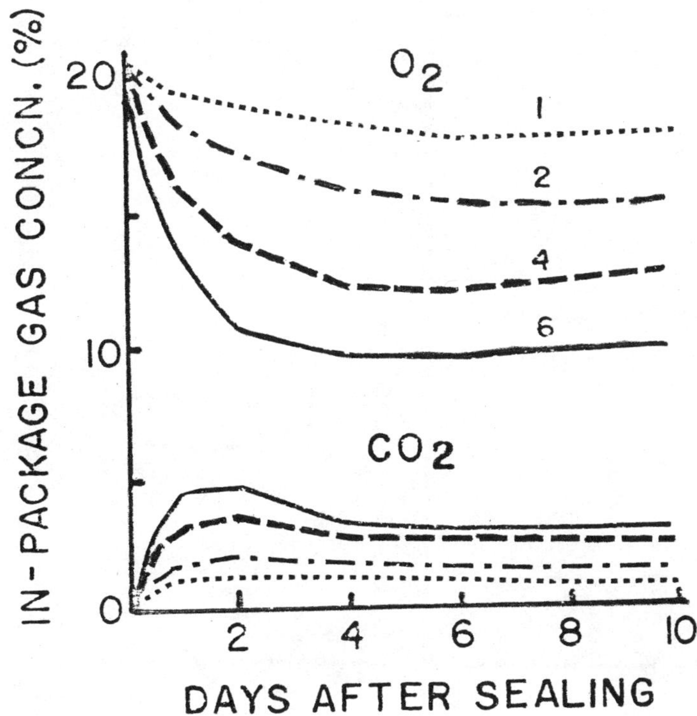

Fig. 3. Effects of number of tomato fruit in a package on in-package O_2 and CO_2 concentrations. From (16).

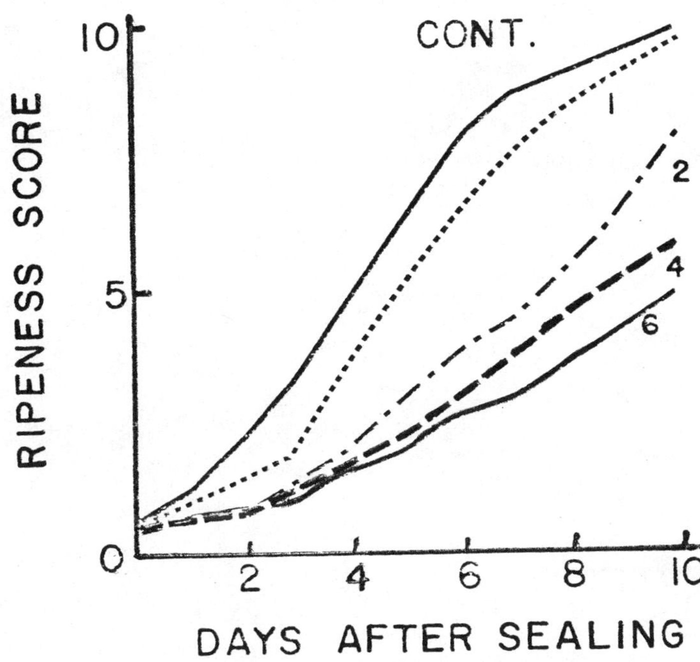

Fig. 4. Effects of number of fruit in a package on ripening of tomatoes packaged in a 0.03 mm PE bag. See Fig. 2 for the ripening score. From (16).

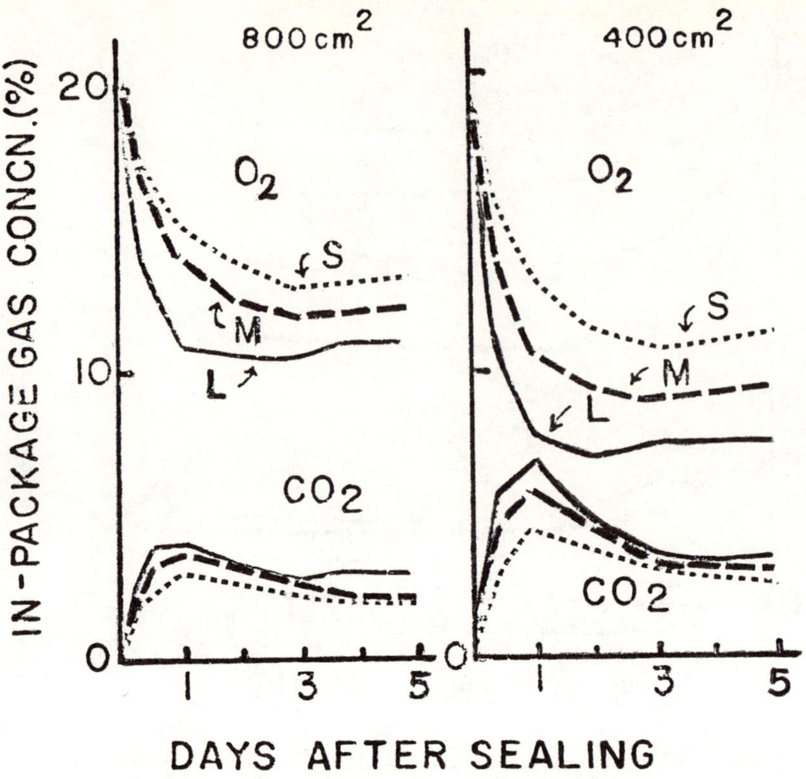

Fig. 5. Effects of PE film surface area and fruit size on in-package O_2 and CO_2 concentrations. From (16).

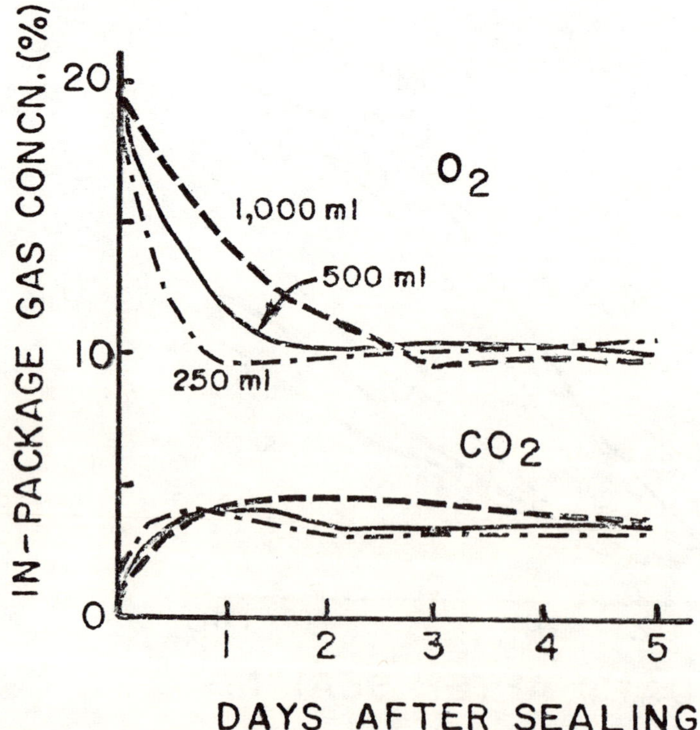

Fig. 6. Effects of initial free air volume on in-package O_2 and CO_2 concentrations. From (16).

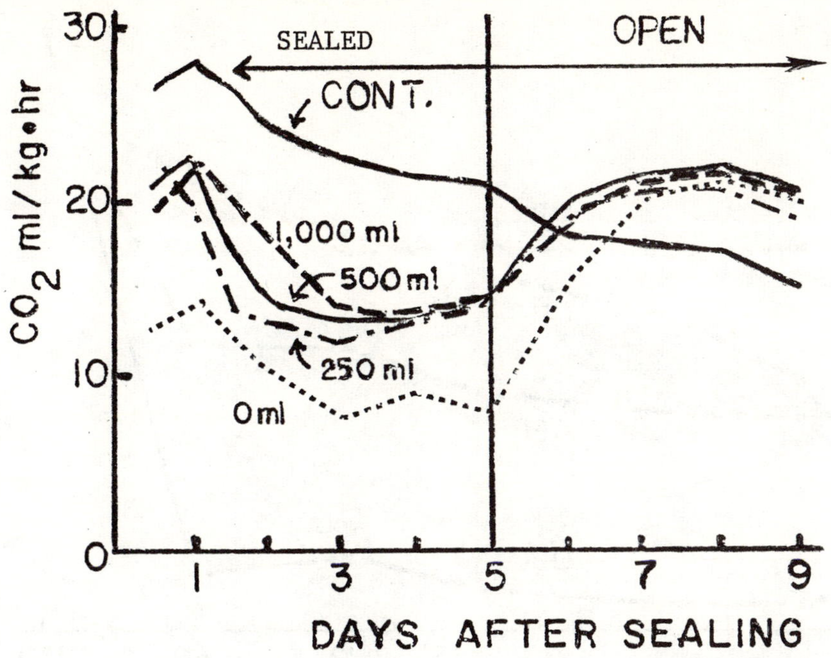

Fig. 7. Effects of initial free air volume on respiration of tomatoes packaged in a PE bag. From (16).

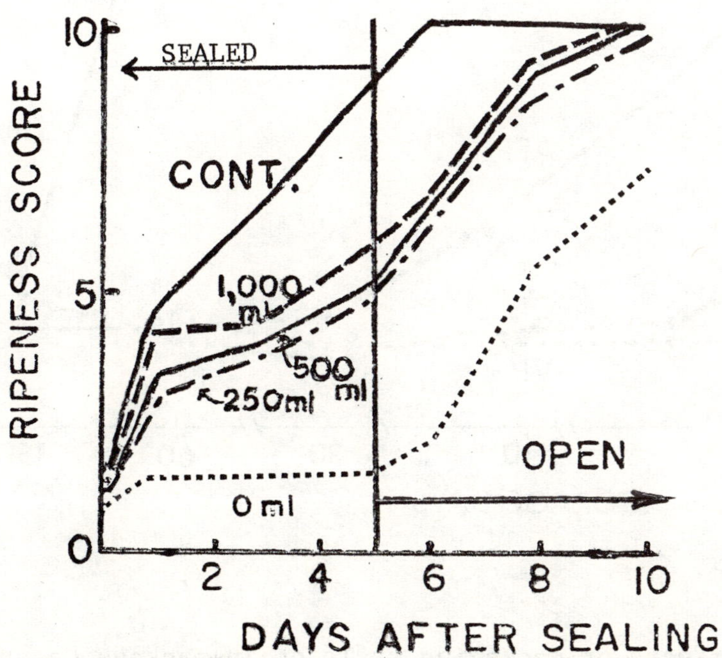

Fig. 8. Effects of initial free air volume on ripening of UNIPACKed tomatoes. See Fig. 2 for the ripening score. From (16).

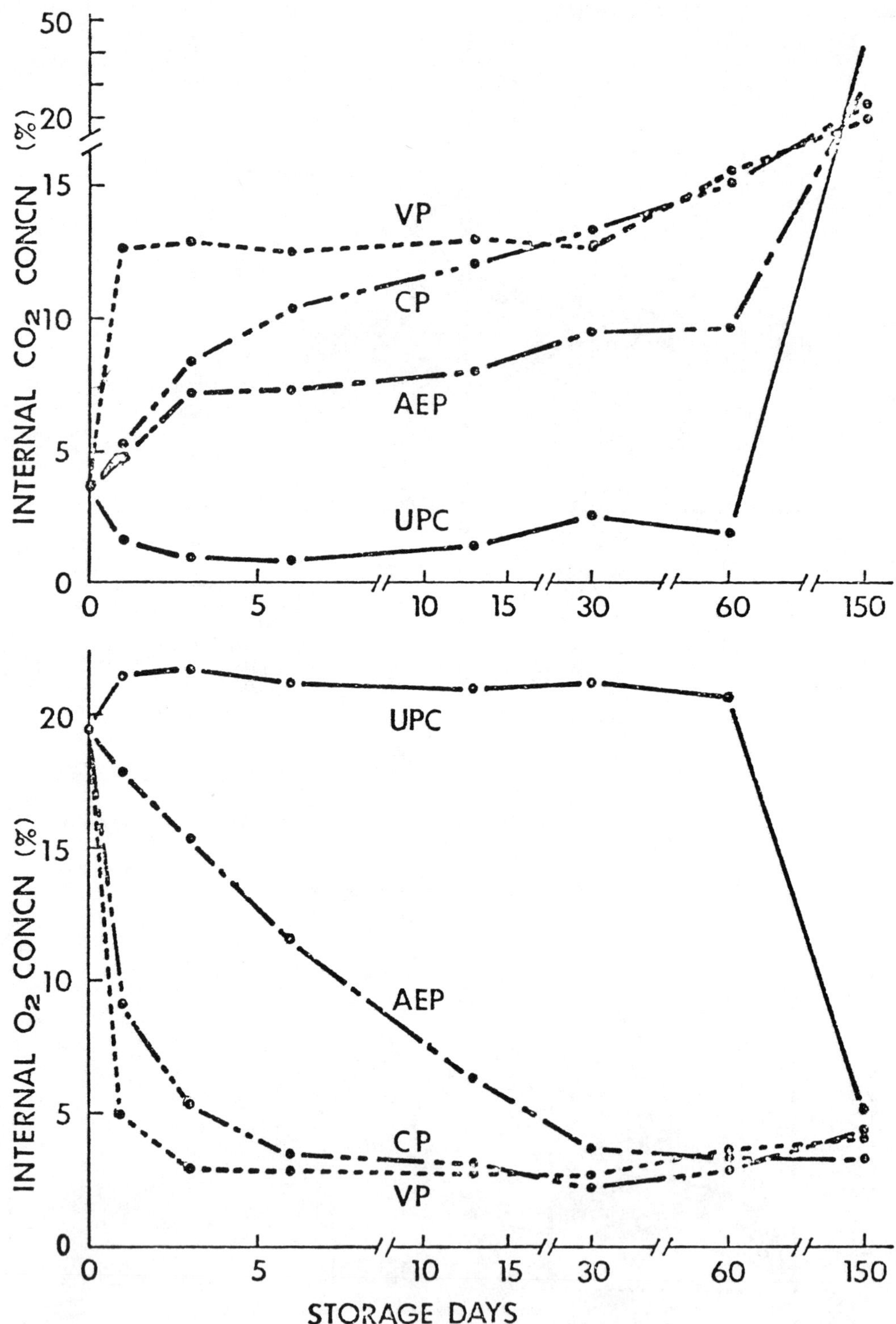

Fig. 9. Effects of packaging style on in-package O_2 and CO_2 concentrations of persimmon UNIPACK. UPC: unpackaged control, AEP: air enclosed packaging, CP: collapsed packaging, and VP; vacuum packaging. From (14).

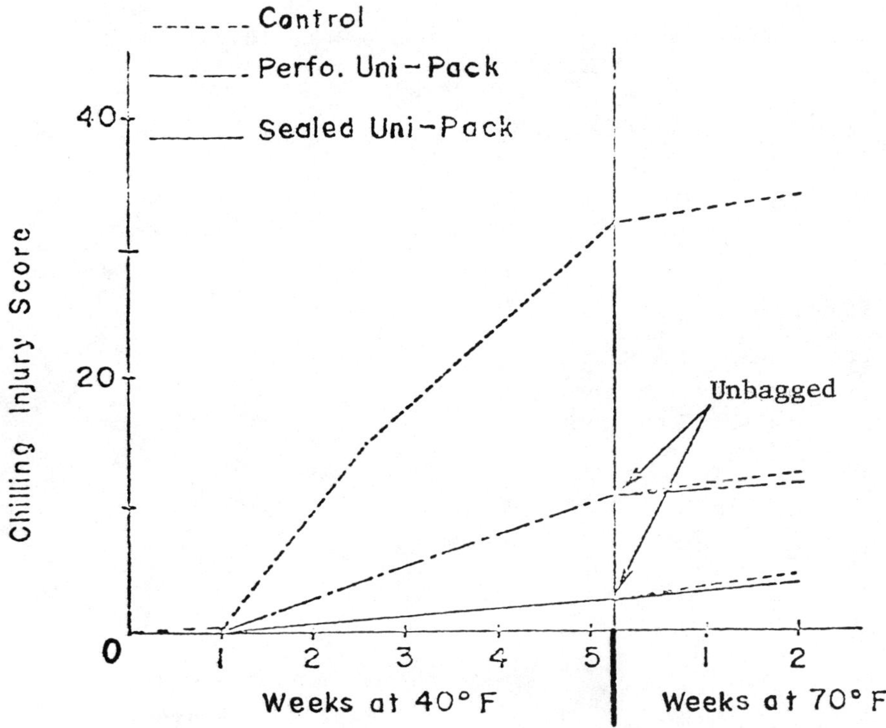

Fig. 10. Chilling injury (pitting) scores (0-100 scale) of early-season Florida 'Marsh' grapefruit as affected by PE film UNIPACK. From (K. Kawada and W. Grierson, unpublished data.)

LOW OXYGEN CA STORAGE OF MCINTOSH APPLES

D. H. Dewey and M. L. Bourne
Department of Horticulture
Michigan State University
East Lansing, Michigan 48824

The controlled atmosphere storage recommendations for cv. McIntosh in Michigan are similar to those employed in New York (3) and Ontario (1), whereby 3% oxygen and 5% carbon dioxide are used at 3°C, with the exception that a temperature of 2.2°C is generally used. Marked flesh softening of this variety in storage even at this lower temperature, has been a common problem particularly for fruit grown in the central and southern producing areas of the state where growing temperatures may be somewhat high. Excessive softening has been a major factor limiting the long-term storage of McIntosh in CA. Since hypobaric studies (2) indicated that low oxygen conditions favor the retention of flesh firmness in apples, studies on the possible benefits and hazards of low oxygen CA storage of McIntosh were undertaken in the fall of 1978.

Low Oxygen and Low Temperatures

The 1978-79 storage season trial incorporated three levels of oxygen, 1.5, 3.0 and 21%, at 0° and 2.2°C, with carbon dioxide maintained below 0.5 in all atmospheres by placement of hydrated lime in the storage chambers. Fruit harvested at the predicted ideal time of harvest from two orchards were used. The patterns of flesh softening for fruit of Orchard 2 at the two temperatures are illustrated in Figs. 1 and 2. Similar results were oftained for fruit of Orchard 1. Flesh softening was markedly retarded in the 1.5% oxygen atmosphere, and more so at 0°C than at 2.2°C. Fruit softening was similar in air and in 3% oxygen. Although there was considerable variation in fruit quality characteristics other than firmness at the conclusion of the 40-week storage period, the apples of Orchard 2 which had been stored at 2.2°C in 1.5% oxygen were superior because of minimum amounts of mealy breakdown, core browning and flesh browning. Fruit injury from low oxygen was of minor incidence and extent.

Encouraged by the previous season's results, the tests were expanded in 1979-80 to include 0.75% oxygen as well as 1.5% with each low oxygen atmosphere compared separately with 3% oxygen CA. Again, temperatures of 0° and 2.2°C were used with low carbon dioxide as maintained by the presence of dry lime in the CA chambers. Apples from 3 orchards were employed.

The effects of temperatures and atmospheres on flesh softening for the 31-week storage period are given in Figs. 3 and 4. There was no softening of fruit in the .75 and 1.5% oxygen atmospheres at either temperature. Contrary to the results of the previous year when similar amounts of softening occurred, the 3% oxygen level markedly reduced softening over air storage in 1979-80. The effects of temperature were similar to the previous year, however, in that the low temperature retarded softening in air and in 3% oxygen, but was of little effect in subnormal oxygen levels.

The development of core browning during 31 weeks of storage in .75 and 1.5% oxygen was minimal at 2.2°C, but considerable at 0°C (see Fig. 5). Relatively good control of this disorder was attained in 3% oxygen at both temperatures. Flesh browning was adequately controlled in CA only in the subnormal oxygen atmosphere at 2.2°C (Fig. 6). This temperature effect was not evident in 3% oxygen, whereas in air the effect inverted with the low temperature retarding the development of the disorder to the extent that it was of negligible amount. There were few symptoms of internal or external suboxidation injury to the fruit at either .75 or 1.5% oxygen.

Low Oxygen with Carbon Dioxide

The 1980-81 tests were conducted exclusively at 2.2°C because of the potential hazards of core and flesh browning in CA storage at 0°C evidenced the previous seasons. This enabled the inclusion of 5% carbon dioxide as well as low carbon dioxide with 1.5 and 3% oxygen atmospheres, for fruits harvested twice from 3 orchards.

The mean changes in flesh firmness, Fig. 7, show that softening was markedly retarded by 1.5% oxygen either with or without carbon dioxide, but only by 3% oxygen in the presence of 5% carbon dioxide. The 3% oxygen atmosphere without carbon dioxide was less effective, but still retarded softening over air storages. It is obvious that an atmosphere of 3% oxygen with 5% carbon dioxide should have been included in the previous trials.

The benefits of low oxygen CA in controlling core browning and flesh browning were verified in 1980-81 (see Fig. 8). Five percent carbon dioxide in both low oxygen atmospheres was of relatively little effect on core browning; its presence with 1.5% oxygen resulted in an increase of flesh browning but with 3% oxygen it decreased flesh browning. The relatively small amounts of browning disorders that developed in 1.5% oxygen would seemingly justify its use for the CA storage of McIntosh, particularly since low oxygen injury was inconsequential. As before, there were marked orchard differences in flesh softening and the development of disorders but only minor differences due to a delay in harvest of one week.

Summary and Conclusion

Lower than normal levels of oxygen (.75 and 1.5%) employed for the CA storage of McIntosh markedly retarded flesh softening over 3% oxygen only when 3% oxygen was used without the normal level of 5% carbon dioxide. However, minimal amounts of core browning and flesh browning were achieved with low oxygen when employed without carbon dioxide at 2.2°C. Low oxygen injury was negligible in all experiments. Orchard differences resulted in marked variations of storage quality; whereas a 1 week delay in time of harvest was of little effect.

Further study of low oxygen for the CA storage of McIntosh is needed before commercial recommendations can be justified.

Literature Cited

1. Andersen, E. T. 1975. Harvesting, storage and packing apples. Ontario Publ. 431. 44 pp.

2. Dilley, D. R. 1977. The hypobaric concept for controlled atmosphere storage. In Horticultural Report #28. Controlled atmospheres for the storage and transport of perishable agricultural commodties. MSU Hort. Dept. 29-37.

3. Smock, R. M. and G. D. Blanpied. 1972. Controlled atmosphere storage of apples. N.Y. (Cornell) Information Bul. 41. 16 pp.

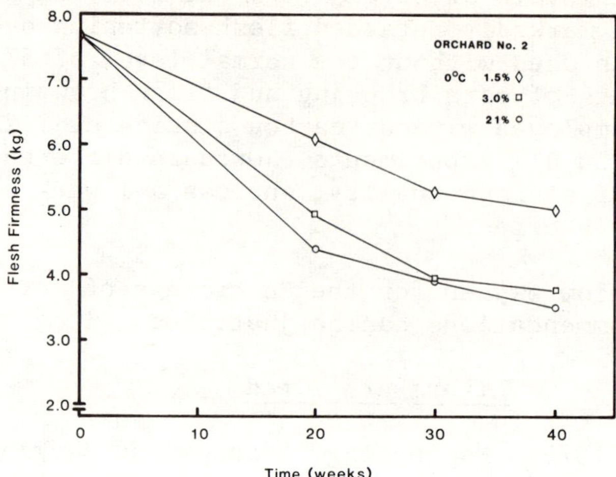

Fig. 1. Mean flesh firmness of McIntosh apples during storage in CA without carbon dioxide and in air at 0°C. (1978-79).

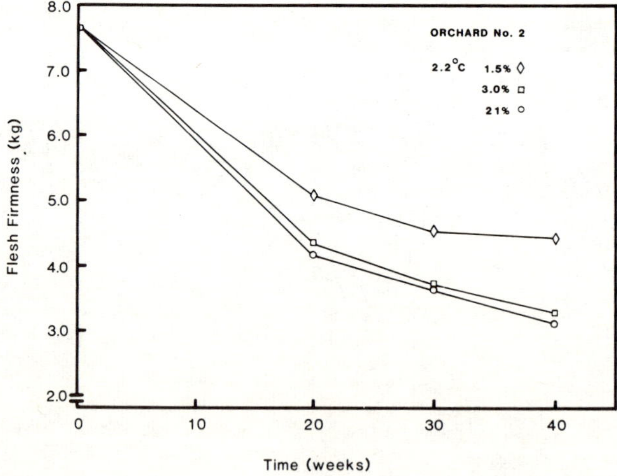

Fig. 2. Mean flesh firmness of McIntosh apples during storage in CA without carbon dioxide and in air at 2.2°C (1978-79).

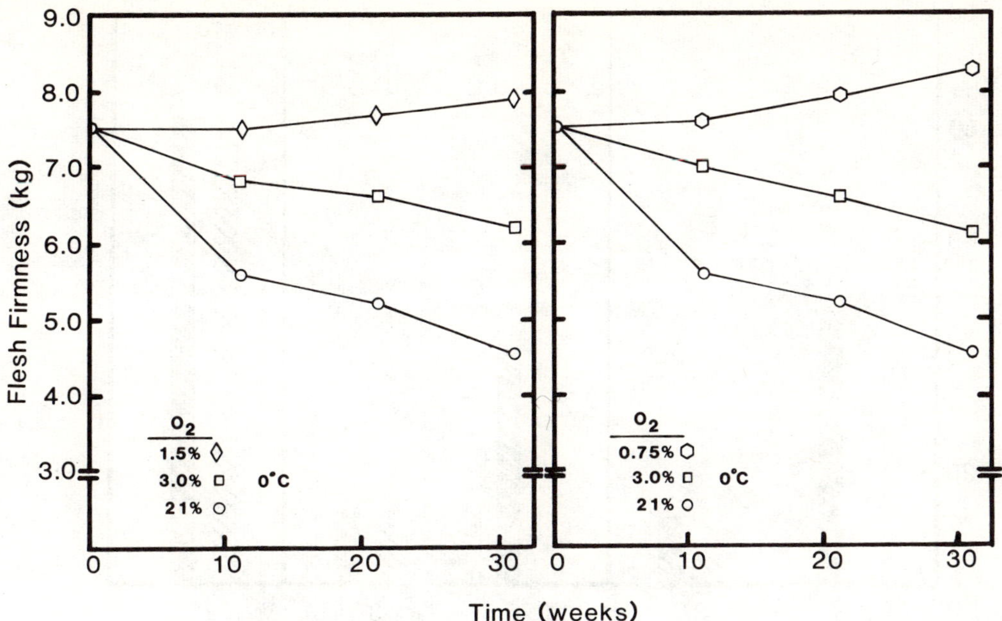

Fig. 3. Mean flesh firmness of McIntosh apples during storage in CA without carbon dioxide and in air at 0°C (1979-80). 1.5% oxygen was compared with 3% oxygen in the experiment at left; .75% oxygen was compared at right.

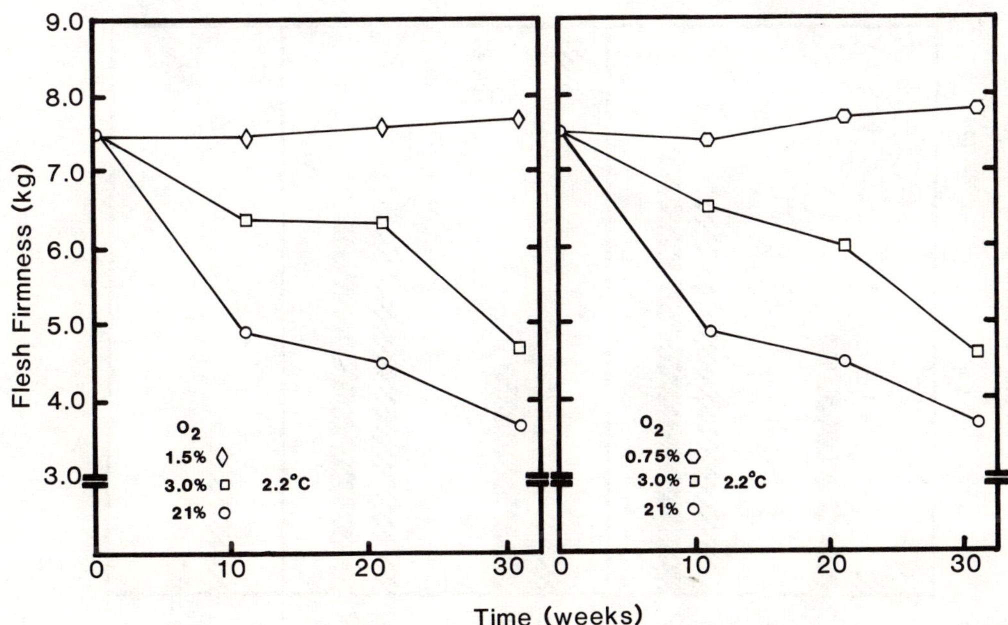

Fig. 4. Mean flesh firmness of McIntosh apples during storage in CA without carbon dioxide and in air at 2.2°C (1979-80). 1.5% oxygen was compared with 3% oxygen in the experiment at left; .75% oxygen was compared at right.

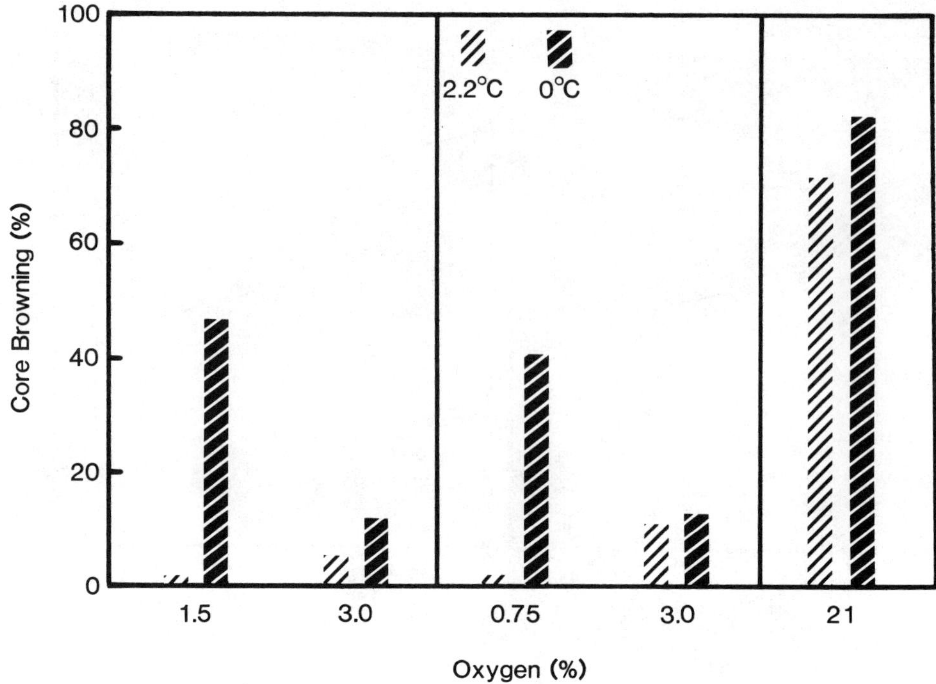

Fig. 5. Mean core browning of McIntosh apples stored in CA without carbon dioxide and in air at 0 and 2.2°C (1979-80).

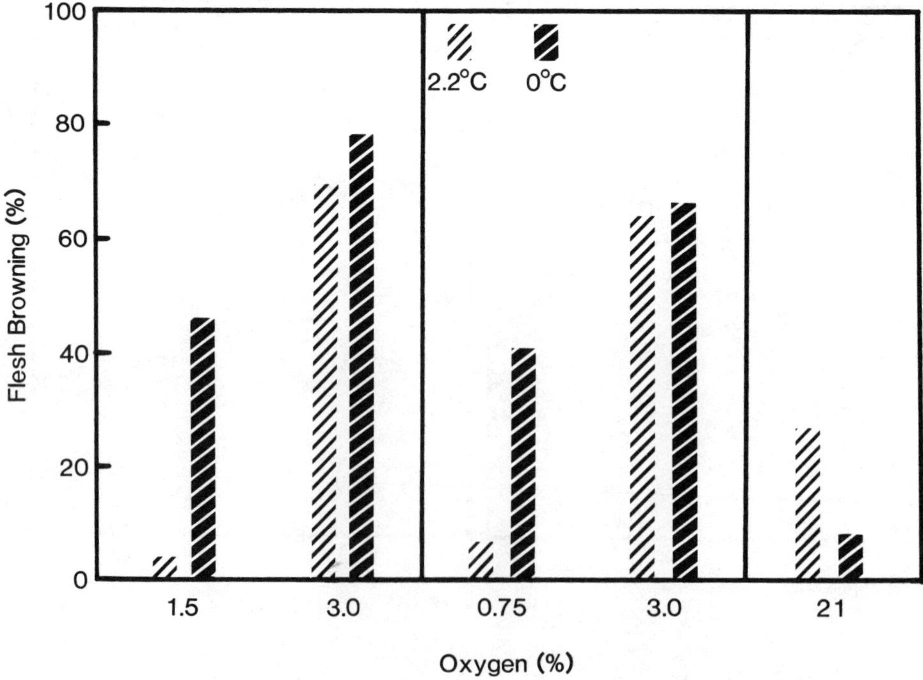

Fig. 6. Mean flesh browning of McIntosh apples in CA without carbon dioxide and in air at 0 and 2.2°C (1979-80).

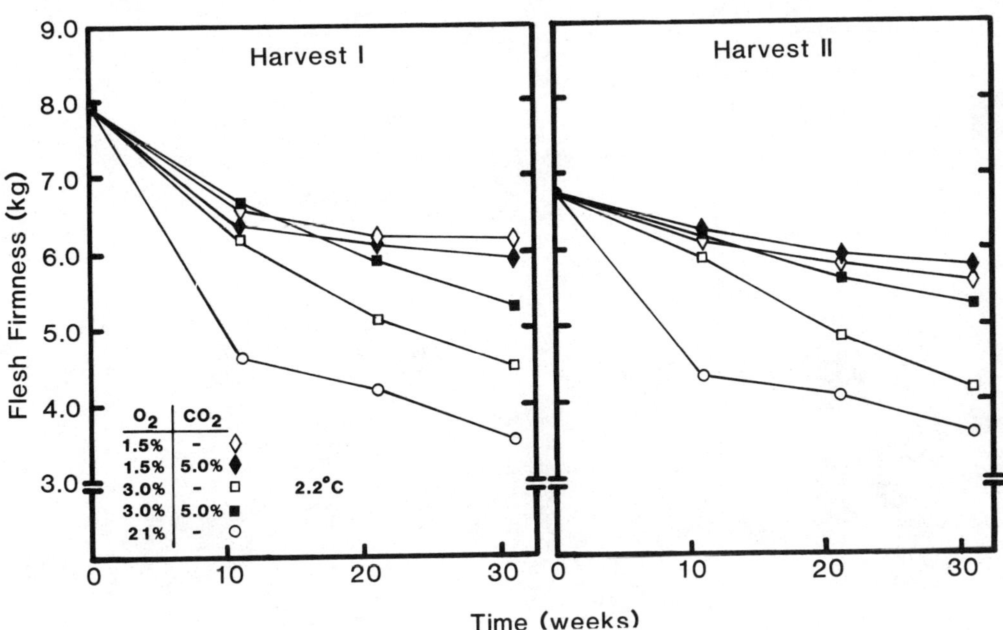

Fig. 7. Mean flesh firmness of McIntosh apples harvested one week apart and stored at 2.2°C (1980-81).

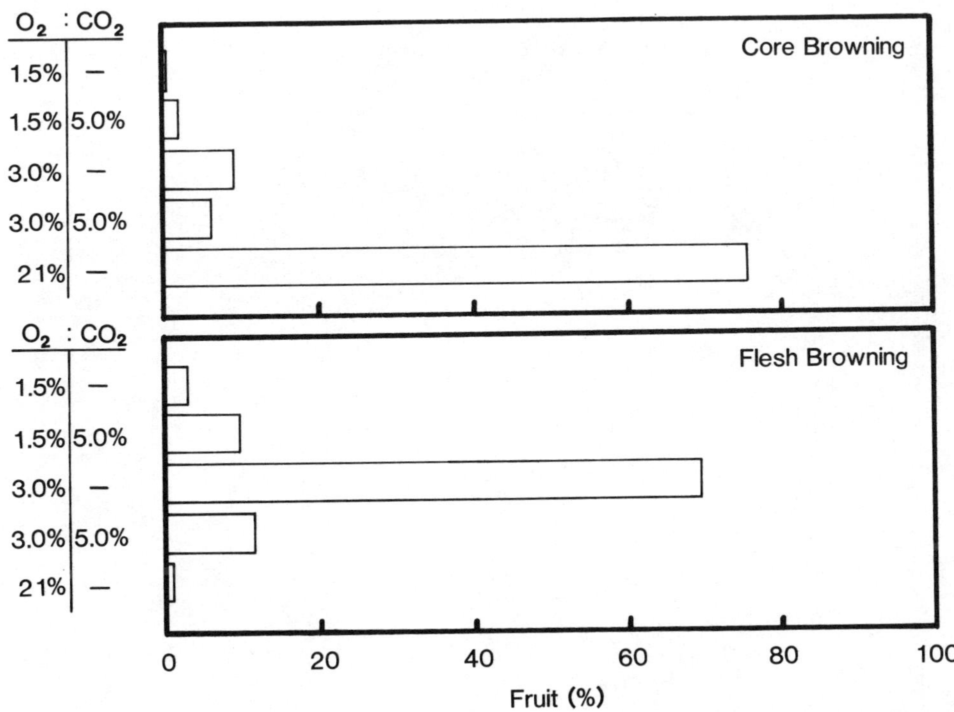

Fig. 8. Mean core browning and flesh browning of McIntosh apples harvested one week apart and stored at 2.2°C (1980-81).

Low Oxygen Atmospheres to Maintain Apple Quality in Storage

P. D. LIDSTER

Agriculture Canada, Kentville, Nova Scotia, Canada

Storage of apples in oxygen atmospheres near 1.0% is the object of renewed research interest (4). The potential beneficial effects of greatly retarding firmness and titratable acid loss in low oxygen storage were observed by previous researchers (1,12) on Delicious and Turley apple cultivars. However the positive effects of low oxygen atmospheres found in experimental storages were considered to be impractical for McIntosh because of fruit injury occurring in existing commercial CA storages in which oxygen levels had accidently dropped to near 1.0% (G. D. Blanpied, E. C. Lougheed personal communication, 10). While the equation of injured fruit in these storages to the low oxygen atmospheres was probably correct, recent evidence supports the conclusion that fruit and storage variables are synergistic with low oxygen atmospheres to result in fruit injury and disorders. The present paper discusses the effects of low oxygen storage on fruit quality retention and the interactions of fruit maturity and storage temperatures on McIntosh and Golden Delicious apple responses to low oxygen atmospheres.

Storage of McIntosh apples in 1.5% CO_2 + 1.0% O_2 atmospheres resulted in significant retention of fruit firmness and titratable acids as compared to similar fruit stored in present recommended CA atmospheres (5% CO_2 + 3.0% O_2) in Nova Scotia (Figs. 1 & 2, Table 1). Firmness losses in 0°C air storage after low O_2 or conventional CA storage were essentially identical when the linear components were compared. Low O_2 storage did not have any residual effect on retarding firmness loss in subsequent air storage but the firmness advantage provided by low O_2 atmospheres over conventional CA were maintained after air storage.

Fruit firmness lost during low oxygen storage was a function of maturity at harvest (6, Fig. 3). The minimum loss of firmness during storage occurred in fruit which had just commenced their climacteric rise in C_2H_4 production which also corresponded to a starch iodine index test (SI) of 3-4 (8). Ethylene production at harvest was determined on the effluent air passed over a 10 fruit sample 24 hr after harvest. Blanpied (3) has determined the critical maturity of McIntosh apples for optimum conventional CA storage to vary depending on crop year. There is evidence that similar yearly fluctuations in optimum harvest maturity may occur for low oxygen storage also. However fruit harvested at maturities corresponding to SI of 2 to 5 respond well to 1.0% O_2 in experimental or commercial trials. McIntosh apples with a SI greater than 5 and stored in 1.0% O_2 appear to be more susceptible to internal browning disorders. The occurrence of internal disorders at advanced fruit maturities will however be mediated by the source of the fruit lot.

The presence of CO_2 in the storage atmosphere was positively correlated to firmness retention in McIntosh apples stored in 3.0% O_2 (Table 2). The effects of CO_2 in maintaining fruit firmness were generally reduced at lower O_2 levels. After the storage period plus a 7 day incubation period at 21°C, CO_2 in the storage atmosphere was again positively related to firmness in apples stored at 3.0% O_2 but tended to be negatively correlated with firmness in apples stored in 0.5, 1.0 or 1.5% O_2 (Table 2). Fruit firmness was negatively correlated with storage O_2 levels. High levels of storage CO_2 and low O_2 tended to be associated with titratable acids retention in apples after 26 weeks of storage (Table 2). Storage O_2 levels of 0.5, 1.0, 1.5 or 3.0% with 5% CO_2 did not modify titratable acids retention.

The effects of low O_2 storage on aromatic volatile production and product acceptability is of critical importance for consumer acceptance of the final product. During long periods of air storage of apples and subsequent shelf durations, the volatiles contributing to off-or atypical flavours i.e. acetaldehyde and ethyl alcohol will increase steadily (2, 9, Fig. 4). Storage of apples in modified atmospheres have been shown to suppress total volatile production; specifically, acetaldehyde, ethanol and ethyl butyrate production are reduced by storage in 5.0% CO_2 + 3.0% O_2 (Fig. 4). However after a 120 day reconditioning period in 0°C air storage ethanol and acetaldehyde production approach normal levels for air storage and the production of ethyl butyrate is stimulated. McIntosh apples stored in 1.0% O_2 (Fig. 4) showed a similar suppression of ethanol, acetaldehyde and ethyl butyrate. However after a reconditioning period in 0° air, ethanol and acetaldehyde levels are reduced whereas ethyl butyrate had been regenerated to near optimum levels. This suggests that ethyl butyrate which contributes to the characteristic McIntosh flavour may be preferentially regenerated to acetaldehyde and ethanol by proper storage management.

Sensory evaluations of 40 duplicate grower lots stored in conventional CA (5.0% CO_2 + 2.8% O_2) and low oxygen storages (1.5% CO_2 + 1.0% O_2) indicated that fruit stored in 1.0% O_2 were perceived to be more acidic (less sweet) and firmer in texture than the conventional CA stored fruit (Table 3). After a 7 day shelf life period at 20°C, the acid and texture differences were maintained but in addition, fruit stored in 1.0% O_2 were perceived to be juicier and more acceptable than comparable fruit stored in conventional CA atmospheres. Differences of skin texture or fruit flavour between the two storage atmospheres were not detected at either examination.

Fruit firmness in McIntosh and Golden Delicious cultivars was negatively correlated to storage temperatures for both conventional and low O_2 atmospheres (Table 4). Lowering storage O_2 levels from 3.0% in conventional atmospheres to 1.5, 1.0 or 0.5% counteracted the decreased firmness response to elevated storage temperature. The major effect of storage temperature on low O_2 storage of McIntosh apples occurred in the temperature ranges of 0-2.5°C and 5.0-20.0°C. In two years of testing, 90-100% of the McIntosh apples stored at 0.5, 1.0 or 1.5% O_2 and 0°C were afflicted with internal browning and coreflush disorders, whereas identical fruit stored in similar atmospheres but at 2.5° or 2.8°C had no or a very low incidence of internal disorders. When red fruit cultivars such as McIntosh or Cortland are stored in 1.0% O_2 atmospheres at temperatures of 10° or 20°C, the red anthocyanin pigments assumed a blue or purple casting. This symptom of oxygen deprivation (11) did not appear in identical fruit held at 2.5 or 5.0°C.

Conclusions:

Storage of McIntosh, or Golden Delicious cultivars in low O_2 (1.0%) retarded loss of firmness and titratable acids as compared to similar fruit stored in conventional CA atmospheres. Low O_2 storage of McIntosh resulted in fruit which were perceived to be firmer, more acidic, juicier and more acceptable by sensory panelists after a 7 day shelf life period than conventionally stored fruit. Low O_2 storage of McIntosh is critically dependent upon optimum fruit maturity and storage temperatures. Pre-climacteric McIntosh with starch-iodine indices of 2-5 have been stored in 1.0% O_2 over a 3 year period in experimental size units (5,6) and for 2 years in commercial applications. The optimum storage temperature for low O_2 storage in Nova Scotia apples is in the range of 2.5°-5.0°C. Successful commercial trials have consisted of 1000 bushels stored at 1.5% CO_2 + 1.0% O_2 in Nova Scotia (1979 crop); 20,000 bushels stored at 1.5% CO_2 + 1.0% O_2 in Nova Scotia (1980 crop); two rooms of approximately 17,000 (total 34,000 bushels) stored at ∿0.8% CO_2 + 1.0% O_2 in Simcoe, Ontario (1980 crop) and approximately 21,000 bushels stored at ∿2.0% CO_2 + 1.0% O_2 in London, Ontario (1980 crop).

The author wishes to credit Dr. George Chu of the Ontario Ministry of Agriculture and Food, Horticultural Research Station in Simcoe, Ontario, for his excellent and necessary cooperation which led to the successful completion of the commercial low O_2 trials in Ontario. Thanks George.

LITERATURE CITED

1. Anderson, R. E. 1967. Experimental storage of Eastern grown 'Delicious' apples in various controlled atmospheres. Proc. Amer. Soc. Hort. Sci. 91: 810-820.

2. Blanpied, G. D., R. M. Smock and L. C. Frank. 1968. Some factors influencing the ethanol content of harvested apple fruits. Proc. Amer. Soc. Hort. Sci. 82: 748-754.

3. Blanpied, G. D. 1969. A study of the relationship between optimum harvest dates for storage and the respiratory climacteric rise in apple fruits. J. Amer. Soc. Hort. Sci. 94: 177-179-

4. Knee, M. 1980. Physiological responses of apple fruits to oxygen concentrations. Ann. Appl. Biol. 96: 243-253.

5. Lidster, P. D., F. R. Forsyth and H. J. Lightfoot. 1980. Low oxygen and carbon dioxide atmospheres for storage of McIntosh apples. Can. J. Plant Sci. 60: 299-301.

6. Lidster, P. D., K. B. McRae and K. A. Sanford. 1981. Responses of McIntosh apples to low oxygen storage. J. Amer. Soc. Hort. Sci. 106: 159-162.

7. Patterson, B. D., S. G. S. Hatfield and M. Knee. 1974. Residual effects of controlled atmosphere storage on the production of volatile compounds by two varieties of apples. J. Sci. Food Agric. 25: 843-849.

8. Poapst, P. A., G. M. Ward and W. R. Phillips. 1959. Maturation of McIntosh apples in relation to starch loss and abscission. Can. J. Plant Sci. 39: 257-263.

9. Smagula, J. M. and W. J. Bramlage. 1977. Acetaldehyde accumulation: Is it a cause of physiological deterioration of fruits? HortScience 12: 200-203.

10. Smock, R. M. 1979. Controlled atmosphere storage of fruits. In: J Janick (ed.) Horticultural Reviews. Avi Pub. Co., Westport, Conn.

11. Smock, R. M. 1977. Nomenclature for internal storage disorders of apples. HortScience 12: 306-308.

12. Workman, M. 1963. Controlled atmosphere studies on Turley apples. Proc. Amer. Soc. Hort. Sci. 83: 126-134.

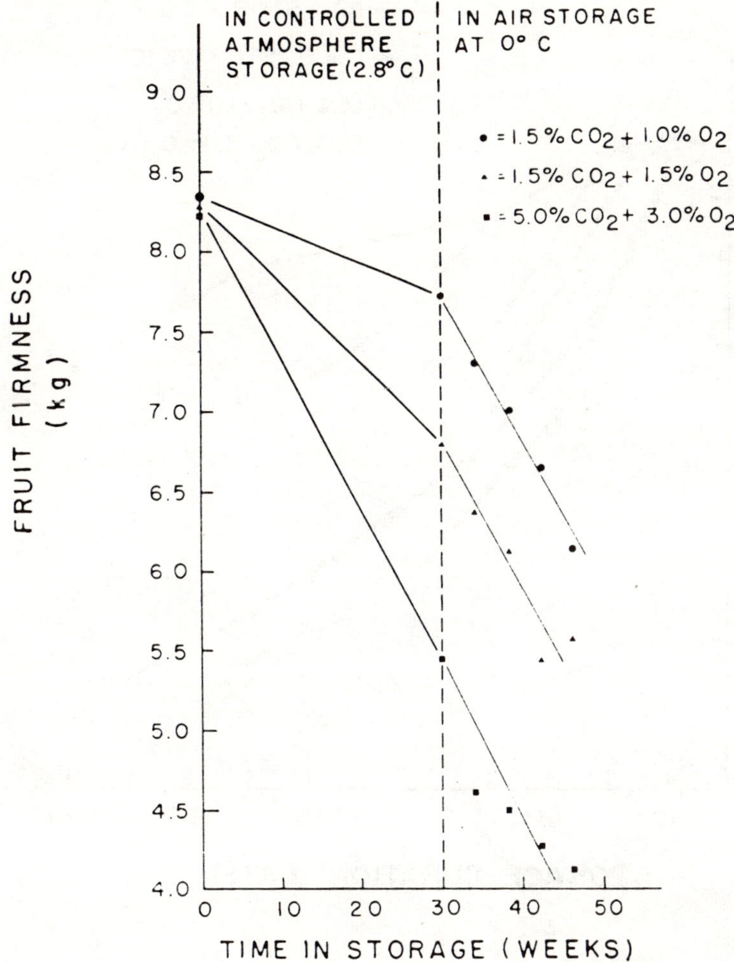

Fig. 1. Firmness loss in McIntosh apples while in low oxygen and controlled atmosphere storage and subsequent loss in 0°C air storage, 1978 crop.

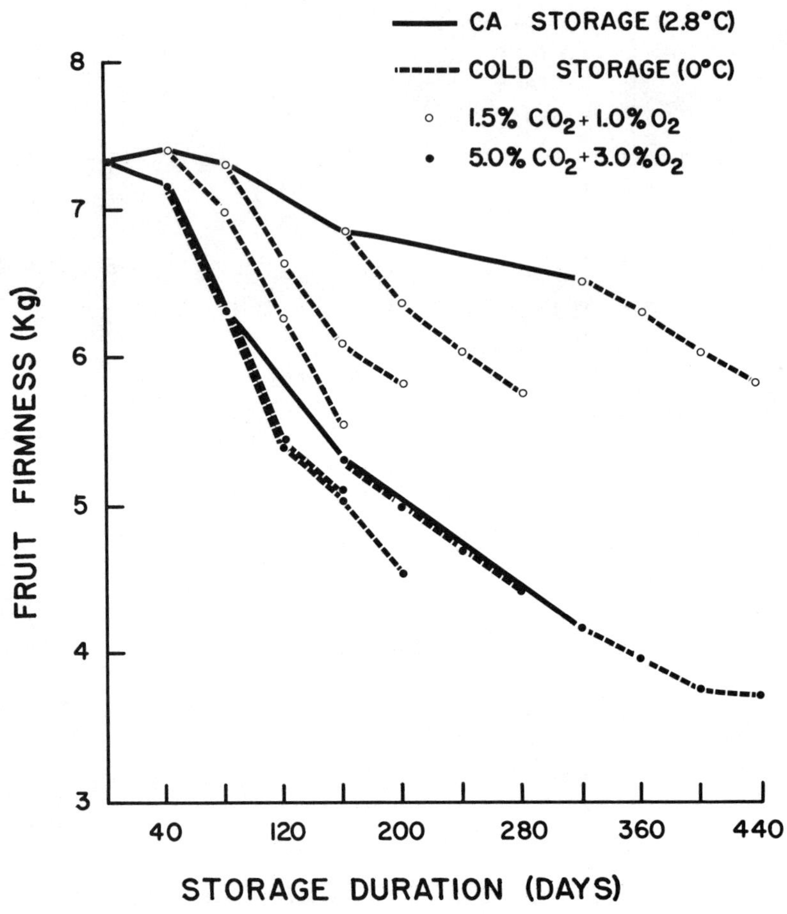

Fig. 2. Loss of firmness in McIntosh apples stored in low oxygen and controlled atmosphere storage and subsequent loss in 0°C air storage, 1979 crop.

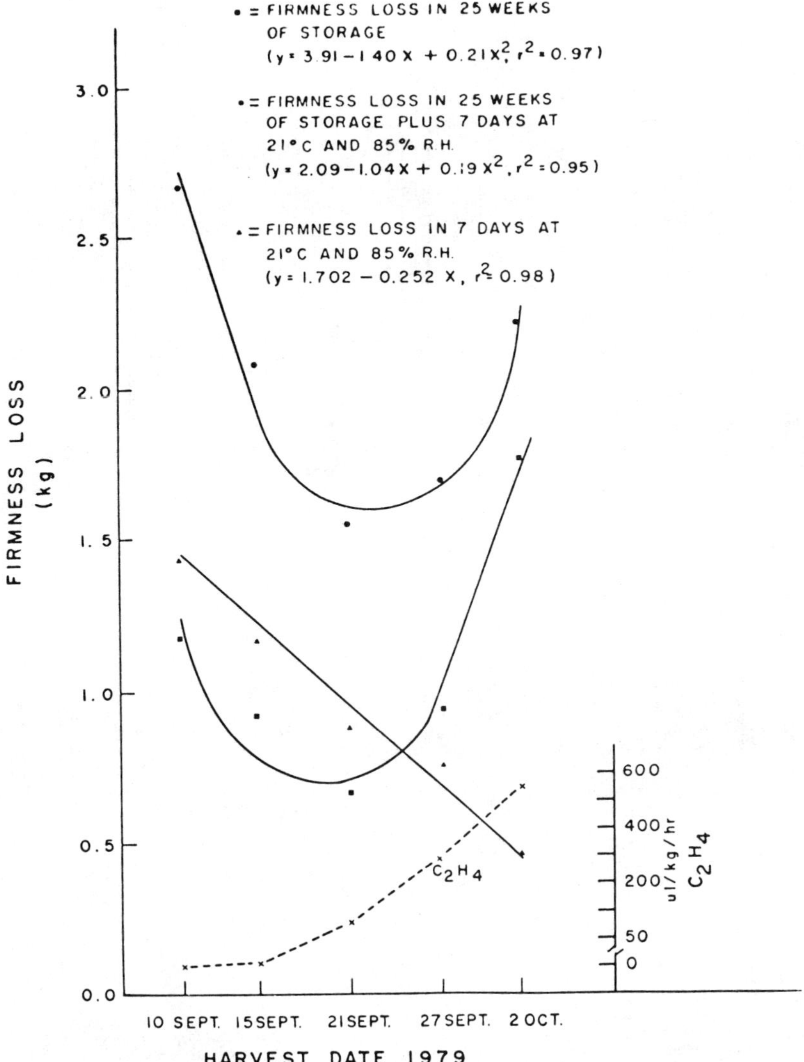

Fig. 3. Effects of maturity on firmness loss in McIntosh apples stored in 1.5% CO_2 + 1.0% O_2 at 2.8°C for 25 weeks. From J. Amer. Soc. Hort. Sci. 106: 159-162.

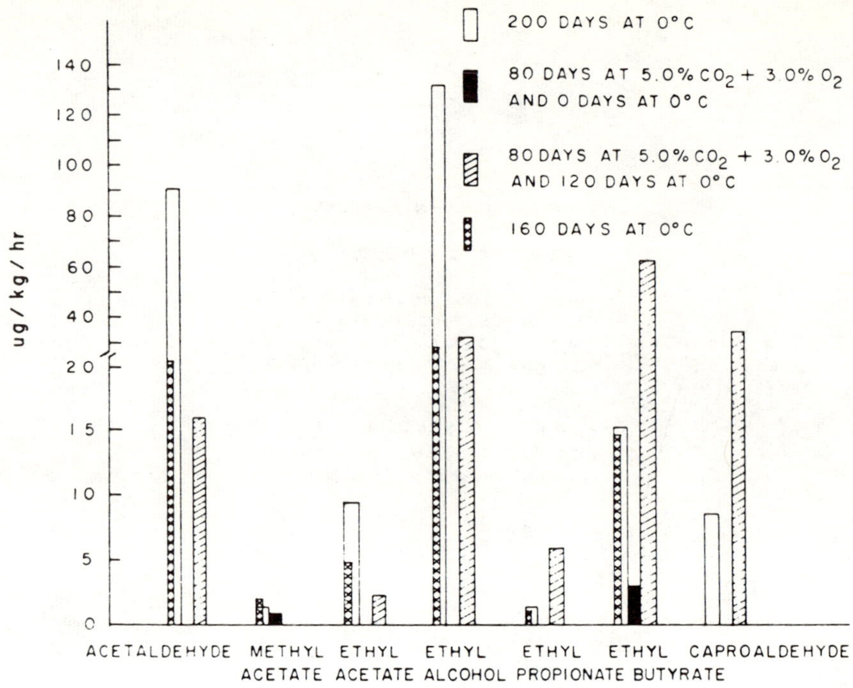

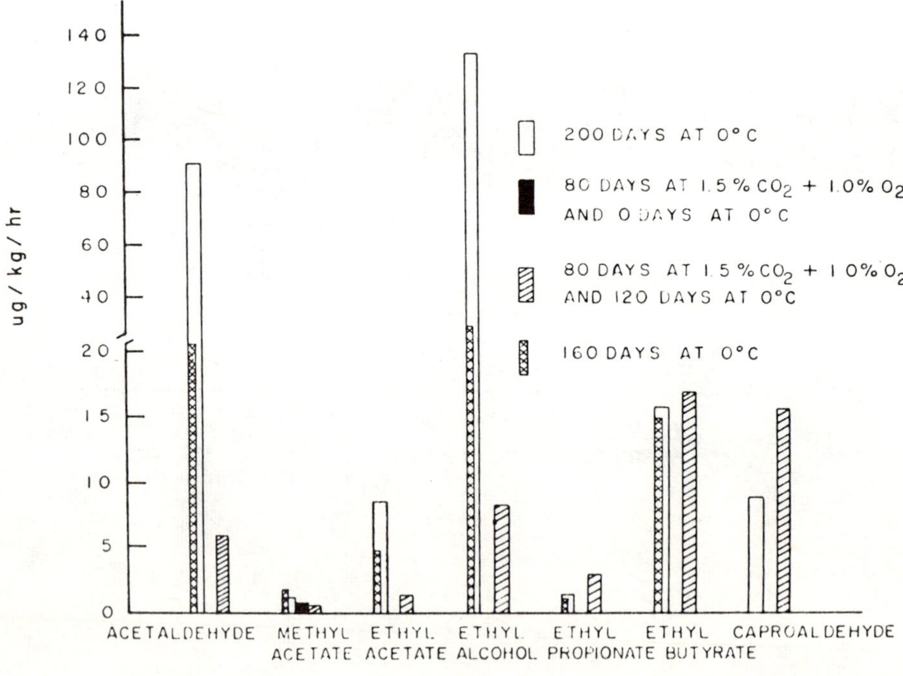

Fig. 4. Headspace volatile production by McIntosh apples removed from CA and cold storage, 1979 crop. Ten apples conditioned for 3 days at 21°C and 85% RH prior to determination. (n = 5).

Table 1. Fruit firmness, titratable acidity and soluble solids contents of 'McIntosh' apples stored in commercial 1.5% CO_2 + 1.0% O_2 and 5.0% CO_2 + 2.8% O_2 atmospheres at 2.8°C for 29 weeks[z].

Storage atmosphere ($\%CO_2 + \%O_2$)	Fruit firmness (kg)		Titratable acidity (mg malic/100 ml juice)		Soluble solids (%)	
	After storage	After storage plus 7 days at 20°C	After storage	After storage plus 7 days at 20°C	After storage	After storage plus 7 days at 20°C
1.5 + 1.0	6.58	5.43	605	553	11.41	11.5
5.0 + 2.8	5.72	4.40	564	498	11.23	11.2
Sig. level[y]	***	***	*	*	NS	*
S.E.D.(n + 40)	0.06	0.05	8	7	0.07	0.06

[z]From: J. Amer. Soc. Hort. Sci. 106: 159-162.

[y]Significance levels: *(P = .05), **(P = .01), ***(P = .001)

Table 2. The effects of storage oxygen and carbon dioxide concentrations on firmness, titratable acids, soluble solids level and internal browning in McIntosh apples stored for 26 weeks at 2.8°C.

Storage [CO_2] (%)	Storage [O_2] (%)	Examined after storage			Examined after storage plus 7 days at 21°C			
		Firmness (kg)	Soluble solids (%)	Titratable acids (mg malic/100 ml juice)	Firmness (kg)	Soluble solids (%)	Titratable acids (mg malic/100 ml juice)	Internal browning (%)
0	0.5	7.2	10.9	541	6.5	10.7	467	0
0	1.0	6.2	10.6	509	6.0	10.5	426	0
0	1.5	6.0	10.7	508	5.5	10.4	386	0
0	3.0	3.7	10.1	366	3.4	10.0	310	80
1.0	0.5	6.9	10.8	559	6.0	10.4	450	0
1.0	1.0	6.6	11.0	521	5.6	10.4	446	0
1.0	1.5	6.0	10.9	520	5.2	10.4	395	0
1.0	3.0	4.1	10.6	443	3.6	10.6	342	62
2.0	0.5	6.8	11.0	529	5.1	10.4	409	0
2.0	1.0	6.3	10.5	478	5.6	10.4	438	0
2.0	1.5	6.3	10.8	518	5.2	10.1	402	0
2.0	3.0	4.1	10.8	452	3.8	10.2	350	0
5.0	0.5	6.9	10.8	543	5.1	10.5	442	1
5.0	1.0	6.9	10.7	553	5.5	10.6	422	2
5.0	1.5	6.3	10.9	536	5.3	10.8	452	3
5.0	3.0	5.4	11.0	526	4.4	10.3	422	8

Table 3. Sensory evaluation of 'McIntosh' apples stored in commercial 1.5% CO_2 + 1.0% O_2 and 5.0% CO_2 + 2.8% O_2 atmospheres at 2.8°C for 29 weeks[z].

Quality attribute	Attribute scale (Keyterms on 1 = 7 scale)	After storage: Sensory rating, 1.5% CO_2 + 1.0% O_2	After storage: Sensory rating, 5.0% CO_2 + 2.8% O_2	After storage: Standard error of mean	After storage plus 7 days at 20°C: Sensory rating, 1.5% CO_2 + 1.0% O_2	After storage plus 7 days at 20°C: Sensory rating, 5.0% CO_2 + 2.8% O_2	After storage plus 7 days at 20°C: Standard error of mean
Skin texture	leathery-friable	4.22	4.10	0.07[NSy]	4.04	4.02	0.07[NSy]
Flavor	no flavor - McIntosh flavor	4.20	4.35	0.08[NS]	4.26	4.14	0.09[NS]
Sweetness/tartness	acidic-sweet	3.68	4.01	0.08***	3.97	4.21	0.06**
Juiciness	dry-juicy	4.88	4.79	0.06[NS]	4.68	4.47	0.06*
Texture	mealy-crisp	4.47	4.18	0.08**	4.33	3.73	0.08***
Overall acceptibility	poor-excellent	4.27	4.38	0.08[NS]	4.36	4.04	0.08***

[z]From: J. Amer. Soc. Hort. Sci. 106: 159-162.

[y]Differences between means within rows and examination date denoted as: NS, nonsignificant at 5% level; *significantly different at 5% level; **1% level or ***0.1% level.

Table 4. Effects of storage atmospheres and temperature on fruit firmness and internal browning disorders in apples held for 203 days, 1980 crop.

Storage atmosphere ($\%CO_2 + \%O_2$)	Storage temperature (°C)	McIntosh[z]		Golden Delicious[z]	
		Firmness (kg)	Internal disorders (%)	Firmness (kg)	Internal disorders (%)
5.0 + 3.0	0.0	6.02	99	6.13	20
	2.5	5.87	43	5.63	12
	5.0	4.33	33	5.27	12
1.5 + 1.5	0.0	6.65	95	6.46	11
	2.5	6.76	1	6.02	16
	5.0	5.23	0	5.77	5
1.5 + 1.0	0.0	7.46	99	7.12	15
	2.5	7.35	2	6.42	7
	5.0	6.77	0	6.51	3
1.5 + 0.5	0.0	7.38	99	7.13	17
	2.5	7.35	4	6.95	13
	5.0	7.04	10	7.09	2

[z]Values are means of 5 grower lots.

LOW-O_2 STORAGE - CAN IT BE?

G. Kapotis, E.C. Lougheed and S.R. Miller
Horticultural Science Department, University of Guelph, Guelph, Ontario
Canada, N1G 2W1, and Agriculture Canada, Smithfield Experimental Farm,
Box 340 Trenton, Ontario, K8V 5R5

For the past 25-30 years most commercial recommendations for CA storage in North America listed O_2 levels not less than 2% (Kader and Morris, 1977; Porritt, 1977; Smock, 1979) and in many cases in the 2.5 to 3.0% range (Andersen, 1975; Blanpied, 1977). Although it was known that in the UK there was some large-scale storage of Cox's Orange Pippin in low-O_2 atmospheres based on work at East Mallying (North et al., 1976 and 1977), the effect of low-O_2 atmospheres did not appear to be as useful for long-term storage of McIntosh Red apples (Bubb, 1977). As well, the occasional occurrence in Ontario, and New York (Blanpied et al., 1975) of complete rooms of apples ruined by low-O_2 atmospheres made the prospect of recommending such atmospheres less than attractive. In some senses this may have been somewhat irrational because in most cases the true O_2 concentration in the room was not known and may have approached zero because of errors involving a leaky sampling line on the analyzer.

Moreover there had been indications in 1969 that atmospheres containing as little as 0.5% O_2 might be used for pears if fruit were carefully selected for maturity (Claypool, 1969), and in 1967 Anderson reported on storing Delicious apples in 1% O_2 (Anderson, 1967). Recent work has confirmed the possibility of using 0.5% O_2 with pears (Mellenthin et al., 1980). In the 1966-'67 and 1967-'68 harvest-storage seasons we had also done work on "low-O_2" atmospheres but with the object of differentiating between low-O_2 and high CO_2 injuries (Lougheed and Franklin, 1967 and 1968). In that work McIntosh was more susceptible to an atmosphere of 1% O_2 than Delicious or probably Northern Spy. McIntosh and Delicious were more susceptible to low-O_2 injury at 3.3°C than 0°C but with Northern Spy this effect was reversed. There were the suggestions that CO_2 levels (3 to 9%) would ameliorate the effect of low-O_2 (1%) levels and that earlier harvested fruit were injured more severely by low-O_2 conditions. There was an obviously large influence of source of fruit on susceptibility to low-O_2 injury. Overall at that time, the variability and unpredictability in response made us disregard low-O_2 storage as a practical possibility and re-affirmed the CA recommendation then in use in Ontario.

As well as the work in the U.K. (North et al., 1976 and 1977) suggesting that low-O_2 storage is a practical possibility, the work with low pressure storage (Bubb, 1975) indicates that apples will withstand low-O_2 atmospheres under certain conditions. Also, work with vegetables (Isenberg, 1979) using atmospheres containing as little as 1% oxygen, indicate the possibility of using such conditions although the dissimilarity in morphology, anatomy and physiology make comparison risky. Probably recent work in low-O_2 storage of apples in North America began some 4 to 5 years ago. This work has resulted in several reports (Bourne and Dewey, 1980; Irwin and Dilley, 1980, Lidster, 1980; Lougheed et al., 1980; Olsen, 1981).

Our work began in 1977 with several pre-conceived ideas:

1) The low-O_2 levels (<2%) should be established slowly to minimize injury. (Wrong).
2) That we should be prepared for considerable alcohol production. (Wrong).
3) That an alcohol detector would be a necessary component of low-O_2 storage equipment. (Wrong).

4) That bringing the O_2 level up to normal air might reduce alcohol-related injuries. (?).
5) That at low-O_2 levels high CO_2 (> 1%) might induce injury. (Wrong).
6) That control of commercial low-O_2 atmospheres would require constant automatic electronic monitoring and maintenance of O_2 levels. (Wrong).
7) That early-harvested apples would be less subject to injury than riper fruit (right?), and would be influenced most beneficially by low O_2 conditions. (?).
8) Few commercial CA rooms were tight enough to allow low-O_2 storage. (?).
9) Commercial rooms should be tight enough that a generator, the use of which could result in wide fluctuations in O_2 levels, would not be required except during initial O_2 pulldown. (Right).

There are also economic considerations; the risk-possible profit ratio, and the question of demand for storage apples on a 12-month basis.

In the first years our results were not good in that with McIntosh we obtained injury without improved firmness; with Spartan and Empire little-or-no injury but no improvement in quality; and in one year we obtained neither injury nor improvement in quality. The lack of results was due to slow O_2 pulldown and/or delayed initiation of CA storage; the injury due to oxygen concentrations lower than 0.5%. But from the results of the first two years and for 1979-'80 there were some conclusions.

1) For a major effect the low-O_2 atmosphere would have to be established quickly.
2) McIntosh are more susceptible to low-O_2 than Spartan or Empire. Field-daminozide application reduced injury in McIntosh apparently because the daminozide-treated fruit were less mature.
3) That an atmosphere of 1 $\pm$ .25% O_2 is desirable and < 0.5% probably harmful although we were not sure of the duration necessary to cause injury.
4) That an atmosphere < 1.5% O_2 was desirable and > 1% could not be compensated for by a higher CO_2 (1.5%) level.

Current Season's Results (1980-'81)

Collation and analyses of the work are not complete but the research will be reported in its entirety in the M.Sc. thesis of the senior author.

Results for fruit firmness of McIntosh are given in Fig. 1-3 for the first removal from storage in February 1981; and for the fruit removed in June, 1981 in Tables 1 and 2. The effectiveness of the 1-0 atmospheres in maintaining the firmness of the fruit is obvious, and the merit of the 1-0 atmosphere versus the 3-5 at 4°C is apparent. Early in the season, storage in the 3-2.5 atmosphere at 0°C was almost as effective as the 1-0 atmosphere at 4°C.

Information which is not given in figure or tabular form is:

1) No low-O_2 injury was observed although the rotting pattern on a few fruits suggested a previous low-O_2 injury.
2) No off-flavors in the low-O_2 fruit were detected.
3) Little alcohol was detected in the chambers although our method of dectection needs refinement.
4) Spartan and Empire were not injured by the low-O_2 atmospheres at 0°C but the benefit derived (data not presented) was not as great as with McIntosh, possibly because there was a delay in imposition of the atmospheres.
5) The on-the-shelf performance of the low-O_2 fruit was excellent and the differences in quality were maintained at 20° for up to a week.

CONCLUSION

Low-O_2 storage of apples, particularly McIntosh, may be effective and practical but it is not for everyone. Only those who choose fruit carefully, who monitor and regulate atmospheres meticulously, and who are prepared to take risks should attempt it. Losing one's nerve in mid-season could be disastrous if there is no CO_2 in the room because at 3.5°C an atmosphere of 3% O_2 plus 0% CO_2 (a "safe" atmosphere) is little better or possibly worse than air storage at 0°C.

Low-O_2 storage of cultivars other than McIntosh <u>may</u> be less risky but probably of less benefit too.

ACKNOWLEDGEMENTS

The authors acknowledge with gratitude the financial assistance of the Ontario Ministry of Agriculture and Food through regular operating and special Wintario grants, the National Science and Engineering Research Council (Canada), and help from the Ontario Ministry of Agriculture and Food, Horticultural Experimental Station, Simcoe, Ontario. The authors also acknowledge the help from papers and people not cited herein.

REFERENCES CITED

Anderson, E.T. (ed), K.A. Clarke, R.W. Davey, E.W. Franklin, H.A. Hughes, E.C. Lougheed, and B.J.E. Teskey. 1975. Harvesting, storing, and packing apples. Ont. Min. Agr. and Food. Pub. 431, Agdex 211/50-60:33-34.

Anderson, R.E. 1967. Experimental storage of eastern-grown Delicious apples in various controlled atmospheres. Proc. Amer. Soc. Hort. Sci. 91: 810-820.

Blanpied, G.D., R.M. Smock and F.W. Liu. 1975. Cornell Fruit Handling and Storage Newsletter, Cornell U. July, 1975:7.

Blanpied, G.D. 1977. Requirements and recommendations for eastern and mid-western apples. Proc. 2nd Nat. CA Res. Conf., Hort. Rpt. #28 (D.H. Dewey Ed.). Mich. State U.:225-230.

Bourne, M.L. and D.H. Dewey. 1980. Very low oxygen for the storage of McIntosh apples. HortScience 15(3), Sec. 2:424.

Bubb, M. 1975. Hypobaric storage. Rpt. East Malling Res. Sta., 1974:81

Bubb, M. 1977. Variety trials. McIntosh Red. Rpt. East Malling Res. Sta., 1977:93.

Claypool, L.L. 1969. Controlled atmosphere storage of pears. Proc. 1st Nat. CA. Res. Conf., Hort. Rpt. #9 (D.H. Dewey, R.C. Herner, and D.R. Dilley Eds.), Mich. State U.:64-65.

Irwin, P.L. and D.R. Dilley. 1980. The effect of hypobaric and low oxygen storage on the ripening of Empire and Idared apple fruits. HortScience 15(3), Sec. 2:424.

Isenberg, F.M.R. 1979. Controlled atmosphere storage of vegetables. Hort. Rev. 1:337-394.

Kader, A.A. and L. L. Morris. 1977. Relative tolerance of fruits and vegetables to elevated CO_2 and reduced O_2 levels. Proc. 2nd Nat. CA. Res. Conf., Hort. Rpt. #28 (D.H. Dewey, Ed). Mich. State U.:260-265.

Lidster, P.D., F.R. Forsyth, and H.J. Lightfoot. 1980. Low oxygen and carbon dioxide atmospheres for storage of McIntosh apples. Can. J. Plant Sci. 60:299-301.

Lougheed, E.C. and E.W. Franklin. 1967. Rpt. of the Comm. on Hort. Res., Can. Hort. Council:154-155.

Lougheed, E.C. and E.W. Franklin. 1968. Rpt. of the Comm. on Hort. Res., Can. Hort. Council:155.

Lougheed, E.C., J.T.A. Proctor and S.R. Miller. 1980. Low-O_2 storage of three apple cultivars. HortScience. 15(3), Sec. 2:424.

Mellenthin, W.M., P.M. Chen, and S.B. Kelly. 1980. Low oxygen effects on dessert quality, scald prevention, and nitrogen metabolism of d'Anjou pear fruit during long-term storage. J. Amer. Soc. Hort. Sci. 105: 522-527.

North, C.J., Bubb, M. and Cockburn, J.T. 1976. Storage of Cox's Orange Pippin apples in low oxygen. Rpt. East Malling Res. Sta., 1975:76-77.

North, C.J., D.J. Chappell, L.E. Sharp, M. Bubb. and J.T. Cockburn. 1977. Semi-commercial scale storage of Cox's Orange Pippin apples in 1.25% oxygen. Rpt. East Malling Res. Sta., 1976:91.

Olsen, K. 1981. Rapid CA, low oxygen storage helps preserve apple quality, but harvest time equally important. The Goodfruit Grower; Jan. 1, 1981: 20-21.

Porritt, S.W. 1977. Conditions and practices used in CA storage of apples in Western United States and B.C. Proc. 2nd Nat. CA Res. Conf., Hort. Rpt. #28 (D.A. Dewey, Ed) Mich. State U.:231-232.

Smock, R.M. 1979. Controlled atmosphere storage of fruits. Hort. Rev. 1: 301-336.

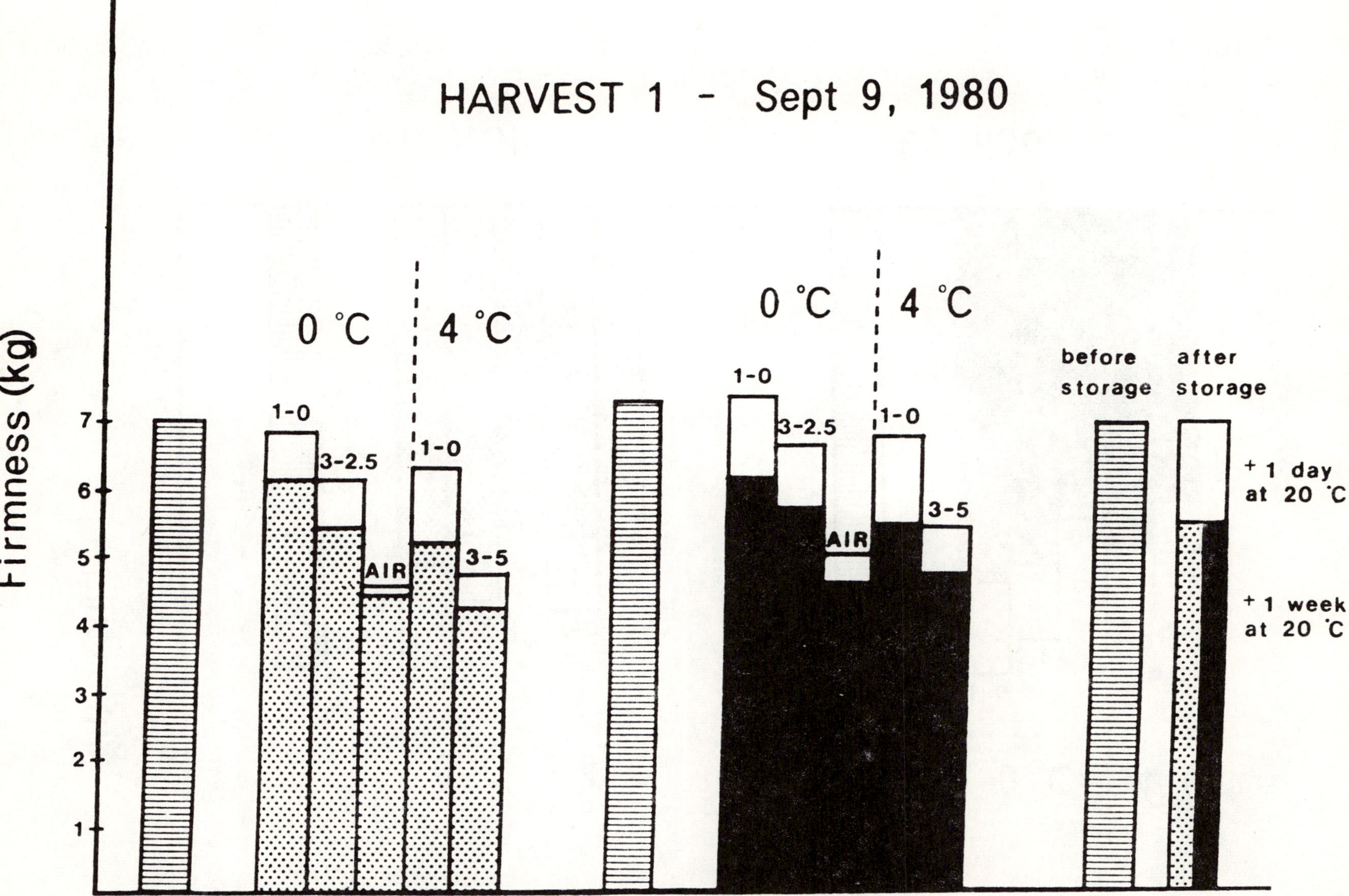

Fig. 1. McIntosh apples evaluated in Feb. 1981. Data are means of 3 replicates, 10 fruits per replicate, 2 readings per fruit. Numbers (eg. 1-0) represent % O_2 + CO_2.

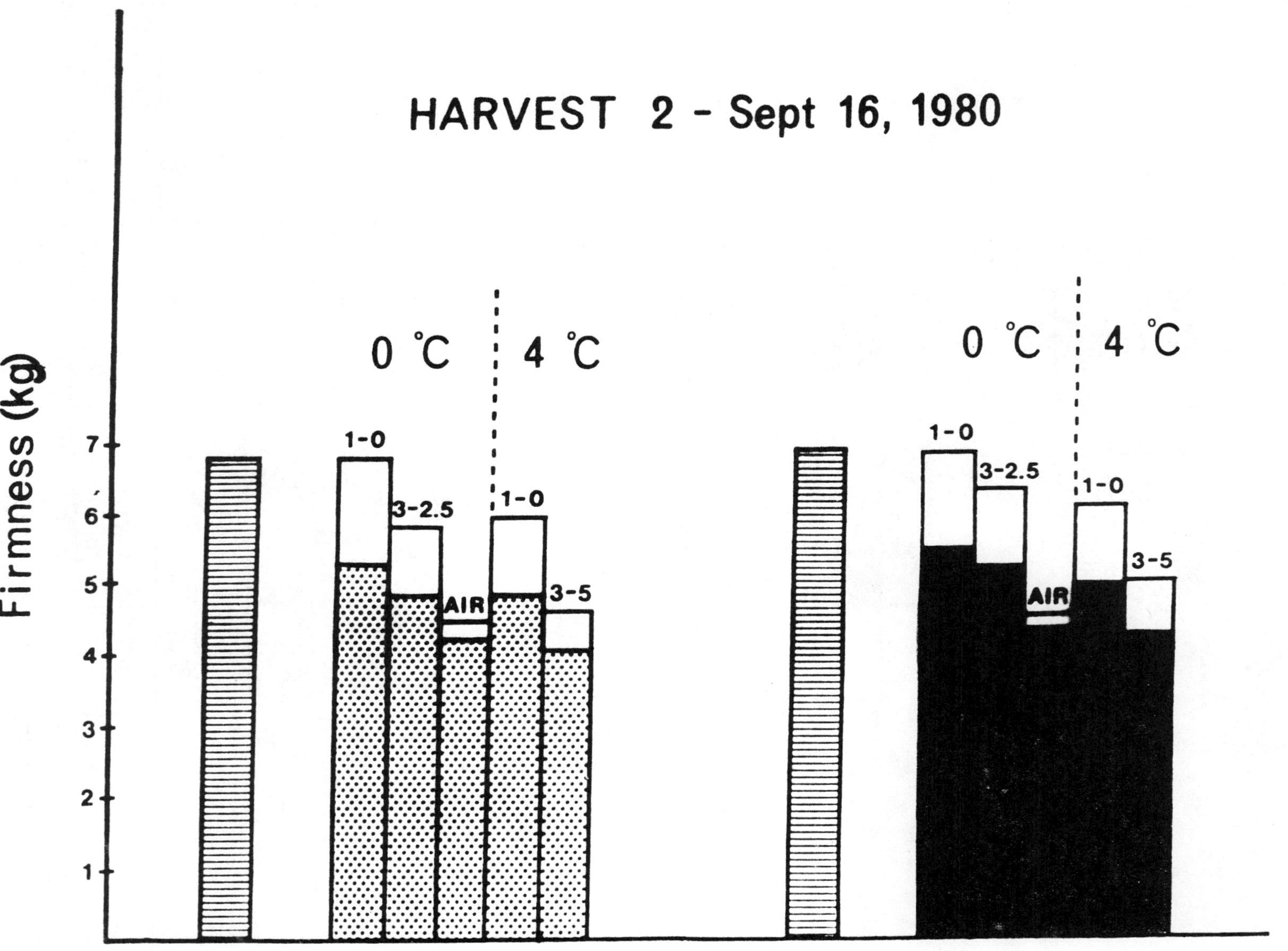

Fig. 2. McIntosh apples evaluated in Feb. 1981. Data are means of 3 replicates, 10 fruits per replicate, 2 readings per fruit. (For explanation of code see Fig. 1)

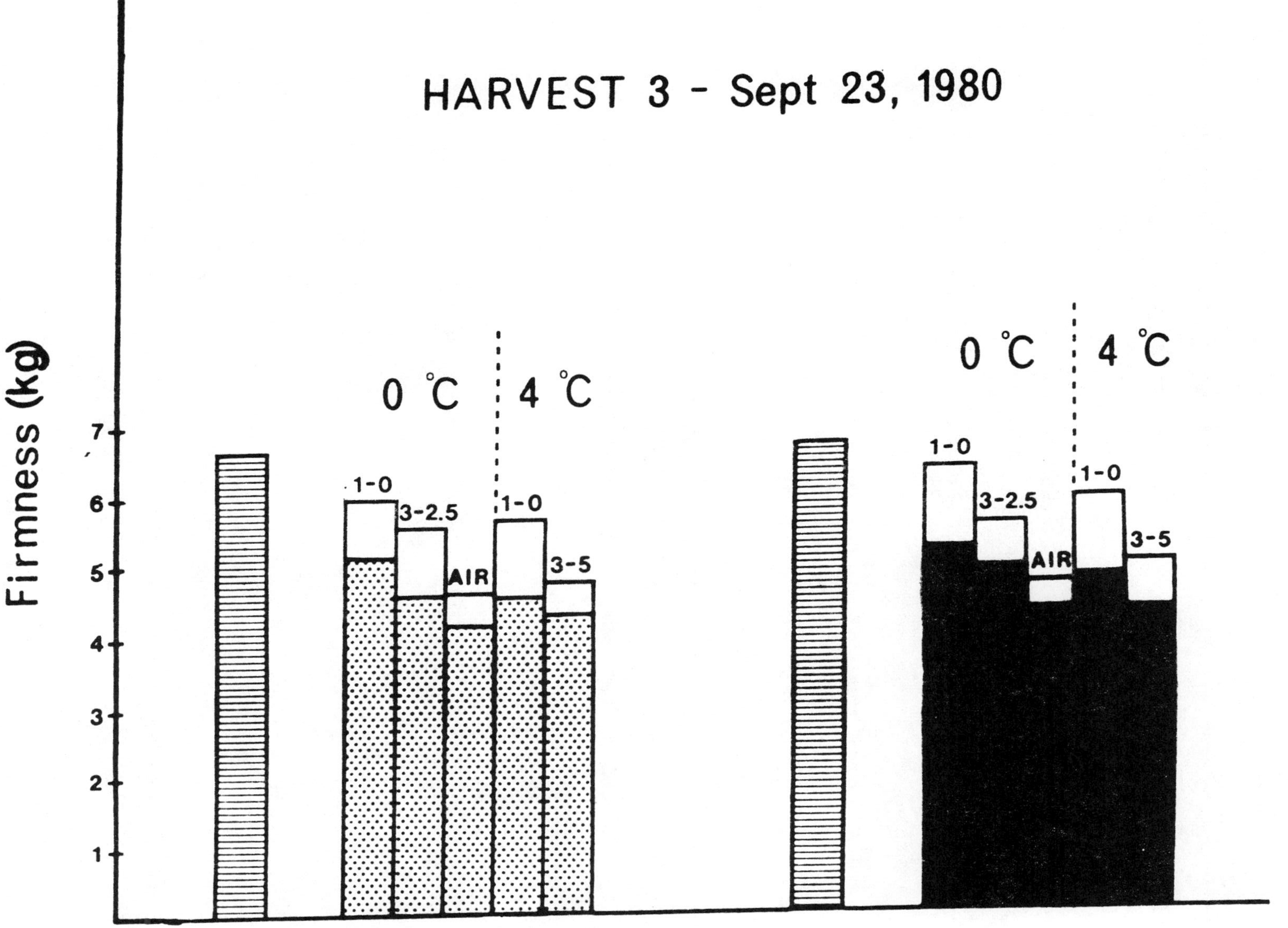

Fig. 3. McIntosh apples evaluated in Feb. 1981. Data are means of 3 replicates, 10 fruits per replicate, 2 readings per fruit. (For explanation of code see Fig. 1).

Table 1: Firmness (kg) of McIntosh apples harvested in Sept. 1980 and stored until June, 1981

Treatments		Harvest date (1980)		
O_2-CO_2(%)	Temp.	Sept. 9	Sept. 16	Sept. 23
1 - 0	0°C	6.9a[z/]	6.6a	6.0[y/]
1 - 0	4°C	6.4b	5.7b	5.4
3 - 2.5	0°C	5.9c	5.6b	5.1
3 - 5.0	4°C	4.4d	4.1c	4.2
Air	0°C	3.8e	3.6d	3.7
Daminozide[x/]				
-		5.3*	4.9*	4.7
+		5.7	5.3	5.0

[z/] Data the mean of 3 replicates, 10 fruit per replicate, 2 readings per fruit. Statistical comparisons only within one harvest date at P = 0.05.

[y/] Significant interaction of daminozide X atmos. temp.; results in next table.

[x/] Difference between - and + daminozide judged significant by F test with 1 degree of freedom.

Table 2: Firmness (kg) of McIntosh apples harvested on Sept. 23 1980 and stored until June, 1981

Treatments		Daminozide	
O_2 - CO_2 (%)	Temp.	-	+
1 - 0	0°C	6.1a[z/]	6.0a
1 - 0	4°C	5.1c	5.6b
3 - 2.5	0°C	4.9d	5.4b
3 - 5.0	4°C	4.1e	4.2e
Air	0°C	3.6g	3.8f

[z/] Data the mean of 3 replicates, 10 fruit per replicate, 2 readings per fruit. Table represents the significant interaction of daminozide treatment X atmos.-temp. Difference between all means judged by Duncan's multiple range test at P = 0.05

EFFECTS OF ULTRA-LOW OXYGEN CONDITIONS ON THE STORAGE QUALITY OF ENGLISH COX'S ORANGE PIPPIN APPLES

R. O. Sharples
Fruit Storage Division
East Malling Research Station
Maidstone, England

The main English apple variety, Cox's Orange Pippin, has excellent dessert qualities but, in common with many other apple varieties grown in northern Europe, it is susceptible to low temperature injury when stored for several months at temperatures below 3°C. Consequently its maximum commercial storage life in air is limited to only 12 weeks and satisfactory marketing of the variety depends on the extensive use of controlled atmosphere (CA) storage. For many years, much of the crop has been stored in 5% CO_2 + 3% O_2 at 3.5-4°C. Although by these means the storage life of the variety can be greatly extended, it is found that after cool growing seasons (Sharples, 1975) core flush (syn. brown core) is often prevalent in Cox stored in this regime due mainly to the relatively high concentration of CO_2. During the 1960's, extensive studies (Table 1) involving samples of Cox from over 150 commercial orchards showed that the incidence of core flush could be greatly reduced by lowering the O_2 concentration to 2% and maintaining CO_2 concentration below 1% by continuous scrubbing of the atmosphere either with an external mechanical scrubber or by placing bags of lime over the tops of the bins of fruit in the store.

Parallel studies (Sharples, 1980) demonstrated the importance of nutritional factors in determining the storage potential of apples grown in different Cox orchards. These findings led to changes in fertilizer recommendations and to the development of orchard sprays which are now used to supplement both calcium and phosphorus concentrations in apples; supplementation of fruit calcium is also provided by post-harvest application of calcium chloride. By combining these measures with storage in 2% O_2 (< 1% CO_2), Cox may now be kept in satisfactory condition for up to 26 weeks. However, in certain seasons, fruit texture becomes soft and mealy towards the end of this storage period and, as market factors now dictate the need for further extension of the storage period of this variety, work during the last 6 years has been concentrated on developing recommendations which will delay ripening changes even more.

In 1974, as part of a study on the interactions of temperature and minimum oxygen requirements, North and Cockburn (1976) found that Cox could be stored for at least 28 weeks in a flow (5 ℓ hr^{-1}) containing 1.0% O_2 in nitrogen as long as the O_2 concentration was maintained within $\pm$ 0.1%. The work also showed that, to avoid anaerobic respiration, it was essential to cool the apples to 4°C before establishing the low O_2 concentrations. Knee (1980) has shown that, when transferring apples from air at 18°C to either air or 2% O_2 at 3°C, the respiration of the fruit continues at a higher rate for at least a day. It is possible that a period of acclimatisation is normally necessary before ultra-low oxygen conditions can be established without the risk of anaerobiosis. In the same year, twenty 18 kg boxes of Cox were stored at 4°C in a CA cabinet equipped for automatic oxygen control. The O_2 level was slowly reduced from air to 1% O_2 over the first 8 days; thereafter conditions were maintained within ± 0.1% O_2. Stored in this way, Cox were consistently firmer, greener and more aicd than similar fruit stored in 2% O_2 (< 0.7% CO_2) or 5% CO_2 + 3% O_2 and they remained in acceptable condition for about 15 weeks longer when stored in ultra-low O_2 instead of the other two CA conditions. Alcohol levels in apples stored in 1% O_2 remained at the same low background level as that generally found in 2% O_2 apples.

Table 1. Core flush (brown core) in Cox's Orange Pippin apples stored for 20 weeks at 3.5-4.0°C (1964-1969) (40-fruit samples from approx. 150 orchards per condition per year).

	Average core flush (%)					
	1964	1965	1966	1967	1968	1969
5% CO_2 + 3% O_2	4	20	4	25	10	7
2% O_2 (< 0.5% CO_2*)	0	3	1	6	1	1

*maintained by continuous scrubbing

In 1976-77, four sets of twenty 18 kg samples of Cox apples were picked just prior to the onset of the climacteric rise in respiration rate and cooled to 4°C. Controlled atmospheres of 1.0, 1.25, 1.5 or 2.0% O_2 were then established and automatically maintained to within ± 0.1%. CO_2 concentrations were kept below 0.7% with continuous scrubbing over lime. Samples of the fruit were inspected on four occasions between January and May. Records were taken immediately ex-store and again after 2 weeks at 10°C. The storage and taste panel data were summarised by averaging the results for the ex-store inspections (Table 2) and those made after a simulated marketing period (Table 3). These data clearly show that as O_2 levels were progressively decreased, so firmness (measured by the penetrometer) and crispness, juiciness and acidity (measured in sensory terms) increased. Ground colour also remained greener at the lower O_2 concentrations but the apples were less sweet and developed progressively less aromatic flavour. Although some recovery of the aroma occurred post-storage, the more pronounced aromatic flavour of Cox stored in 2% O_2 was not realised even after 14 days at 10°C. The improvement in texture and acidity persisted throughout marketing, particularly after storage in 1.0% O_2. Similar findings have been reported for McIntosh stored in 1.5% CO_2 + 1% O_2 (Lidster et al., 1980). These clear-cut effects on aroma and texture have been found in all subsequent work on Cox in which ultra-low O_2 has been compared with storage in 2% O_2.

Table 2. The influence of oxygen concentration on the quality of Cox apples measured 1 day after removal from CA storage (1976/77) (Means of 4 inspections on 17 Jan, 28 Feb, 12 Apr and 9 May)

	Oxygen concentrations (%)*			
	1.0	1.25	1.50	2.0
Ground colour (0=green, 15=yellow)	7.4	8.2	8.5	9.8
Firmness (kg, 8 mm dia. probe)	3.70	3.61	3.51	3.43
Sensory scores: (1=low, 5=high)				
Juiciness	3.03	2.83	2.73	2.43
Crispness	3.15	2.68	2.13	1.95
Acidity	2.65	2.35	2.10	1.93
Sweetness	2.35	2.73	2.63	2.90
Aroma	2.07	2.27	2.69	2.87

*CO_2 concentration maintained below 0.7%

Similar differences in eating quality were detected in a comsumer survey organised in early April 1977 and involving 200 respondents living in different parts of the UK. Cox apples from a 17.5 tonne store operated at 1.25% O_2 were given higher ratings for texture and juiciness but lower ones for appearance and aroma than fruit stored in 2% O_2. Overall acceptability of the fruit from 1.25% O_2 was slightly greater than that of the 2% O_2 fruit, but this preference was more marked among younger people and those living in the south of England.

It should be noted that this consumer study was made while the texture of the 2% O_2 fruit was still acceptable; evidence from taste panel assessments made after a further 5 weeks storage of fruit from the same trial indicated that the overall acceptability of the Cox from ultra-low O_2 storage had become markedly superior.

Table 3. The influence of oxygen concentration on the eating quality of Cox apples held for 14 days at 10°C following CA storage (1976/77) (Means of 4 assessments on 31 Jan, 14 Mar, 25 Apr and 23 May)

	Oxygen concentration (%)*			
	1.0	1.25	1.50	2.0
Sensory scores: (1=low, 5=high)				
Juiciness	3.00	2.53	2.23	2.25
Crispness	2.78	2.28	2.05	2.00
Acidity	2.23	1.90	1.63	1.68
Sweetness	2.80	2.78	3.00	2.73
Aroma	2.30	2.39	2.52	2.64

*CO_2 concentration maintained below 0.7%

Recent work (Knee and Sharples, 1981) suggested that storage in low O_2 atmospheres may limit the availability of the alcohols from which the aromatic esters are derived thereby depressing production of the important flavour volatiles, hexyl and butyl acetates. It seems possible that, of necessity, CA storage reduces aroma production just as it retards softening, loss of acidity and other ripening changes. Thus the optimum O_2 concentration for any variety will depend on the balance which is required between its aromatic and textural attributes. This is likely to vary according to local marketing requirements and, in the case of English Cos, recommendations have been made which are based on a mean O_2 concentration of 1.25% to allow slightly better aroma development without seriously affecting the improved retention of texture.

In 1975, North, Bubb and Cockburn (1976) investigated the effect of time of picking on the storage of Cox in 1.25% O_2. Fruit samples were picked and immediately cooled to 4°C on the 9th and 24th September and the 15th October. The O_2 concentration was allowed to fall to about 5% over the first 7 days before being reduced to 12.5% by displacing the atmosphere with nitrogen. Although the fruit (which had not received a post-harvest fungicide treatment) from the third pick developed severe rotting by February, none of the samples showed any build-up of alcohol throughout 30 weeks storage. However, an internal disorder, which appeared as discrete areas of corky tissue drying out to form a diffuse speckled pattern throughout the cortex, was recorded. The disorder tended to appear earlier and more frequently in fruit from the later picks. A second picking date experiment was done on the 1978 crop but no adverse effects of storage in ultra-low O_2 were found even when picking was delayed until mid-October. The latter result appears to contradict the findings of Lidster et al. (1981) who showed that, when stored in 1.5% CO_2 + 1% O_2, late-picked McIntosh apples were more sensitive to internal senescent breakdown than those picked earlier.

During the 1976/77 season, a semi-commercial scale trial was done in which Cox apples from six different orchards were harvested in bulk bins, drenched in 0.1% thiophanate-methyl, and then randomly distributed between two, 17.5 tonne stores, one of which was operated at 1.25% O_2 (< 0.7% CO_2) and the other at 2% O_2 (< 0.7% CO_2). The condition of the fruit after storage until late March differed in the same way as that found in the laboratory trials (eg Table 2), but this large-scale trial confirmed the feasibility of extending the technique to apples stored in bulk bins and handled on a commercial basis. Details of the type of control equipment used are given by Jameson (1981). No differences were found in the concentration of O_2 or CO_2 in any part of the store and no significant build-up of alcohol occurred in the fruit. The late-storage corking disorder was recorded by the end of March in samples of fruit from four of the six orchards. Where it occurred, it was consistently more prevalent in apples stored in 1.25% O_2 rather than 2% O_2. In the most severe case, 24.5% of the fruit were affected in ultra-low O_2 compared with only 0.8% of fruit in 2% O_2.

The disorder was again evident at the end of May 1979 in samples of Cox obtained from the 1978 crops in fourteen commercial orchards (two from Germany, four from Holland and eight from S.E. England). The fruit was stored either in 1.25% O_2 (< 0.4% CO_2) or 2% O_2 (< 0.4% CO_2) when it was assessed for both classical bitter pit symptoms and late-storage corking. Some difficulty was experienced in separating the symptoms since late-storage corking has a similar texture and bears a superficial resemblance to bitter pit. However, the lesions of the corking disorder become much more extensive than those of typical bitter pit often forming complete sectors of affected tissue up to 1 cm in diameter. It will be seen (Table 4) that, whilst some of the orchards (eg sites D, J, L, N) with a high potential for bitter pit (as evidenced by the development of typical symptoms during 12 weeks air storage at 3°C) also developed a considerable amount of both bitter pit and late-storage corking in 1.25% O_2, no correlations were apparent between the incidence of pit in air storage and that of pit or the corking disorder in samples from other orchards (eg sites G and K). Both bitter pit and late-storage corking were much less prevalent in 2% O_2. Although the cause of the latter disorder is not yet understood, it is being recommended, as an interim measure, that apples which when analysed at harvest contain less than 5 mg 100 g^{-1} of calcium or over 150 mg 100 g^{-1} of potassium should be excluded from storage in O_2 concentrations of less than 2%.

In 1978/9, Johnson et al. (1980) found that Cox stored in ultra-low O_2 was slightly more sensitive to low temperature injury (LTB). Although, when Cox are stored at 4°C, the incidence of this disorder in 1.25% O_2 is generally of no commercial signifi-

Table 4. Effects of O_2 concentration on development of bitter pit and "late-storage corking" in Cox apples (1978/79)

Site†	Bitter pit (%)			Late-storage corking (%)	
	12 wk Air 3°C	36 wk 2% O_2*	36 wk 1.25% O_2*	36 wk 2% O_2*	36 wk 1.25% O_2*
A	2.5	0	1.7	0	0
B	0	0	0	0	0
C	5.0	0	6.3	0	0
D	21.1	1.7	7.2	0	11.1
E	0	0.9	1.3	0.9	0
F	3.4	4.9	5.7	0	0
G	17.5	0	0	0	1.3
H	12.7	5.2	12.8	0	2.6
I	3.4	1.6	2.6	0	4.1
J	36.3	5.3	25.4	1.3	14.4
K	0	0	11.5	0	16.7
L	26.7	26.5	17.0	1.4	28.5
M	0	1.8	1.9	0	3.6
N	18.2	0	35.0	3.7	31.5

† A–D, S.W. Holland; E–L, S.E. England; M & N, Rhineland, Germany

* Stored at 3.8°C with CO_2 kept below 0.4%

cance, data obtained in 1980/81 (Table 5) for Cox stored at 3.7°C and 2.7°C demonstrated that sensitivity to LTB is markedly increased by storage in ultra-low O_2. It seems probable that the additional stress which is caused to the cortical tissues by storage at limiting O_2 concentrations increases sensitivity to membrane damage at storage temperatures which are safe for storage at higher O_2 levels.

Current UK recommendations for Cox limit the storage period in 1.25% O_2 (< 1% CO_2) until the end of April and exclude any fruit which does not meet the required mineral composition standards.

It is also recommended that the store should be held at 2% O_2 for at least one week before the O_2 concentration is reduced further. Although the normal storage temperature is 3.5-4.0°C, a minimum temperature of 4°C should be adopted when storing fruit harvested after cool summers to minimise the risk of LTB. Where these guidelines have been followed, no LTB or late-storage corking have been reported in commercial ultra-low O_2 stores and, following the very cool summer of 1980, several thousand tonnes of Cox were stored in 1.25% O_2 with marketing continuing until the first week of May. In most cases, O_2 was kept within ± 0.1% and the fruit temperature was maintained at 3.9 ± 0.3°C. Thus, in a year when fruit texture was generally suspect, ultra-low O_2 conditions enabled the English fruit industry to maintain fruit quality and extend its marketing season for 6 weeks longer than normal.

Table 5. Low temperature breakdown (LTB) in Cox apples stored for 35 weeks at 3.7°C or 2.7°C in either 2% O_2 or 1.25% O_2 (1980/81) (Means for samples from 25 orchards in S.E. England)

	2% O_2 (<0.7% CO_2)		1.25% O_2 (<0.7% CO_2)	
	3.7°C	2.7°C	3.7°C	2.7°C
LTB (%)	0.3	3.7	3.3	24.5
LTB, severity index (max 75)	0.2	0.6	0.5	4.6

Current work is concerned with the influence of different levels of CO_2 on Cox stored in 1%, 2% and 3% O_2. This investigation became necessary since most mechanical scrubbers cannot maintain CO_2 levels below 1% without introducing relatively large volumes of air which then make it impossible to obtain ultra-low O_2 conditions. Hence, in practice, it would be more convenient to operate at somewhat higher CO_2 concentrations. Further clarification is also needed of the relationship between the rate of establishment of ultra-low O_2 conditions and subsequent storage quality while more information is required on maturity at harvest in relation to minimum O_2 requirements. Practical methods for restoring the aromatic qualities of fruit and of controlling the late-sotrage corking disorder will also be necessary before the use of ultra-low oxygen storage can be fully exploited.

I am indebted to many colleagues, past and present, for much of the research on which this paper is based. Particular thanks are due to Mr M Bubb and Mrs J P Perkins for operating the small-scale CA cabinets, Mr P Gull for constructing the automatic O_2 controller and Mr D S Johnson for help in identifying the various types of storage disorders.

References

Jameson, J. (1981). Automated control of CA conditions. Proceedings of 1981 National C.A. Research Conference, Oregon State Univ. Corvallis, Oregon, USA, pp 43-53.

Johnson, D.S., Perring, M.A. & Pearson, K. (1980). The effects of ultra-low O_2 storage on Cox from different commercial orchards. Ann. Rep. E. Malling Res. Stn for 1979, p 155.

Knee, M. (1980). Physiological responses of apple fruits to oxygen concentrations. Ann. appl. Biol., 96, 243-53.

Knee, M. & Sharples, R.O. (1981). The influence of controlled atmosphere storage on the ripening of apples in relation to quality. In 'Quality in stored and processed vegetables and fruit', Eds. Goodenough, P.W. & Atkins, R.K. Academic Press, London & New York, pp 341-52.

Lidster, P.D., Forsyth, F.R. and Lightfoot, H.T. (1980). Low oxygen and carbon dioxide atmospheres for storage of McIntosh apples. Can. J. Plant Sci., 60, 299-301.

Lidster, P.D., McRae, K.D. & Sandford, K.A. (1981). Some effects of maturity and cultural origin on response of McIntosh apples to low oxygen storage (in press).

North, C.J. & Cockburn, J.T. (1976). The tolerance of Cox's Orange Pippin apples to low levels of oxygen. Ann. Rep. E. Malling Res. Stn for 1975, p 76.

North, C.J., Cockburn, J.T. & Bubb, M. (1977). Storage of Cox's Orange Pippin apples in low levels of oxygen: effects of picking date. Ann. Rep. E. Malling Res Stn for 1976, p 90-91.

Sharples, R.O. (1975). Influence of climate on the storage quality of pome fruits. In 'Climate and the Orchard' (Ed. by H.C. Pereira), Res. Rev. Commonw. Bur. Hort. Plantn Crops, No. 5, pp 113-124.

Sharples, R.O. (1980). The influence of orchard nutrition on the storage quality of apples and pears grown in the United Kingdom. In 'Mineral Nutrition of Fruit Trees', eds. D. Atkinson, J.E. Jackson, R.O. Sharples and W.M. Waller, Butterworths, Sevenoaks, U.K., pp 17-28.

STORAGE BEHAVIOR OF 'd'ANJOU' PEARS IN LOW OXYGEN AND AIR

P. M. Chen and W. M. Mellenthin
Mid-Columbia Experiment Station
Oregon State University
Hood River, OR 97031

Controlled atmosphere (CA) storage of 'd'Anjou' pears results in a greater reduction of respiratory activity and quality deterioration than is obtained by conventional air storage. Commercial CA levels for 'd'Anjous' are 2.0-2.5% O_2 with 0.8-1.0% CO_2 (7). Increased levels of CO_2 have resulted in internal injury (brown core) of fruit after long periods in CA storage. Mature 'd'Anjou' pears grown in a cool season are more susceptible to CO_2 injury than those grown in a warm season (6). Furthermore, commercial CA will not eliminate the disorder of superficial scald of 'd'Anjou' fruit. Low O_2 concentration at 0.5-1.0% during storage can benefit storage life and dessert quality of certain climacteric fruit (4, 5, 9, 13, 14). 'd'Anjou' pears in atmospheres of 1% O_2 with 0.03% CO_2 not only maintained high dessert quality, but also prevented scald disorder and internal brown core after long-term cold storage (3, 10). This report makes a comparative analysis of storage behavior between 1% O_2-stored and air-stored 'd'Anjou' fruit.

Free Amino Acids and Proteins

In general, fruit stored at 1% O_2 had higher concentrations of free amino acids with lower total proteins as compared with those stored in air for 5 months. The fruit in 1% O_2 had a slower rate of decrease in amino acids and increase in proteins than those in air throughout 8 months of storage (Fig. 1). Li and Hansen (8) reported that active transamination reactions would tend to maintain a lower organic acid and higher protein content as observed in pears stored in air. Total proteins synthesized in pears during long-term cold storage would be from the free amino acids pool and/or from the transamination of organic acids. Since fruit stored in air did not ripen normally with good dessert quality after 8 months storage, the higher accumulation of proteins in air-stored fruit might inhibit the further synthesis of required enzymes for the normal ripening activities (16).

Organic Acids and Sugars

Fruit stored at 1% O_2 had higher amounts of titratable acids (TA) after 5 months storage than those stored in air. TA in 1% O_2-stored fruit decreased at a slower rate from 300 mg/100 ml juice to 200 mg after 8 months while that in air-stored fruit dropped rapidly from 250 mg/100 ml juice to 170 mg (Fig. 2). Soluble solids (SS), on the other hand, were relatively similar between two storage treatments and increased a little after 8 months storage (Fig. 2). The slower rate of organic acid decrease in 1% O_2-stored fruit might be due to slower rates of overall metabolic activities. Since SS are mainly soluble sugars, the decrease in TA without changing SS during 8 months storage indicated that the energy and carbon sources in fruit for the maintenance of living activities during long-term cold storage must be directly contributed by organic acids rather than sugars. The majority of organic acids in 'd'Anjou' pears is malic acid (Table 1). Malic acid can be directly incorporated into tricarboxylic acids (TCA) cycle to generate energy or decarboxylated by malic enzyme and pyruvic decarboxylase (15). The faster rate of organic acid decrease in air-stored fruit indicated that those fruit had higher activities of TCA enzymes, malic enzyme, and decarboxylase than 1% O_2-stored fruit.

Ripening Activities, Dessert Quality, and Juice Binding Capacity

After 8 months storage, changes in internal C_2H_4 in 1% O_2- and air-stored fruit at 20°C (Fig. 3) appeared to be similar to the climacteric ethylene production and also to the climacteric respiration (Fig. 4). Thus, both 1% O_2- and air-stored fruit apparently underwent simultaneous ripening. However, softening of 1% O_2-stored fruit was much faster than that of air-stored fruit during ripening process (Fig. 5). This difference led to a distinct difference in dessert quality. The 1% O_2-stored fruit passed through the "buttery and juicy" texture during days 8-10, changed to "soft but juicy" texture during days 11-13, and finally became "mealy and dry" on days 14-15. The air-stored fruit, on the other hand, tasted "coarse and dry" on days 8-10, and directly changed to "mealy and dry" during days 11-15. The "buttery and juicy" pulp of 1% O_2-stored fruit had the same water content as the "coarse and dry" pulp of air-stored fruit (Table 2). However, the amount of juice that could be extracted from the "buttery and juice" pulp was significantly less than that from the "coarse and dry" pulp (Table 2). Apparently, the "buttery and juicy" pulp of 1% O_2-stored fruit had a greater juice binding capacity than the "coarse and dry" pulp of air-stored fruit. The differences in fruit softening and texture change between 1% O_2- and air-stored 'd'Anjou' pears during ripening may be due to the difference enzymic activities of pectinesterase and polygalacturonase (1, 2, 17). In ripened 'd'Anjou' fruit, solubilized pectin (11), in combination with acids and sugars released from the vacuole (12) and other sugars released from the breakdown of cell walls (1, 2), might form a distinct gel matrix with water. This might alter juice binding capacity between the "buttery and juicy" pulp of 1% O_2-stored fruit and the "coarse and dry" pulp of air-stored fruit.

Literature Cited

1. Ahmed, A. E., and J. M. Labavitch. 1980. Cell wall metabolism in ripening fruit. II. Changes in carbohydrate-degrading enzymes in ripening 'Bartlett' pears. Plant Physiol. 65:1014-1016.
2. Ben-Arie, R., L. Sonego, and C. Frenkel. 1979. Changes in pectic substances in ripening pears. J. Amer. Soc. Hort. Sci. 104:500-505.
3. Chen, P. M., R. A. Spotts, and W. M. Mellenthin. 1981. Further studies on low oxygen storage of 'd'Anjou' pears. J. Amer. Soc. Hort. Sci. (In print).
4. Claypool, L. L. 1973. Further studies on controlled atmosphere storage of 'Bartlett' pears. J. Amer. Soc. Hort. Sci. 98:289-293.
5. Hansen, E. 1957. Reaction of Anjou pears to carbon dioxide and oxygen content of the storage atmosphere. Proc. Amer. Soc. Hort. Sci. 69:110-115.
6. Hansen, E., and W. M. Mellenthin. 1962. Factors affecting susceptibility of pears to carbon dioxide injury. Proc. Amer. Soc. Hort. Sci. 80:146-153.

7. Hansen, E., and W. M. Mellenthin. 1979. Commercial handling and storage practices for winter pears. Oregon State Univ., Agric. Expt. Sta. Special Report 550.
8. Li, P. H., and E. Hansen. 1964. Effects of modified atmosphere storage on organic acid and protein metabolism of pears. Proc. Amer. Soc. Hort. Sci. 85:100-111.
9. Lidster, P. D., F. R. Forsyth, and H. J. Lightfoot. 1980. Low oxygen and carbon dioxide atmospheres for storage of McIntosh apples. Can. J. Plant Sci. 60:299-301.
10. Mellenthin, W. M., P. M. Chen, and S. B. Kelly. 1980. Low oxygen effects on dessert quality, scald prevention, and nitrogen metabolism of 'd'Anjou' pear fruit during long-term storage. J. Amer. Soc. Hort. Sci. 105:522-527.
11. Pilnik, W., and A.G.J. Voragen. 1970. Pectic substances and other uronides. pp. 53-87. In A. C. Hulme (ed.) The Biochemistry of Fruit and their Products. Vol. 1. Academic Press, New York.
12. Sacher, J. A. 1966. Permeability characteristics and amino acid incorporation during senescence (ripening) of banana tissue. Plant Physiol. 41:701-708.
13. Salunkhe, D. K., and M. T. Wu. 1973. Effects of low oxygen atmosphere storage on ripening and associated biochemical changes of tomato fruits. J. Amer. Soc. Hort. Sci. 98:12-14.
14. Stow, J. R. 1978. Controlled atmosphere storage of Conference pears. P. 156. In East Malling Research Station Rpt. for 1978. Maidstone, Kent, U.K.
15. Ulrich, R. 1970. Organic acids. pp. 89-118. In A. C. Hulme (ed.) The Biochemistry of Fruits and their Products. Vol. 1. Academic Press, New York.
16. Wang, C. Y., and W. M. Mellenthin. 1975. Effect of short-term high CO_2 treatments on storage of d'Anjou pears. J. Amer. Soc. Hort. Sci. 100:492-495.
17. Zauberman, G., and M. Schiffmann-Nadel. 1972. Pectinmethylesterase and polygalacturonase in avocado fruit at various stages of development. Plant Physiol. 49:864-865.

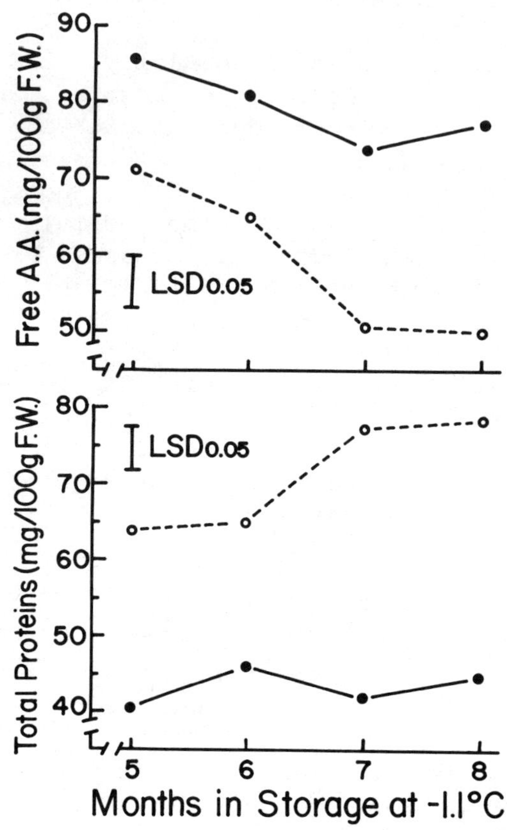

Fig. 1. Changes in free amino acids and total proteins in pulp tissue of 'd'Anjou' pears stored in 1% O_2 (●) and in air (○) at -1.1°C for 8 months.

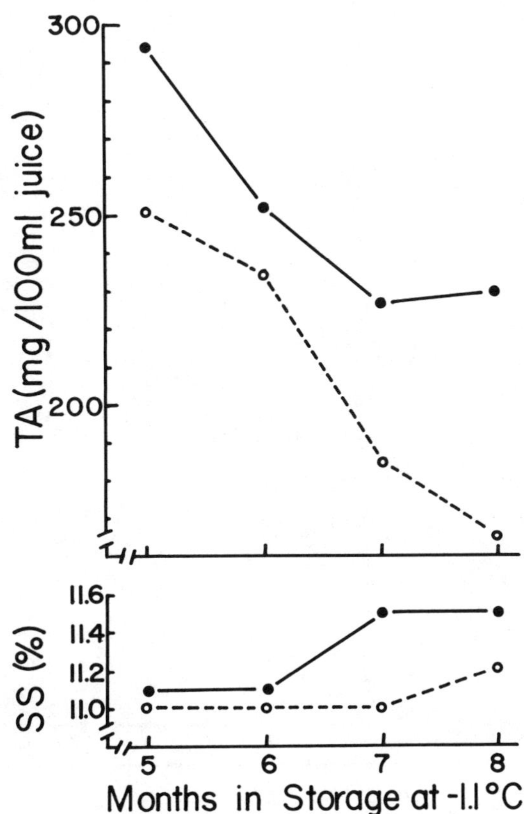

Fig. 2. Changes in titratable acids (TA) and soluble solids (SS) in pulp tissue of 'd'Anjou' pears stored in 1% O_2 (●) and in air (○) at -1.1°C for 8 months.

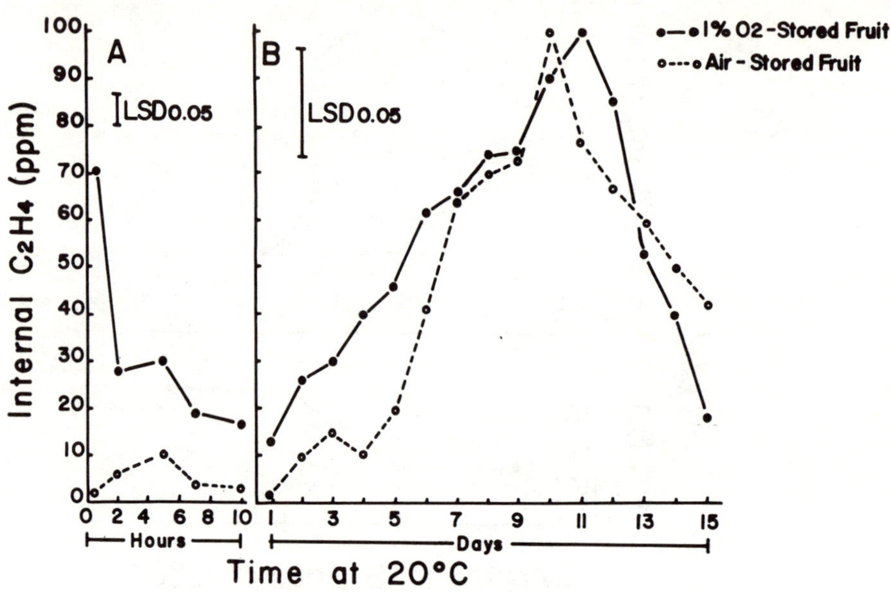

Fig. 3. Changes in internal C_2H_4 concentrations in 1% O_2- (●) and air-stored (○) 'd'Anjou' pears during ripening at 20°C. The fruit were previously stored at -1.1°C for 8 months.

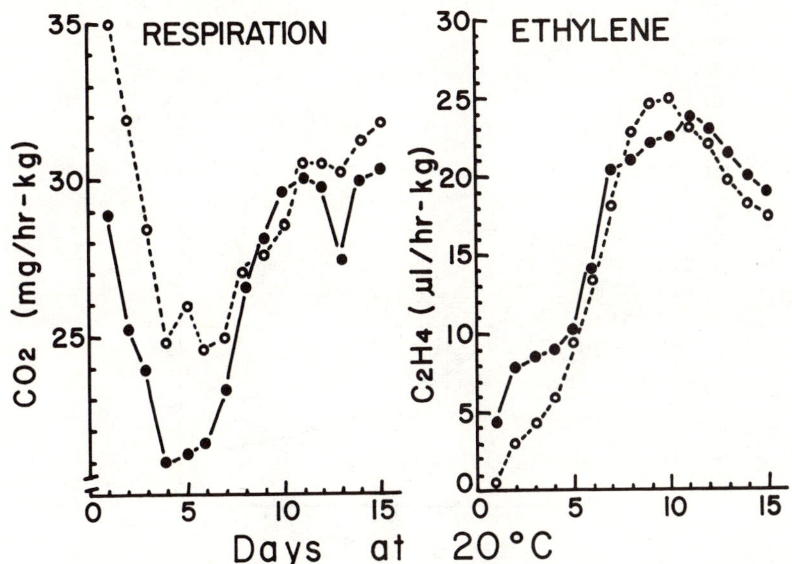

Fig. 4. Respiration and ethylene production of 1% O_2- (●) and air-stored (○) 'd'Anjou' pears during ripening at 20°C. The fruit were previously stored at -1.1°C for 8 months.

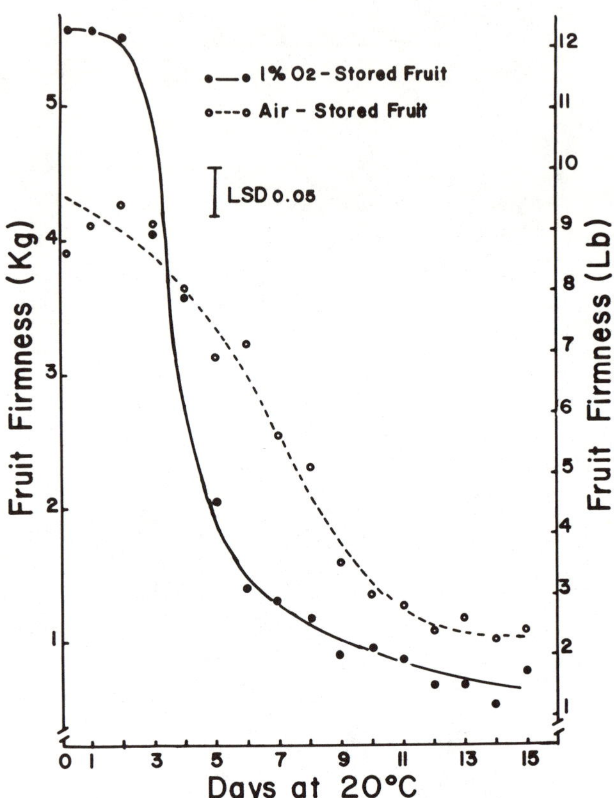

Fig. 5. Changes in fruit firmness of 1% O_2- (●) and air-stored (○) 'd'Anjou' pears during ripening at 20°C. The fruit were previously stored at -1.1°C for 8 months.

Table 1. Organic acid content of 'd'Anjou' pears stored in 1% O_2 or in air[z]

Type of storage	Fruit condition	Mg/100 g fresh weight [Y]						
		Shikimic acid	Quinic acid	Succinic acid	Chlorogenic acid	Oxaloacetic acid	Malic acid	Fumaric acid
1% O_2	Unripened	9	11	68	11	18	310	0.9
	Ripened	5	10	33	14	11	180	0.2
Air	Unripened	2	7	20	15	9	212	0.5
	Ripened	3	2	17	10	15	124	0

[z] The fruit had been stored at -1.1°C for 8 months and then ripened at 20°C for 8 days.

[Y] The quantity of each acid was estimated by the external standard method based on a single injection.

Table 2. Differences in water content and juice binding capacity of pear pulp in unripened and ripened 'd'Anjou' fruit stored in either 1% O_2 or air at -1.1°C for 8 months and then ripened at 20°C.

Storage atmosphere	Days at 20°C	Pulp texture and juiciness	Water content of pear pulp (% of fresh weight)	Amount of juice (ml) extracted from 100 g fresh pear pulp
1% O_2	0	(Unripened)	86.3[z]	74.0 ab[y]
	8	Buttery, juicy	86.0	63.7 cd
	9	Buttery, juicy	85.9	58.9 d
	10	Buttery, juicy	86.0	57.9 d
Air	0	(Unripened)	86.6	73.0 ab
	8	Coarse, dry	85.2	72.3 ab
	9	Coarse, dry	86.2	68.9 abc
	10	Coarse, dry	85.8	68.7 bc

[z] Values in the column are not significantly different according to the F tests at 5% level.

[y] Mean separation in the column by Duncan's multiple range test, 5% level.

CA STORAGE OF CHERRIES

Max E Patterson[1]
Department of Horticulture and Landscape Architecture
Washington State University
Pullman, WA 99164

Introduction

The Pacific West Coast is the leading center for production of fresh market sweet cherries. Production is dominated by the single cultivar, Bing, and marketed by a relatively small group of shippers. Marketing is dependent upon long distance shipments to domestic and export markets. The industry is favorably located for shipments to the Orient, and shipments to Taiwan, Hong Kong and Japan now constitute a significant part of the market.

The largest crop was produced in 1979 when Washington, Oregon, Idaho, Montana, Utah and California shipped 76,231 tons. In 1980 fresh shipments per state were as follows.

State	Tons
Washington	34,312
California	30,470
Oregon	5,260
Idaho	1,814
Utah	1,759
Montana	364
	73,979

Total production in Washington in 1979 was 45,735 tons. That was the highest production year in an increasing trend since 1970 when production was only 17,314 tons (1). Since 1979, adverse weather has reduced crop size, but potential production is for even larger crops than in 1979 due to acreage and increased growth of young trees.

Most can be gained from controlled atmospheres by spreading peak shipments in large crop years and by application to transit containers going to distant markets. It can also be used to advantage anytime retention of premium fruit quality will give premium returns.

Developments in Storage and Transit

Relatively little true cherry storage is practiced. However, the ability to extend the marketing period with harvest-fresh quality would be particularly valuable in specific cases and beneficial to the entire industry. Refrigerated cold rooms are used primarily as temporary holding rooms in what is essentially a pick, pack and ship operation. The importance of

---

[1]This report is an overview of several years work performed with major contributions by: G. Apel, C. L. Chu, G. Johnson, J. L. Melstad, K. Patten and A. Yazdaniha. Work was supported, in part, by the Columbia River Orchards Foundation and the Washington State Tree Fruit Research Commission.

rapid heat removal and a pulp temperature of -1.1 to 0°C are beginning to be accepted. There has been a large increase in hydrocooling but there is no uniformity of box fill temperature, and additional heat removal of boxed fruit is not properly emphasized.

Some commercial controlled atmosphere storage has occurred since the report of the last conference. In the 1979 peak production year, it is estimated from unofficial reports that 120,000 boxes were put in a 20% CO_2 CA atmosphere.

This year, 1981, all the export fruit contracted to Japan from the Yakima area were shipped in transit containers which were charged with CO_2, allowed to equilibrate to about 20% and maintained near that percentage until arrival.

At approximately the time of this conference the Dormavac Division of Grumman Allied Industries is scheduled to deliver the first load of Washington cherries to Sweden in a hypobaric container.

Although the cherry responds very favorably to various forms of atmosphere control, low pressure, low O_2 or high CO_2, no atmosphere control should even be contemplated until pulp temperatures are in the -1.1 to +1.1°C range and can be maintained at 0°C or below.

Importance of Pre-CA Requirements

Experience has demonstrated the major pre-CA factors are: (1) large firm fruits, (2) freedom from damage, and (3) prevention of moisture loss. Fruit should be large and firm with high soluble solids and free of rain cracking and rain softening. Picking, hauling, and packing operations should attempt to minimize physical damage since damage promotes pitting and bruising. Assuming only good fruit, free of damage, is selected for CA, then fruit moisture content must be preserved from the moment of harvest in order to retain freshness.

Temperature and humidity conditions which accentuate aging and moisture loss prior to a CA environment will result in loss of freshness and quality that become increasingly visible as the storage duration increases. Some markets routinely require 3 weeks or more for delivery after harvest. Any transit environment recommendations must consider delivery time plus storage and display time in order to maximize the treatment effect.

Proximate vapor diffusion resistance values developed by Gary Apel in our laboratory (unpublished) show that sweet cherry has one of the poorest barriers to water vapor diffusion within a group of assorted commodities. Comparing values with Delicious and Golden Delicious apples reveals a potential for moisture loss from 7 to 14 times greater than these apple cultivars due to the poor vapor diffusion barrier.

Therefore, every effort must be made from the moment of detachment to provide temperature and relative humidity environments that will minimize vapor pressure deficits between fruit internal atmospheres and external atmospheres. We have observed freshly harvest fruit lose weight at the rate of 1% per hour under the drying conditions of the laboratory. Under the same conditions, stems may lose 4% per hour. Field and packing house conditions may result in much greater losses. Vapor pressure losses will be minimal when fruit internal and external environments equilibrate just above the freezing point of the tissue and at humidities approaching 100%.

Materials & Methods

Cherries are commercially packed in 12 or 20 pound fiberboard boxes within polyethylene film liners. Boxes are frequently stacked 66 to 99 to a pallet. Heat removal at this time is very slow. The boxing and palletizing system creates atmosphere and heat exchange barriers that may result in widely different room atmosphere and box atmosphere gas composition for extended periods before equilibration occurs. Until more rapid, uniform and precise temperature control of boxed fruit is attained, varying respiration rates and fruit-generated box atmospheres limit the imposed CA to proximate median composition values between the most effective for fruit longevity and extremes causing toxicity. Fortunately the cherry is quite tolerant to a range of atmospheric gas levels and has wide limits between maximum response and toxicity.

We have had Bing cherries in some form of experimental modified atmosphere treatments on an annual basis since 1971. Treatments have included low pressures, various CO_2 and O_2 combinations, static systems and flowing systems.

Most of our treatments have been with flowing systems of premixed gases, metered to small lots of fruit in chambers in cold rooms with no refrigeration limitations to rapid heat removal. Mixtures were prepared by adding CO_2 and/or N_2 to air to provide an analyzed composition within ± 5% of stated values. Fruit characterization parameters were generally evaluated at 0, 2, 4, 6, and 8 weeks.

Results

Firmness Table 1 is typical of mean firmness data obtained at 2, 4, 6, and 8 weeks with a Durometer hardness tester modified to prevent puncturing the fruit. Firmness is usually increased above harvest values by refrigeration. It remains greater than initial values even after fruit is returned to warm temperatures. Softening, which occurs with extended storage time, is accelerated by temperatures above $0^{\circ}C$. Oxygen levels of 10% or below also appear to accelerate loss of firmness. Contrasts between 20% CO_2 and 1% O_2 during storage are given in Table 2 which shows better retention of firmness with 20% CO_2. Similar data have been obtained other years with other pressure testing devices.

Table 1. Mean Durometer values after 2, 4, 6, and 8 weeks storage.

Firmness

$^{\circ}C$	Atm.	Mean
--	initial	31.9 bc
0	air	35.1 a
1.7	air	32.4 b
1.7	30% CO_2	34.9 a
1.7	20% CO_2, 10% O_2	31.3 bc
1.7	20% CO_2	34.6 a
1.7	1% O_2	30.5 c

Table 2. Durometer values during storage. (Each value a mean of 4 single tree replications of 40 fruits per tree.)

		Firmness			
°C	Atm.	2	4	6	8 wks
1.7	air	31.4	32.0	34.8	31.3
1.7	20% CO_2	35.3	34.8	35.8	32.6
1.7	1% O_2	34.8	32.8	29.1	25.3

Soluble Solids At 0°C there is little change in soluble solids over 8 weeks of storage (Table 3). However, at 1.7°C there are losses in air which are retarded somewhat by low oxygen or more so by elevated CO_2. Elevated CO_2 (20%) is slightly more effective in reducing the loss of soluble solids than other atmospheres when temperatures exceed 0°C.

Table 3. Mean soluble solids changes during 8 weeks storage.

Soluble Solids - 1979

°C	Atm.	Mean (%)
--	-- (initial)	16.4 a
0	air	16.0 ab
1.7	air	13.5 c
1.7	30% CO_2	17.0 a
1.7	20% CO_2, 10% O_2	15.1 b
1.7	20% CO_2	16.9 a
1.7	1% O_2	15.1 b

Decay Elevated CO_2 levels below fruit toxicity levels have consistently reduced the precentage of decayed fruit as well as the amount of decay per fruit. Much greater suppression is obtained at near 0°C than above (Table 4).

Table 4. Percent decayed fruit during Storage at 1.1 and 5.0°C.

% Decay (1.1°C)

Atm.	2	4	6	8 wks
air	0.8	3.8	9.8	--
40 mm Hg	0	0.8	0	--
20% CO_2	1.5	2.3	2.3	0.8
40% CO_2	0	2.3	2.3	2.5

% Decay (5.0°C)

air	6.7	21.7	52.3	--
40 mm Hg	3.5	3.3	25.0	--
20% CO_2	0	8.1	5.8	--
40% CO_2	0	4.1	24.0	--

Respiration Respiration rates following 10 days of controlled atmospheres at 0°C shows some residual suppression by 20% CO_2. Rate reductions continued for 3 days at 20°C following treatment before returning to the level normal for air stored fruit. Mean rates for 4 color classes in each atmosphere for 3 days at 20°C after treatment are given in Table 5.

Table 5. Atmosphere effects on respiration rates after storage.

Post Storage Respiration

Storage 1°C Atm. 10 d	ml CO_2/Kg/hr mean
2.5 O_2 + 1% CO_2	25.9 a
2.5% O_2	25.1 ab
air	24.5 ab
20% CO_2	22.1 b

Color Post harvest low temperatures, low oxygen, low pressure and high CO_2 have often been observed to suppress further red coloration in in proportional amounts to factor levels. Color ratings of random samples seldom reveal the suppression in anthocyanin synthesis due to initial color variability. A better indication is given in Table 6 showing pigment differences after 5 weeks in samples selected for color uniformity before storage at 0°C in treatment atmospheres. Differences are apparent even at this temperature, with the most pigment synthesis inhibition by 20% CO_2 and the lower O_2 levels.

Table 6. Anthocyanin pigment in cherries after 5 weeks in CA at 0°C.

Anthocyanin (7/10-8/15)

Atm.	nm/g
air	272
5% O_2	260
1% O_2	206
.5% O_2	239
N_2	220
20% CO_2	190

Discussions and Conclusions

Because of increasing production trends and expanding distant markets, there is an increasing need for storage and transit environments that maintain harvest quality condition and freshness. A poor moisture barrier makes the cherry very susceptible to moisture loss from the moment of detachment. Appropriate environments can reduce moisture loss and retard senescent changes. Even short durations of adverse environments become damaging after extended storage periods. Experience has shown that large crisp fruit handle, store and ship better than smaller or softer fruit. Fruit for extended transit or storage in controlled atmospheres must be selected for size and firmness, handled carefully to prevent bruising and pitting and cooled and humidified rapidly to slow moisture loss and senescence changes before placing in a CA environment. If these conditions are met the cherry responds favorably to low O_2, low pressure, and high CO_2. Temperature and humidity remain the fundamental physical factors affecting postharvest longevity. Durations of exposure to high temperature and low relative humidity as short as 2 hours can markedly reduce freshness and quality. Intermediate temperatures and humidities are equally bad when the time factor is extended. In-store room temperature displays can overcome weeks of the best care in a few hours.

Excellent responses to CA are possible with good fruit that has been promptly cooled to near $0^{o}C$. Greatest response is obtained when fruit temperatures are maintained at -1.1 to $0^{o}C$ during the CA period.

Variable decay control measures, variable intervals between harvest and packaging and variable pre-and-post packaging temperatures require some compromise between the most effective atmospheres and those causing toxicity. Our work indicates air enriched with 20% CO_2 has been a safe and effective atmosphere for the broad range of variables encountered. It has been very effective in suppressing decay, maintaining firmness and soluble solids, reducing respiration and suppressing pigment synthesis. Properly used CA for cherries has a bright future for bringing a better quality fruit to more people.

Literature Cited

1. Anonymous. 1981. Sweet cherry crop expected to fall short of '80 production. The Goodfruit Grower. 32(10):3.

GAS EXCHANGE BETWEEN ATMOSPHERES OF DIFFERENT OXYGEN CONCENTRATIONS AND "COX'S ORANGE PIPPIN" APPLES

Fernando Lalaguna and Stuart Thorne
Queen Elizabeth College
University of London
Campden Hill Road
London, W8 7AH
England

INTRODUCTION

Controlled atmosphere storage of apples was first introduced, in England, in the late 1920s; by 1972, the capacity for such storage had grown to 339,000 tons in the United States and to 223,000 tons in Great Britain (Ryall & Pentzer(1974)). This growth has been accompanied by many studies of the metabolism of apples stored in modified atmospheres, but, until recently, the precision of respiration experiments has been limited by the techniques available for gas analysis.

Fidler & North (1951), working with Sturmer Pippin and Bramley's Seedling apples, reported a quantitative agreement between the loss of carbohydrate plus acid and the carbon dioxide produced plus alcohol- in both air and nitrogen at temperatures between 3C and 20C. However, the same authors (1967b) found that the loss of respirable carbohydrates exceeded the loss of carbon as carbon dioxide and the theoretical oxygen consumption required was greater than observed; this was for Cox's Orange Pippin apples at 0C and 3.3C in air and in several carbon dioxide and oxygen concentrations. Between these two sets of observations, Fidler and North had changed from a traditional, volumetric gas analyser to an infra-red carbon dioxide analyser and it is possible that the inherent non-linearity and the considerable zero supression necessary to provide adequate resolution with the latter instrument accounted for some of their unexpected results. Fidler and North (1968) later suggested that accumulation of sorbitol accounted for some of the lost carbohydrate. But their observed respiratory quotients (ratio of carbon dioxide produced to oxygen consumed) of up to 1.8 were not satisfactorily explained and, indeed, are not readily accounted for by any likely metabolic pathways. As they reported that respiratory quotients were constant at any temperature, irrespective of atmospheric composition, high values could not be explained in terms of anaerobic metabolism.

The present work is an attempt to repeat some of Fidler and North's observations using improved gas analysers. Differential non-dispersive infra-red carbon dioxide analysers and differential paramagnetic oxygen analysers have been used. These overcome the need for great zero supression with absolute instruments and eliminate the necessity of observing inlet and outlet concentrations of a chamber sequentially.

MATERIALS AND METHODS

Apples cv. Cox's Orange Pippin, grown at the East Malling Research Station, Maidstone, Kent, England, were used throughout the work. Fruits were stored in a flow-through system in 4.2litre jars, each of which contained 1.9kg of apples of mean weight 103g. Six jars were used for each of four different treatments; these treatments were 1%, 2%, 5% and 10% oxygen, each with 0.5% carbon dioxide, the balance being nitrogen. Gases for blending were taken from bottles of high purity gas and reduced to 4 bar pressure with conventional regulators. Flow control for each gas for each treatment was effected with a diaphragm type controller (Porter Instruments Inc type 1000VCD) and the resultant flow rate was measured as the pressure drop across a standard capilliary, observed with a U-tube manometer. The total flow rate for each treatment was set at 100ml/min, this being a compromise between keeping the concentration change an experiment small and ensuring that it was sufficiently large to be measured with adequate resolution. In the worst cases, concentration changes were of the order of 0.25%, which could be resolved with a resolution of 1 part in 50.

Gas exchange was measured differentially using a non-dispersive infra-red carbon dioxide analyser (Analytical Development Company Ltd) and a paramagnetic oxygen analyser (Taylor-Servomex Ltd). Both instruments could also operate in an absolute mode and were used to record absolute concentrations at the storage chamber inlets. The resolution of the carbon dioxide analyser was far greater than needed; that of the oxygen analyser was about 0.005% oxygen difference and limited the precision of our measurements.

RESULTS AND CONCLUSIONS

Results for the 1979-1980 and 1980-1981 seasons were similar, though the latter were rather more consistent due to equipment improvements. Rates of carbon dioxide production and oxygen consumption for the second season are given in Fig 1. Respiratory quotients derived from these data are given in Fig 2. Fig 3 shows the mean respiratory quotients for consecutive periods of 50 days for each of the treatments.

In general, there was a fall in gas exchange rates during the first 50 days of storage, though this was much more marked at 1% and 2% oxygen than at the higher concentrations. After this, there was a slow increase in gas exchange rates.

Respiratory quotients for the apples stored at 5% and 10% oxygen were almost constant throughout 180 days storage and did not differ significantly from unity, suggesting that respiration was through a

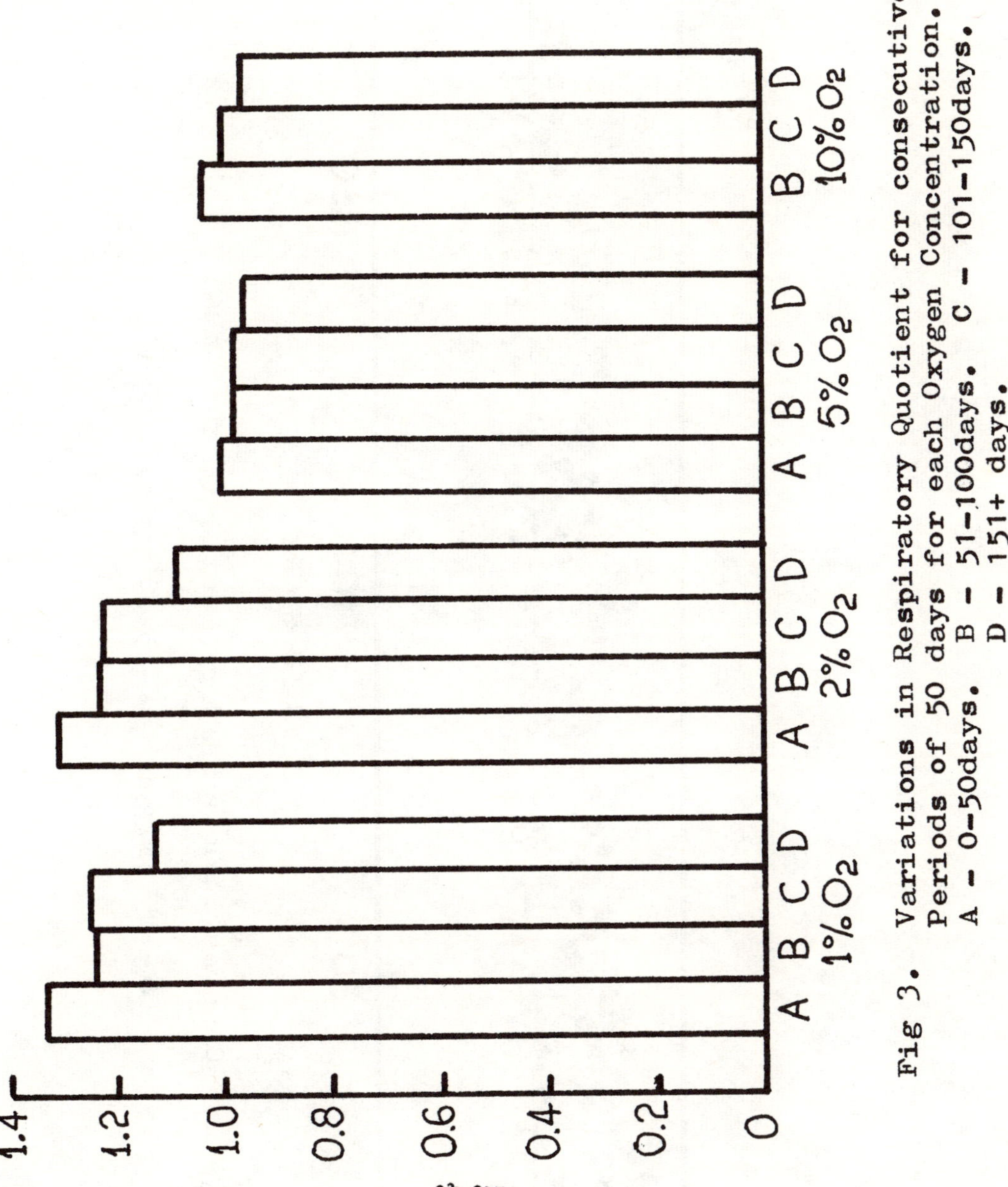

Fig 3. Variations in Respiratory Quotient for consecutive Periods of 50 days for each Oxygen Concentration. A - 0-50days. B - 51-100days. C - 101-150days. D - 151+ days.

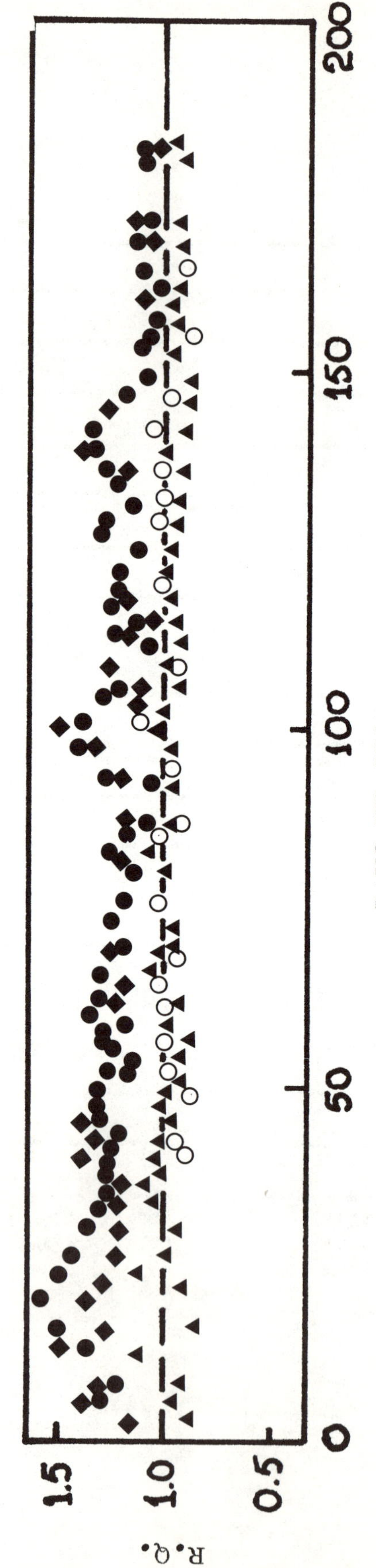

Fig 2. Changes in Respiratory Quotient, 1980-1981

●- 1% Oxygen. ◆- 2% Oxygen. ▲- 5% Oxygen. ○- 10% Oxygen.

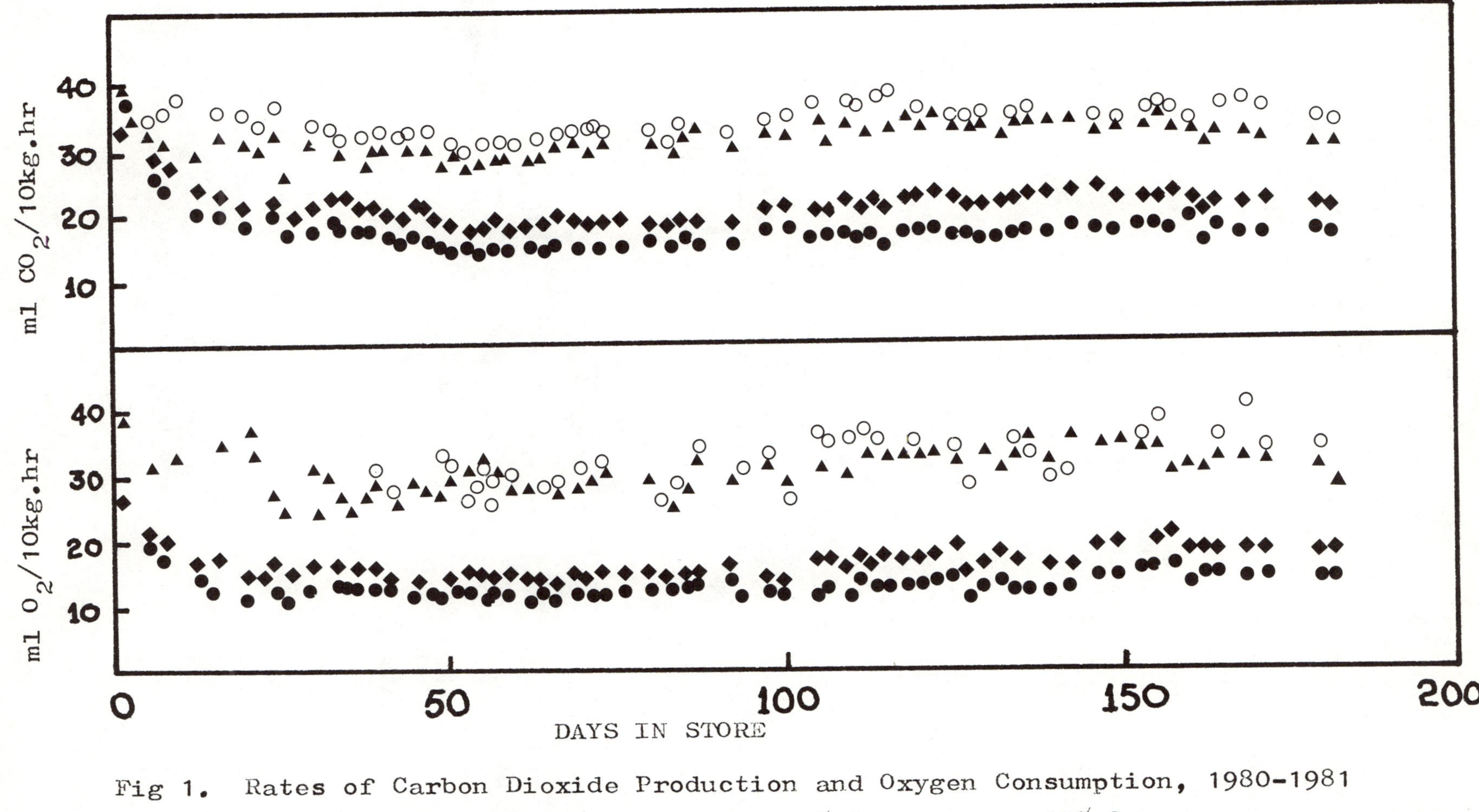

Fig 1. Rates of Carbon Dioxide Production and Oxygen Consumption, 1980-1981
● - 1% Oxygen. ◆ - 2% Oxygen. ▲ - 5% Oxygen. ○ - 10% Oxygen.

normal, aerobic pathway. Respiratory quotients in 1% and 2% oxygen were considerably higher but fell continuously during storage, until they were but little higher than in high oxygen concentrations. Anaerobic metabolism is an inefficient, emergency mechanism and it is suggested that it is adopted as a reaction to low oxygen concentration, but that normal, aerobic respiration is resumed when it is found possible. Gas exchange rates at the two lowest oxygen concentrations remained throughout at about one half those at the two higher concentrations.

Studies of substrate changes are still incomplete, but initial calculations suggest that the higher respiratory quotients in low oxygen concentrations can be accounted for by the anaerobic production of ethanol; the concentrations involved are insufficient to be detected by taste.

ACKNOWLEDGEMENT

We are grateful to Dr R.O. Sharples of the East Malling Research Station for help and advice and for supplying apples.

REFERENCES

Fidler J.C. (1951) J.Exp.Bot. 11 41-64
Fidler J.C. & North C.J. (1967a) J.Hort.Sci. 42 189-206
Fidler J.C. & North C.J. (1967b) J.Hort.Sci. 42 207-221
Fidler J.C. & North C.J. (1968) J.Hort.Sci. 43 429-439
Ryall A.L. & Pentzer W.T. (1974) "Handling, Transportation and Storage of Fruits and Vegetables", Vol II, AVI, p342

EFFECTS OF LOW OXYGEN CONCENTRATION ON FRUIT RESPIRATION: NATURE OF RESPIRATORY DIMINUTION

Theophanes Solomos
Department of Horticulture
University of Maryland
College Park, Maryland 20742

Introduction

Kidd and West [19] were probably the first to show that apple fruit respiration decreases with decreasing external oxygen concentrations. In addition, this same phenomenon has been demonstrated with fruits other than apples [4, 5]. It is observed with both preclimacteric and climacteric fruits [4, 5]. The nature of this inhibition remains to be fathomed, however. Measurements of the respiratory quotients, as well as of the carbon balance, show that in preclimacteric fruits, the respiratory substrate is carbohydrate [4, 5, 15, 22, 24]. Even in fruits like avocado, which contain as much as 15% fats, carbohydrates appear to be the respiratory substrate. On the one hand, fat content does not decrease with ripening and the fruits lack the enzymes involved in the mobilization of triglycerides [6]. On the other, the $^{13}CO_2/^{12}CO_2$ stable carbon isotope of respiratory CO_2 resembles that of cellular "carbohydrates" [Solomos and Laties, unpublished]. Several lines of evidence indicate that the bulk of reducing equivalents which are transferred to oxygen are mediated by respiratory electron carriers [14]. Hence, the substantial decrease in the rate of oxygen uptake with decreasing external oxygen concentrations is probably connected with the enzymes involved in the respiratory machinery. It is unlikely that this decrease is an indirect effect of a decrease in the general metabolism of the tissue. For one reason, the decrease is rapid. Thus, when green peas are transferred to nitrogen, the new steady state of glycolysis is established within 10-15 min. [1]. In the case of fruits, there are no precise measurements of the rate of such a decrease, but the slopes of the published data indicate a rapid adjustment [3, 4, 5]. For another reason, this decrease occurs in tissues whose physiology differs greatly. Besides fruits, it has been observed in storage organs as well as fully developed leaves [18, 20]. In theory, O_2 can affect the initial steps of glucose breakdown or the terminal electron carriers. There are no data to indicate that molecular oxygen is a positive modulator of the enzymes involved in the early stages of glucose oxidation. Moreover, the lack of oxygen leads to the enhancement of glycolytic enzymes. Thus, oxygen must limit the terminal respiratory electron acceptors, namely, cytochrome oxidase or alternate oxidase, in tissues where this enzyme is present and functioning. It is also possible that oxygen may limit "oxidases" with relatively low affinity for oxygen [14].

Gas diffusion in apples

We shall first attempt to examine the physical parameters which control the rate of oxygen diffusion in fruits and other storage organs. It has been established by previous workers that the diffusion of oxygen into fruits follows Fick's law. Moreover, oxygen, as well as other gases, diffuses into and out of fruits through channels filled with air [7, 16]. The fruit we shall use as a model is McIntosh apples, whose permeability characteristics have been studied extensively by Burg and Burg [7] and whose geometry approaches a sphere.

For a sphere the change in concentration with time is given by the equation of Fick's second law of diffusion:

$$\frac{\partial c}{\partial t} = D\left(\frac{\partial^2 c}{\partial R^2} + \frac{2\partial c}{R\partial R}\right) - V \qquad (1)$$

where c = the concentration of oxygen in μmoles/cm^3, R = the radius of the fruit or cell in centimeters, V = the rate of oxygen uptake in μmoles/cm^3/sec and D = the diffusion coefficient in cm^2/sec. For oxygen to diffuse into a cell at the center of an apple, it has to cross the skin and the intercellular spaces of the pulp. Furthermore, O_2 must diffuse to the center of the cell. We shall examine first the diffusion into the cell and shall attempt to calculate the oxygen concentration of the intercellular spaces required to give a concentration of oxygen at the center of the cell sufficient to saturate cytochrome oxidase. The $K_m^{O_2}$ of cytochrome oxidase is very small and the values quoted in the literature vary between 10^{-7} and 10^{-8} M [11]. We shall assume that the $K_m^{O_2}$ of the cytochrome oxidase of the apples is 0.15 μM. Hence, if the oxygen concentration decreases below 0.3 μM, then cytochrome oxidase will not be saturated. We shall make the following assumptions regarding the size of the apple and individual cells:

(a) weight 150g;
(b) density 0.85 g/cm^3;
(c) 40 x 10^6 cells per fruit;
(d) the porosity of the skin 10^{-6} to 2 x 10^{-6}, i.e. only 10^{-6} to 2 x 10^{-6} of the total surface is permeable to gases;
(e) the length of the lenticel is 2.5 x 10^{-3} cm;
(f) the volume of the intercellular spaces is 25% of the total fruit volume [7, 12].

Burg and Burg [7], in their critical studies of gas diffusion in fruits, reported a 30-35% intercellular space volume. Further, we assume that the fruit respiration at 0°C is 4 μl/g/h. At that temperature the critical tables give a diffusivity of O_2 in water $\left(D_w^{O_2}\right)$ of 1.95 x 10^{-5} cm^2/sec and in air $\left(D_{air}^{O_2}\right)$ = 0.178 cm^2/sec. In addition, the molarity of O_2 in water at 0°C and 0.21 atm. partial pressure is 460 μm. The shape of the cells as well as that of the fruit is taken to approximate a sphere.

On the basis of the above assumptions, the radius of the fruit is R_f = 3.48 cm, that of the cell is R_c = 9.24 x 10^{-3}, the rate of oxygen consumption is 4.22 x 10^{-5} μmoles/cm^3 fruit/sec, or 5.64 x 10^{-5} μmoles/cm^3 cell/sec. A small correction for the solubility of oxygen should have been introduced, since its solubility will be decreased because of the solutes present in the cell sap. However, these corrections will not be expected to appreciably influence the present calculations.

The steady state solution of equation (1), i.e., when the concentration of oxygen does not change with time, which is the case for stored apples, is given by the equation [13]:

$$C = \frac{V}{6D} R^2 - \frac{A}{R_{cell}} + B \qquad (2)$$

Since we require that the concentration of oxygen at the center of the cell have a finite value, A must be zero. If the concentration at $R_{cell} = R_1$ is C_1, then B can be calculated. Hence, the concentration gradient between the center of the cell and its outer surface is given by:

$$C_o = C_i + \frac{V}{6D} R^2_{cell} \qquad (3)$$

where C_o is the concentration of oxygen at the outer surface of the cell and C_i the concentration at the center. We assume that diffusivity of O_2 through the cell is the same as that in water. If the concentration at the center of the cell is 0.3×10^{-3} μmoles/cm^3, as is required, then its concentration at the surface will be $C_o = 0.341 \times 10^{-3}$ μmoles/cm^3 cellular sap. Hence, at 0°C and one atmospheric pressure, the partial pressure of O_2 in the adjacent intercellular spaces must be 1.56×10^{-4} atm, i.e. 0.0156% O_2. In other words, at that partial pressure of oxygen in the intercellular spaces the oxygen concentration at the center of the cell will be sufficient to saturate cytochrome oxidase.

Equation (3) can be used to calculate the concentration of oxygen in the intercellular spaces just below the skin which will maintain a p_{O_2} of 1.56×10^{-4} atm. at the center of the fruit. To calculate this, the magnitude of the value of D for the diffusion of oxygen must be calculated. Burg and Burg [7] could detect no gradient for either CO_2 or ethylene between the center of the apple and the skin. The results of Cameron and Reid [10] also indicate that the resistance to diffusion of the apple flesh is negligible. Burton [9] observed that in potato tubers, the diffusion coefficient of O_2 in air was decreased by about 600 due to the tortuosity of the intercellular spaces. It should be noted that the volume of the intercellular spaces in potatoes is 1.3%, as opposed to 25% in McIntosh apples. We have thus decreased the diffusivity in the intercellular spaces by either 20 or 100 fold. It can be seen from Table 1 that the p_{O_2} under the skin is 9.372×10^{-3} and 1.227×10^{-3} atm., corresponding to a 20 and 100 fold decrease in the value of $D^{O_2}_{air}$, respectively. In order to calculate the concentration of oxygen in the ambient atmosphere which will give these concentrations inside the skin, we proceed as follows: We assume that the gas diffuses through a hollow sphere

Table 1. Oxygen concentrations in the intercellular spaces just below the apple skin.

Diffusion coefficient in intercellular spaces (cm^2/sec)	µmoles O_2/ $cm^3 \times 10^{-3}$	p_{O_2} in atm. $\times 10^{-3}$	% O_2 at 1 atm.
$0.178 \times \frac{1}{20}$	16.62	0.372	0.037
0.178×10^{-2}	54.81	1.227	0.123

whose outside diameter is that of the fruit, 3.48 cm, while that of the inside, R_i, is the difference between the outside and the depth of the lenticel, i.e. $3.48 - 2.5 \times 10^{-3} \sim 3.48$ cm. Further, we assume that the walls of the lenticels do not utilize oxygen. Under these conditions it can be shown [13] that the flux of oxygen per unit time through the surface available for diffusion is:

$$J = 4\pi D \times K \frac{R_o R_i}{R_o - R_i} (C_o - C_i) \text{ µmoles/unit time.} \tag{4}$$

K is a constant by which the total surface of the sphere must be multiplied to give the surface of the lenticels. The porosity of the apple skin can vary from 10^{-6} to 2×10^{-6} [7, 12]. If we know the flux, then equation (4) can be solved for C_o. At the steady state the total flux must numerically equal the rate of oxygen uptake of the whole fruit, i.e. 7.44×10^{-3} µmoles/sec. Table 2 shows the external concentrations of oxygen which will give the

Table 2. Oxygen concentrations of the ambient atmosphere

Porosity	Diffusion coefficient in intercellular spaces cm^2/sec.	µmoles O_2 /cm^3	p_{O_2} atm $\times 10^{-3}$ in ambient atmosphere	% O_2 in ambient atmosphere at 1 atm.
1×10^{-6}	$0.178 \times \frac{1}{20}$	0.70	15.60	1.56
1.5×10^{-6}	$0.178 \times \frac{1}{20}$	0.47	10.53	1.05
2.0×10^{-6}	$0.178 \times \frac{1}{20}$	0.36	8.07	0.81
1×10^{-6}	0.178×10^{-2}	0.74	16.62	1.66
1.5×10^{-6}	0.178×10^{-2}	0.51	11.42	1.15
2.0×10^{-6}	0.178×10^{-2}	0.40	8.93	0.89

desired concentrations just under the skin in order to attain the required concentrations quoted in Table 1. It can be seen from the data of Table 2 that for the cytochrome oxidase to be limited by O_2 at the center of the fruit the external oxygen concentration should be decreased below 2%. The data of Fidler et al. [17] show that the rate of respiration of Cox's Orange Pippin kept at 0°C begins to fall at 10-11% external oxygen concentration. Therefore, the curtailment of respiration at O_2 concentrations below 10% should be due to the diminution of the activities of "oxidases" other than cytochrome oxidase, and whose $K_m^{O_2}$'s are at least five-fold higher than that of the cytochrome oxidase.

Effects of low O_2 concentration on banana fruit respiration

Figure 1 shows that the application of 3% oxygen reduces by about 30% the rate of respiration of preclimacteric bananas. When 3% oxygen is applied to banana fruits which are halfway to the climacteric peak, the rate of CO_2 output decreases sharply and its peak value is about 50% less than that of the control (Figure 2). This decrease in the peak values, however, can be attributed to the decrease in the activity of ethylene resulting from low O_2 concentrations [8]. Nevertheless, the rate of CO_2 at 3% oxygen is about four-fold higher than that of the preclimacteric fruit. It thus appears that the rate of oxygen diffusion at 3% O_2 concentration is sufficient to sustain rates of respiration much higher than those observed in the preclimacteric fruits. It should be pointed out that the increase in respiration by ethylene at 3% oxygen cannot be attributed either to the activation of the cytochrome oxidase by ethylene or the synthesis of new cytochrome oxidase. On the one hand, ethylene is not known to be a positive modulator of cytochrome oxidase [14], and on the other, the capacity of the cytochrome oxidase of preclimacteric banana fruits is sufficient to sustain rates of respiration observed at the climacteric peak [23].

Effects of low O_2 concentrations on the respiration of sweet potato roots

In order to avoid the complication of the interaction between oxygen and ethylene we studied the effects of O_2 concentrations on the rate of CO_2 output of sweet potato roots. Figure 3 shows that a gradual decrease in O_2 concentration results in a stepwise decrease in the respiration until the concentration of oxygen reaches about 2%, where the rate of CO_2 output begins to rise as the oxygen concentration falls to zero. If the roots are kept at 1.45% O_2 for 12 hours, lactate accumulates (Table 3), which in turn indicates that the tissue experiences partial anaerobiosis. If the roots are treated at 10% oxygen with 100 ppm ethylene, the rate of respiration doubles that of the air value. Thus, the decrease in respiration at 10% O_2 cannot be attributed to the fact that the rate of oxygen diffusion limits the activity of the cytochrome oxidase. In addition, this increase may not be the result of the synthesis of new cytochrome oxidase since its capacity far exceeds the rate of respiration of roots treated with ethylene [Solomos, unpublished].

Table 3. Effect of O_2 concentration on the level of lactate in intact sweet potato roots.

	Lactate nmoles/g FW
Air	< 10
10% O_2	< 10
1.4% O_2 12 h	260

It is well established that oxidative phosphorylation is one of the most important factors involved in the regulation of plant respiration. The addition of uncouplers of oxidative phosphorylation or anoxia invariably leads to the enhancement of the respiratory breakdown of glucose [2]. Only in tissues such as climacteric fruits, where the full respiratory potential is realized, does the addition of uncouplers or anoxia fail to elicit a further increase in the rate of glycolysis [21]. Nevertheless, it would be anticipated that a partial restriction of the cytochrome path will result in an increase in the rate of production of ethanol and/or lactate. Control atmospheres with oxygen con-concentrations as low as 1.5% for apples do not appreciably increase the rate of ethanol accumulation.

On the basis of the present knowledge concerning the regulation of plant respiration, the above data indicate that the observed decrease in the rate of respiration in response to the decrease in external oxygen concentration is not the result of cytochrome oxidase being restricted by the diffusion of oxygen: rather, it stems from the diminution of the activity of "oxidases" other than cytochrome oxidase, and whose affinity for oxygen may be five to six times lower than that of the latter enzyme. Further, these "oxidases" must not play as important a role in the generation of ATP as does cytochrome oxidase. Mapson and Burton [20] postulated that in potato tubers there are two oxidases with different affinities for oxygen. Alternatively, plant tissues may possess an as yet unknown mechanism to compensate for a partial restriction of the cytochrome oxidase.

Acknowledgements

This is scientific article no. A-3032, contribution no. 6095 of the Maryland Agricultural Experiment Station and the Department of Horticulture, University of Maryland, College Park.

References

1. Barker, J., Khan, A.A., and Solomos, T., 1967, The mechanism of the Pasteur effect in peas, New Phytol., 66:577.
2. Beevers, H., 1974, Conceptual developments in metabolic control, 1924-1974, Plant Physiol., 54:437.
3. Biale, J.B., 1946, Effect of oxygen concentration on respiration of Fuerte avocado fruit, Amer. J. Botan., 33:363.
4. Biale, J.B., 1960, Respiration of fruits, in "Encyclopedia of plant physiology," vol. XII/2, Springer Verlag, Berlin.

5. Biale, J.B., 1960, The post-harvest biochemistry of tropical and subtropical fruits, Adv. Food Res., 10:293.
6. Biale, J.B., and Young, R.E., 1970, The avocado pear, in "The Biochemistry of fruits and their products," vol. 2, A.C. Hulme, ed., Academic Press, New York.
7. Burg, S.P., and Burg, E.A., 1965, Gas exchange in fruits, Physiol. Plant, 18:870.
8. Burg, S.P., and Burg, E.A., 1967, Molecular requirements for the biological activity of ethylene, Plant Physiol., 42:144.
9. Burton, W.G., 1950, Studies on the dormancy and sprouting of potatoes. I. The oxygen content of the potato tuber, New Phytol., 49:121.
10. Cameron, A.C., and Reid, M., 1981, Diffusive resistance. Measurements in CA storage, in these Proceedings, pp. 171-180.
11. Chevillotte, P., 1973, Relationship between the reaction cytochrome oxidase-oxygen and oxygen uptake in cells in vivo, J. Theor. Biol., 39:277.
12. Clements, H.F., 1935, Morphology and physiology of the pome lenticels of Pyrus malus, Bot. Gaz., 97:101.
13. Crank, J., 1970, "The mathematics of diffusion," Oxford U.P.
14. Day, D.A., Arron, J.P., and Laties, G.G., 1980, Nature and control of respiratory pathways in plants. The interaction of cyanide-resistant respiration with cyanide-insensitive pathways, in "The Biochemistry of plants. A comprehensive treatise," vol. 2, D.D. Davies, ed., Academic Press, New York.
15. Fidler, J., and North, C.J., 1967, The effects of storage on the respiration of apples. I. The effects of temperature and concentrations of carbon dioxide and oxygen on the production of carbon dioxide and the uptake of oxygen, J. Hort. Sci., 42:189.
16. Fidler, J.C., and North, C.J., 1971, The effect of conditions of storage on the respiration of apples. V. The relationship between temperature, rate of respiration and composition of the internal atmosphere of the fruit, J. Hort. Sci., 46:229.
17. Fidler, J.C., Wilkinson, B.G., Edney, K.L., and Sharples, R.O., 1973, The biology of apple and pear storage, Research Review #3, Commonwealth Agricultural Bureaux.
18. James, W.O., 1953, "Plant respiration," Oxford U.P.
19. Kidd, F., and West, C., 1927, Gas storage of fruit, Spec. Rep. Fd. Invest. D.S.I.R., 30.
20. Mapson, L.W., and Burton, W.G., 1962, The terminal oxidases of potato tuber, Bioch. J., 82:19.
21. Millerd, A., Bonner, J., and Biale, J.B., 1963, The climacteric rise in respiration as controlled by phosphorylative coupling, Plant Physiol., 28:521.
22. Solomos, T., and Laties, G.G., 1976, Induction by ethylene of cyanide-resistant respiration, Biochem. Bioph. Res. Commun., 70:663.
23. Theologis, A., and Laties, G.G., 1978, Respiratory contribution of the alternate path during various stages of ripening of avocado and banana fruits, Plant Physiol., 62:255.
24. Thorne, S., 1981, Respiration and metabolism of apples stored under CA conditions, in these Proceedings, pp. 155-160.

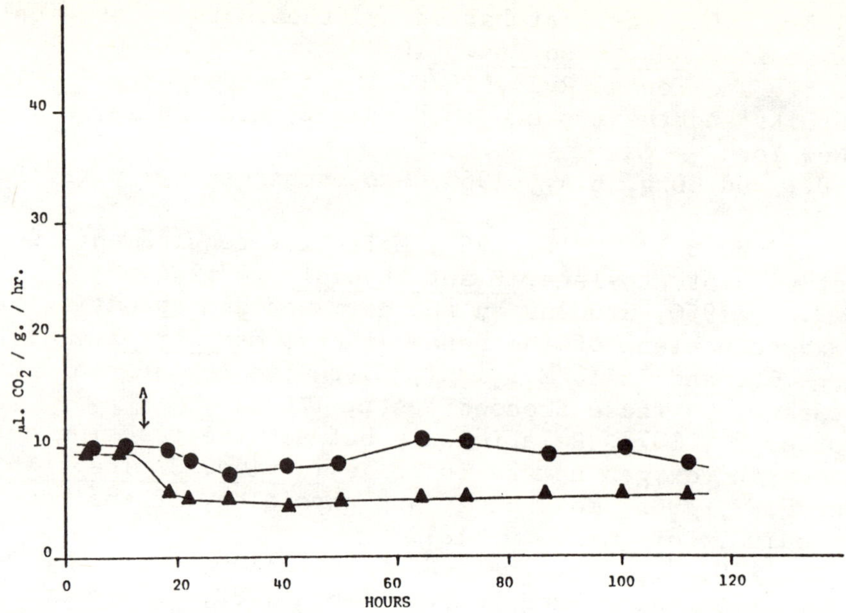

Figure 1. Effect of 3% O_2 on the rate of CO_2 output of preclimacteric bananas.

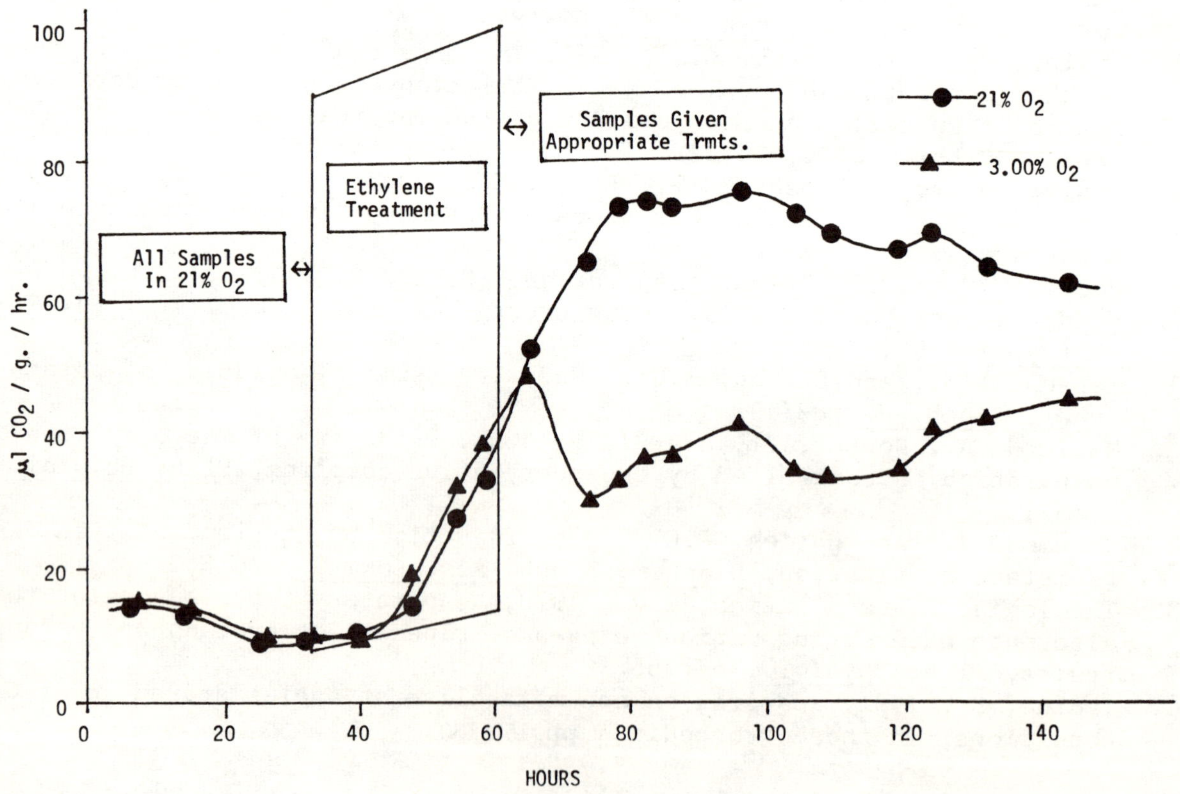

Figure 2. Effect of 3% O_2 on the rate of CO_2 output of climacteric bananas. The fruits were transferred to 3% at the arrow.

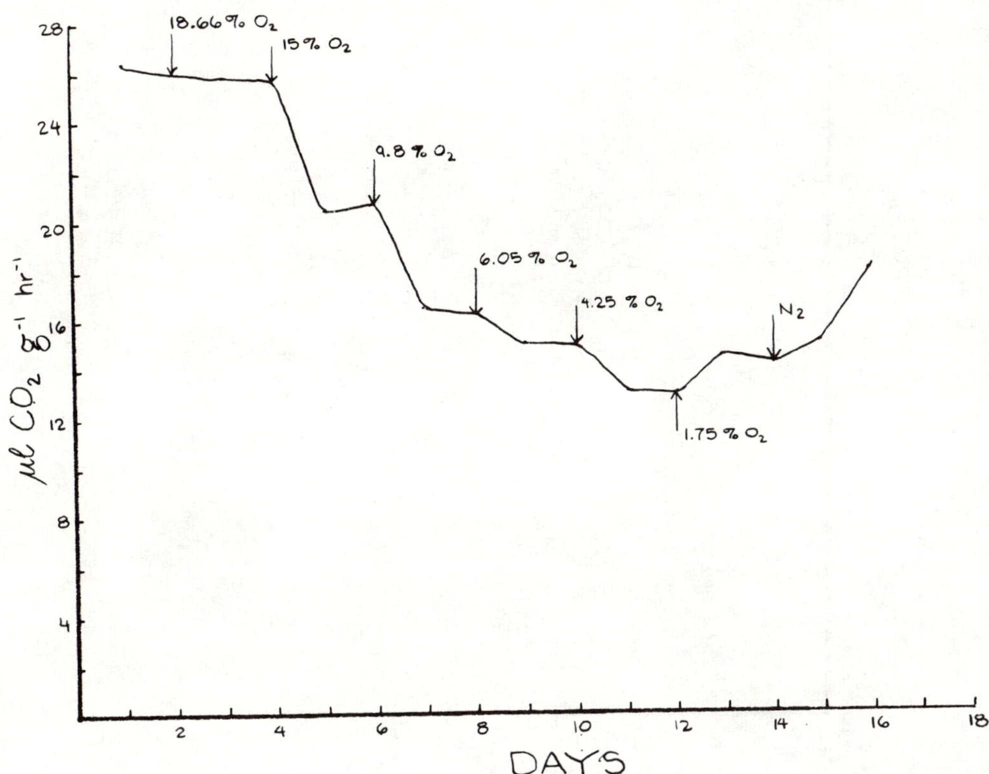

Figure 3. Effect of the gradual decrease in external oxygen concentrations on the rate of CO_2 output of sweet potato roots.

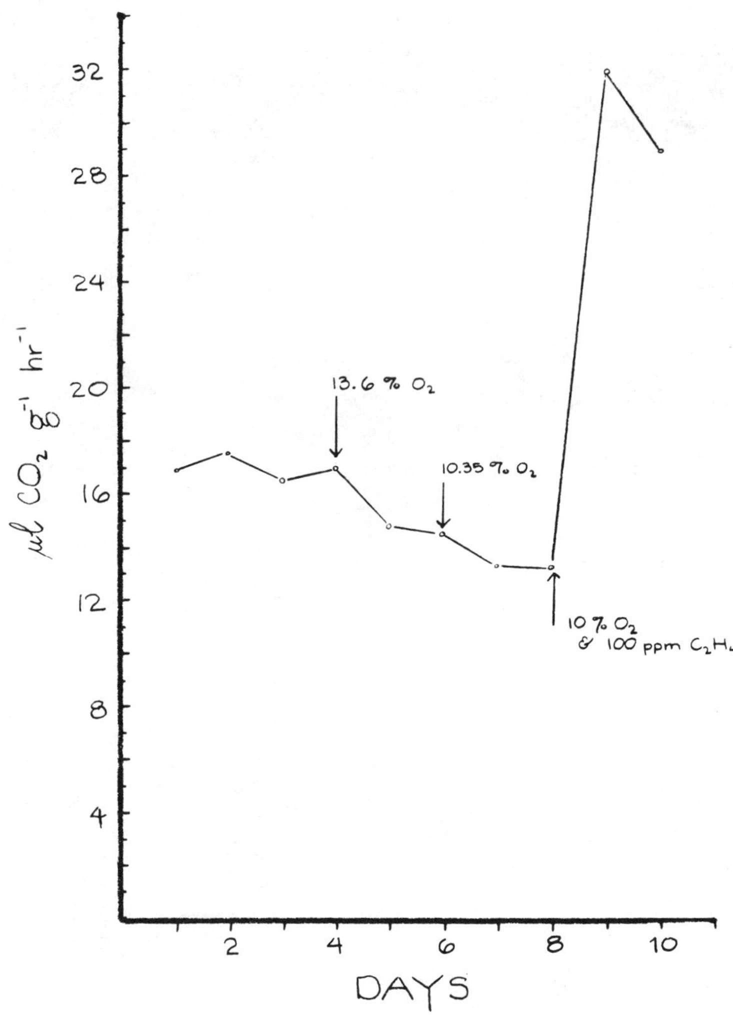

Figure 4. Effect of ethylene on the rate of CO_2 output of sweet potato roots kept under 10% O_2.

Diffusive Resistance: Importance and Measurement in Controlled Atmosphere Storage

Arthur C. Cameron and Michael S. Reid
Department of Environmental Horticulture
University of California
Davis, CA 95616

The concentration limits for oxygen reduction and carbon dioxide enrichment in controlled atmosphere (CA) storage depend on a number of factors which include the respiration rate and the "tolerance" of the tissue to lowered oxygen or increased carbon dioxide. Tolerance, an operational term, is a function of a number of physiological and physical characteristics of the tissue. One physical factor governing tolerance which is often not taken into consideration but is nonetheless of major importance is the resistance of the skin to diffusion. For example, it has been found empirically for many fruits that the limit to which oxygen can be lowered in CA storage without subsequent injury varies from 1 to 5% (12). However, direct work with the enzymes involved in oxygen utilization during respiration has demonstrated that only 0.1% oxygen is required for half of maximal activity (see 8). This suggests that aerobic respiration by the cells can proceed at concentrations of oxygen much lower than what is known to be required by the intact organ. The discrepancy between oxygen requirements for the storage atmosphere and for the enzymes of respiration implies that there exists a gradient between the external atmosphere and the cytoplasm of the cells of the stored plant organs which can range between 1 and 5%. The magnitude of this gradient is a function of the physical characteristics of the skin, or more specifically, the resistance of the skin of the fruit to oxygen diffusion; the greater the resistance, the greater the gradient for any given rate of respiration. These same considerations apply to tolerance to carbon dioxide although there is evidence that a wider range of physiological tolerances exist than for oxygen. The resistance characteristics nevertheless dictate the upper limit of carbon dioxide which can be present in the storage atmosphere since they control the gradient which exists between the internal atmosphere of the tissue and its external environment.

In order to gain a clearer understanding of the importance of the diffusive resistance characteristics of an organ to its tolerance to lower levels of oxygen or increased levels of carbon dioxide, it is desirable to investigate the laws which govern gas movements. Within limits, these laws apply to all gases and are summarized by Fick's First Law of Diffusion which states that the movement or flux of a gas in or out of a plant tissue depends on the concentration drop across the barrier involved, the surface area of the barrier, and the resistance of the barrier to diffusion. A simplified version of Fick's Law can be written as follows:

$$J_j = \frac{A \cdot \Delta c_j}{R_j} \quad \text{[Eq. 1]}$$

where the total flux of species j (ie oxygen, carbon dioxide, water vapor, ethylene, etc.) per unit time (J_j cm^3 sec^{-1}) is moving across a barrier which has surface area A (cm^2) and resistance to diffusion of species j of R_j (sec cm^{-1}). The driving force for the movement depends on the concentration gradient across the barrier (Δc_j) and for the purposes of this discussion, we assume that Δc_j is a good approximation of the gradient in chemical potential which is the true driving force. The simplified form of Fick's Law is valid for plant organs as long as the resistance of the tissue is insignificant compared to the resistance of the skin, an assumption which has been verified for a number of fruits (3,6,7,9), and that the thickness of the skin is an insignificant part of the fruit diameter. If resistance within the fruit becomes significant, for instance when a fruit softens and becomes waterlogged, the equation becomes invalid, since concentration gradients will then be established between the skin and the center of the fruit.

Equation 1 reveals the direct role that R_j plays in limiting the flux of gas across the boundary. For example, if R_j is large, the flux will be relatively small for any given set of conditions compared to a barrier with a small R_j. The analogy between resistance to gas movement and resistance in electrical circuits has been recognized and applied to the study of fluxes accompanying photosynthesis (see 18). We felt that it it would be useful to also apply this approach to the study of gas diffusion in stored fruits and vegetables. The resistance network approach, as it is called, is straightforward in application and is demonstrated in Fig. 1 which shows the various barriers to diffusion encountered by ethylene as it leaves a nonstomatal fruit which is waxed, wrapped in plastic, and placed in a carton. The total resistance can be calculated as for an electrical circuit:

$$R_j^{total} = R_j^{tissue} + \frac{R_j^{skin} + R_j^{stem\ scar}}{R_j^{skin} \cdot R_j^{stem\ scar}} + R_j^{wax} + R_j^{plastic} + R_j^{carton} \qquad \text{[Eq. 2]}$$

Note that the net resistance for a system in parallel, such as the stem scar and the skin (Fig. 1) is calculated exactly as for two resistors in parallel in an electrical circuit. The advantage of the resistance network approach is that it permits the segregation of the total resistance into its many components, which allows investigation of how a change in the resistance of any one conponent might affect the total resistance. An important consequence of the resistance network approach is an understanding of the concentration profile of gas species j from the atmosphere to the tissue at steady state. The steady state condition for any gas species is met when the mass flux of that species is constant throughout the entire system under study. For example, when diffusion of oxygen is at steady state, it is utilized at a constant rate of x moles sec^{-1}, precisely x moles moves across the plastic film every sec, x moles per sec crosses the wax barrier, and x moles per sec crosses the stem scar and skin. However, whereas the flux of oxygen is constant, the concentration of oxygen drops from the atmosphere to the site of respiration. The concentration difference across any barrier for a given steady state flux will depend only on the resistance of the barrier and its surface area. A rearrangement of Equation 1 reveals that:

$$\Delta c_j = \frac{J_j \cdot R_j}{A} \qquad \text{[Eq. 3]}$$

Equation 3 shows that driving a given flux across a boundary of increased resistance requires a greater concentration gradient. If the area over which diffusion can occur is increased, however, the concentration gradient required to drive the same flux will be lowered. For the oxygen example, the oxygen concentration in the tissue will ultimately be controlled by the combined resistances of the tissue, skin, stem scar, wax and so on as calculated in Equation 2. The internal oxygen in the fruit tissue at the steady state can be expressed as:

$$[O_2]_{tissue} = [O_2]_{atmosphere} - \frac{J_{O_2} \cdot R^{total}_{O_2}}{A} \qquad \text{[Eq. 4]}$$

where at steady state J_{O_2} equals the rate of utilization of oxygen (ml O_2 sec^{-1}) and where $[O_2]_{atmosphere}$ (ml O_2/ml air) can vary according to CA storage conditions. Equation 4 demonstrates how respiration rate, resistance coefficient and surface area combine to limit the level to which oxygen can be decreased in the atmosphere surrounding the tissue without establishment of anaerobic conditions within the tissue.

A similar approach can be used for the study of all other gases which move in and out of the stored plant tissue. In the case of carbon dioxide, which is produced within the tissue and moves outward, the equation which describes internal concentration is as follows:

$$[CO_2]_{tissue} = [CO_2]_{atmosphere} + \frac{J_{CO_2} \; R^{total}_{CO_2}}{A} \qquad \text{[Eq. 5]}$$

Here it is seen that any increase in the concentration of atmospheric carbon dioxide will have a direct additive effect on the concentration of carbon dioxide within the tissue.

Measurement of R_j

The commercially available leaf porometer provides a simple method for the measurement of resistance to water vapor diffusion. The measurements are based on the assumption that internal water vapor is at 100% relative humidity; the water vapor concentration can be calculated from a knowledge of the temperature of the tissue or organ under examination. The leaf porometer measures the time for a predetermined amount of water vapor to diffuse across a known surface area and is calibrated against plates of known resistance. Porometer measurements of R_{H_2O} have been made for a number of bulky spherical organs (19).

The resistance of tissues and organs to carbon dioxide, oxygen, and ethylene gases has normally been investigated using the steady state approach (4,5,7,8,10,13). For each of the gases, the approach is the same; the production rate (or utilization rate) of the gas by the organ under investigation is measured, the concentrations of the gas in the internal and external atmospheres are experimentally determined, and the resistance is calculated as follows:

$$R_j = \frac{\text{concentration gradient}}{\text{production rate}} \quad \text{[Eq. 6]}$$

Since both the concentration gradient and the production rate can be expressed in a number of different ways, resistance values have been reported in a wide array of units and are generally difficult to interpret and compare. A particular problem develops when production or consumption rates are expressed on the commonly accepted per weight basis, since the surface area of a tissue or organ, not the weight or volume, is the factor governing diffusion. Thus, resistance factors (based on weight) can vary with a change in the surface to volume ratio, even for one tissue or organ. The steady state approach can be used to obtain values for resistance in acceptable units for all gases if the flux of the gas is measured in units consistent with those used for measurement of concentration. For instance, if the concentration difference across the skin for ethylene is given in nl cm^{-3}, then flux should be expressed as nl sec^{-1} and the surface area as cm^2. For oxygen, where concentration is expressed as ml O_2/ml air, flux should be expressed as the total ml of O_2 consumed per second by the organ. For the steady state approach, R_{O_2} would be calculated as follows:

$$R_{O_2} = \frac{A\ (cm^2)\quad \Delta c_{O_2}\ (ml\ O_2/ml\ air)}{J_{O_2}\ (ml\ sec^{-1})} \quad \text{[Eq. 7]}$$

and would have units of sec cm^{-1}. These units are convenient in studying plant systems and permit the direct comparison of resistance values of carbon dioxide, oxygen, water vapor and ethylene for different tissues.

Although the steady state method can be used to obtain information on a wide variety of gases and tissues, it has some serious drawbacks. By definition, the system must be in steady state, with the flux of gas throughout the system constant. This might become a limiting factor if the resistance of a fruit to oxygen was being measured during the climacteric rise, a time when fluxes would most likely not be at steady state. The need for the steady gas flux would also prevent the measurement of resistance following wounding or other pertubations which cause nonsteady state fluxes. The tissue must also produce significant levels of the gas of interest. For example, both production rates and internal concentrations of ethylene are often extremely difficult to detect in preclimacteric fruits or in vegetative tissues. Accurate measurement of the internal concentration of gases is often difficult. Although internal samples are easily withdrawn from fruits with internal cavities such as cantaloupes, extraction methods used to determine the internal atmospheres of bulky organs such as avocado or potato can often yield inconsistent or misleading results.

In order to overcome these problems, Cameron and Yang (9) developed an approach to the quantitative measurement of R_j based on the kinetic analysis of the efflux of preloaded gases from plant organs which does not require the flux to be at the steady state. Ethane was chosen for preloading in the first study since it is neither produced nor metabolized to a significant degree by the tissue, has properties similar to ethylene, and can be quickly and accurately determined at low concentrations by gas chromatography. In essence, the method involves the equilibration of the tissue with 500-1000 ppm ethane, the transfer of the tissue to a second, ethane free chamber, and the measurement of the increase in ethane concentration in the second chamber as the ethane escapes from the tissue. It is apparent that if there is a high resistance to ethane movement, the measured rate of ethane increase in the jar will be slow and vice versa. Equations were developed from Fick's First Law of Diffusion which allowed the calculation of $R_{C_2H_6}$ from the efflux data (9). Values obtained by this method compared well with those obtained for ethylene by the steady state method and it is anticipated that this method will permit the estimation of R_j's which have hitherto been virtually impossible to obtain. The primary limitation of the efflux approach in the determination of R_j for other physiologically important gases is that suitable analogs are difficult to find.

Reported Resistance Values

In Table 1, we have gathered resistance values for carbon dioxide, oxygen, ethylene and water vapor for a number of fruit and vegetative organs. All values of resistance were converted to units of sec cm^{-1} to enable direct comparisons. The surface area to weight ratio (Table 1) (essentially the surface to volume ratio if the density is assumed to be 1.0 g cm^{-3}) was estimated either from data presented in the individual papers or by assuming resonable sizes and shapes for each of the fruits. This factor enabled conversion of fluxes from a weight basis to an area basis. Although the absolute values of resistance presented in Table 1 may be somewhat in error as a result of assumptions made during their calculation, we feel that they at least give relative trends.

One trend that appears in all organs for which resistances are presented is that R_{H_2O} values are consistently much lower than R_{CO_2}, R_{O_2} and $R_{C_2H_4}$ (Table 1). It has been suggested that carbon dioxide, oxygen and ethylene move primarily through holes in the skin of the fruits (stomates, lenticels, etc.) rather than through the cuticle (5, 17). Considerable evidence has been presented suggesting that waxes in the cuticle are the primary constituent which limits water vapor diffusion (20, 21). Perhaps water vapor actually moves through both holes and cuticle while the other gases are for some reason restricted primarily to the holes. The resistance network analysis may provide an approach for investigation of this phenomemon.

Apples appear to have relatively high resistances when compared with most of the other fruits listed (Table 1). Cultivar differences have not been examined to any extent, though such information could be important in explaining observed cultivar differences in sensitivity to CA conditions. In addition, only limited work has been conducted to measure seasonal changes in resistances to gas diffusion for fruits (17) even though it is well known that there are seasonal variations in sensitivity to low oxygen. Resistance values for apples can be substituted in Equation 4 to demonstrate the concentration gradients which might be expected during storage. For preclimacteric apples utilizing oxygen at a rate of 5 ml kg^{-1} hr^{-1} (1.4×10^{-6} cm^3 cm^{-2} sec^{-1}) with an R_{O_2} of 30,000 sec cm^{-1}:

$$\Delta C_{O_2} = \left(\frac{J_{O_2}}{A}\right) \cdot R_{O_2} = (1.4 \times 10^{-6} cm^3 cm^{-2} sec^{-1}) \cdot (30{,}000 \text{ sec } cm^{-1})$$

$$\Delta C_{O_2} = 0.042 = 4.2\%$$

Thus, anaerobic conditions could develop in the tissue of these fruits if they continued to respire at this rate when held in CA of less than 4% oxygen. Although there would be a drop in respiration rate in response to the shortage of oxygen, a potential for low oxygen injury would remain. It is equally clear in this example how an increase in resistance to oxygen diffusion or an increase in respiration rate might affect the concentration gradient. For instance, if the respiration rate doubled, as during the climacteric, the concentration gradient would rise to 8% in this example!

Leonard and Wardlaw (14) measured carbon dioxide production rates and internal concentration of both carbon dioxide and oxygen of bananas in a number of situations. Their data is summarized in Table 1. A temperature change of 8 C had little effect on resistance, even though production rates and internal concentrations changed substantially. This might be expected since resistance to diffusion is a physical parameter and hence should be relatively unaffected by temperature. Note also the marked increase in banana resistance as the fruits entered final senescence. This probably reflects an increase in the resistance of the pulp to diffusion as the intercellular air spaces became filled with water (diffusion in water is approximately 10,000 times slower than in air).

In tomatoes, the stem scar has been demonstrated to be the primary site of gas exchange (5,9), and Cameron and Yang (9), using the ethane efflux method, were able to show that the resistance of the skin was over 100 times that of the stem scar per unit surface area. This explains early observations which noted that waxing of the stem scar considerably retarded ripening rates (3). The increased resistance caused by blocking the stem scar would immediately cause increaseed concentration gradients for both carbon dioxide and oxygen, resulting in "controlled atmosphere" conditions within the tissue.

Summary and Conclusions

Carbon dioxide toxicity and low oxygen breakdown often accompany CA storage of fruits and vegetables and limit the extent to which the levels of oxygen can be reduced and carbon dioxide can be increased in the storage atmosphere. Optimum storage concentrations of these two gases depend on a variety of factors and we have attempted to stress that it is the concentration of oxygen and carbon dioxide within the tissue that is of vital importance. We have shown, assuming that the pulp of the fruit offers relatively little resistance compared to the skin and that the skin has an insignificant thickness relative to the radius of the fruit, that the internal concentration of oxygen and carbon dioxide is dependent not only on the external concentration, but also on the production or utilization rate, the surface area, and the resistance of the skin to diffusion (Equations 4 and 5). These three factors thus play a decisive role in limiting oxygen and carbon dioxide levels which can be safely used for CA storage.

Table 1 presents, in common units, resistance values for a number of fruits. Generally, due to the lack of attention which gas diffusion in fruits has received, the inconsistencies seen in Table 1 have not been addressed and are not easily explained. For instance, the differences between the resistances to carbon dioxide, oxygen, and ethylene in cranberries are particularly perplexing, as are the differences in resistance values between oranges and lemons, two closely related fruits. The consistently lower resistance values for water vapor also present an intriguing problem. Cultivar differences in resistance to gas diffusion may be important in determining sensitivity to CA limits. With the steady state method as described in Equation 7 or the ethane efflux method of Cameron and Yang (9), the means to determine these values are readily available. Given the importance of gas diffusion in successful CA storage of fruits and vegetables, it would seem that more effort will be devoted to these questions in the future.

Literature Cited

1. Ben-Yehoshua, S. and I.L. Eaks. 1969. Mode of action of abscisic acid in inducing selective abscission of citrus fruit, *Citrus sinensis*, Osbeck. J. Amer. Soc. Hort. Sci. 94:292-298.
2. Ben-Yehoshua, S. and I.L. Eaks. 1970. Ethylene production and abscission of fruits and leaves of orange. Botanical Gazette. 131:144-150.
3. Brooks, C. 1937. Some effects of waxing tomatoes. Proc. Am. Soc. Hort. Sci. 35:720.
4. Burg, S.P. and E.A. Burg. 1962. Role of ethylene in fruit ripening. Plant Physiol. 37:179-189.
5. Burg, S.P. and E.A. Burg. 1965. Gas exchange in fruits. Physiol. Plant. 18:870-884.
6. Burton, W.G. 1950. Studies on the dormancy and sprouting of potatoes. I. The oxygen content of the potato tuber. New Phytol. 49: 121-134.
7. Burton, W.G. 1974. Some biophysical principles underlying the controlled atmosphere storage of plant material. Ann. Appl. Biol. 78:149-168
8. Burton, W.G. 1978. Biochemical and physiological effects of modified atmospheres and their role in quality maintenance. *In* Postharvest Biology and Technology. H.O. Hultin and M. Milner, Eds. Food and Nutrition Press. Westport, Conn. 97-110.
9. Cameron, A.C. and S.F. Yang. 1981. A simple method for the determination of resistance to gas diffusion. Submitted to Plant Physiol.
10. Forsyth, F.R., I.V. Hall and H.J Lightfoot. 1973. Diffusion of carbon dioxide, oxygen, and ethylene in cranberry fruit. HortScience. 8:45-46.

11. Horrocks, R.L. 1964. Wax and the water vapor permeability of apple cuticle. Nature. 203:547.
12. Kader, A.A. 1980. Prevention of ripening in fruits by use of controlled atmospheres. Food Tech. 34(3):51-54.
13. Kidd, F. and C. West. 1949. Resistance of the skin of the apple fruit to gaseous exchange. Report of the Food Investigation Board, London, for 1939, 59-64.
14. Leonard, E.R and C.W. Wardlaw. 1941. Studies in tropical fruits. XII The respiration of bananas during storage at 53°F and ripening at controlled temperatures. Ann. Bot. NS. 5:379-423.
15. Lyons, J.M., W.B. McGlasson and H.K. Pratt. 1962. Ethylene production, respiration and internal gas concentrations in cantaloupe at various stages of maturity. Plant Physiol. 37:31-36.
16. Lyons, J.M. and H.K. Pratt. 1964. Effect of stage of maturity and ethylene treatment on respiration and ripening of tomato fruits. Proc. Amer. Soc. Hort. Sci. 84:491-500.
17. Marcellin, P. 1974. Conditions physiques de la circulation des gaz respiratoires a travers la masse des fruits et maturation. In, Colloques internationaux CNRS, No 238, Facteurs et Regulation de la Maturation des Fruits. 241-251.
18. Nobel, P.S. 1974. Introduction to Biophysical Plant Physiology. W.H. Freeman and Company. San Francisco. 488 pp.
19. Nobel, P.S. 1975. Effective thickness and resistance of the air boundary layer adjacent to spherical plant parts. Jour. Expt. Bot. 26:120-130.
20. Schonherr, J. 1976. Water permeability of isolated cuticular membranes. The effect of waxes on diffusion of water. Planta. 131:159-164.
21. Soliday, C.L., P.E. Kolattukudy and R.W. Davis. 1979. Chemical and ultrastructural evidence that waxes associated with the suberin polymer constitute the major diffusion barrier to water vapor in potato tuber (*Solanum tuberosum*, L.). Planta. 146:607-614.
22. Vendrall, M. 1970. Relationship between internal distribution of exogenous auxins and accelerated ripening of banana fruits. Aust. J. Biol. 23:1133-1142.

Table 1. Values of R_j's after conversion into sec/cm.

Tissue	$cm^2\ g^{-1}$	R_{CO_2} sec/cm	R_{O_2} sec/cm	$R_{C_2H_4}$ sec/cm	R_{H_2O} sec/cm	Source
APPLE						
Cox's Orange	1.0	21,200	30,000			13
McIntosh Red	1.0	18,000	30,000			13
Golden Delicious	-			30,000		*
McIntosh	1.0			35,000		5
Granny Smith	1.0				700	11
Granny Smith	-				7,000	19
Gravenstein	-				1,700	19
BANANA						
12 C	1.0	1,300	2,500		40	14
20 C	1.0	1,000	2,000		40	14
climacteric	1.0	800	1,500			14
postclimacteric	1.0	500	300			14
final senescence	1.0	5,000	4,800			14
preclimacteric	1.0			3,000		22
final senescence	1.0			17,000		22
CANTALOUPE						
preclimacteric	0.6	4,200	3,800			15
postclimacteric	0.6	5,000	3,000	6,900		15
CRANBERRIES						
whole fruit	2.0	70,000	70,000	400		10
LEMON						
whole fruit	1.5			10,000		4
whole fruit	-				240	19
ORANGE						
navel	1.0			3,000		1
navel	1.0			3,700		2
navel	-				1,000	19
valencia	1.0			14,000		4
valencia	-				1,500	19
PEACH						
whole fruit	1.0			19,000		4
whole fruit	1.0				26	19
POTATO						
whole tuber	1.2		5,000			8
whole tuber	0.9		5,500			8
whole tuber	-			36,000		*
TOMATO						
whole fruit	1.0	11,000	11,000	18,000		17
whole fruit	-			12,500	100	9
skin	-			27,000		9
skin	-				280	19
stem scar	-			260		9

* Cameron and Reid, unpublished.

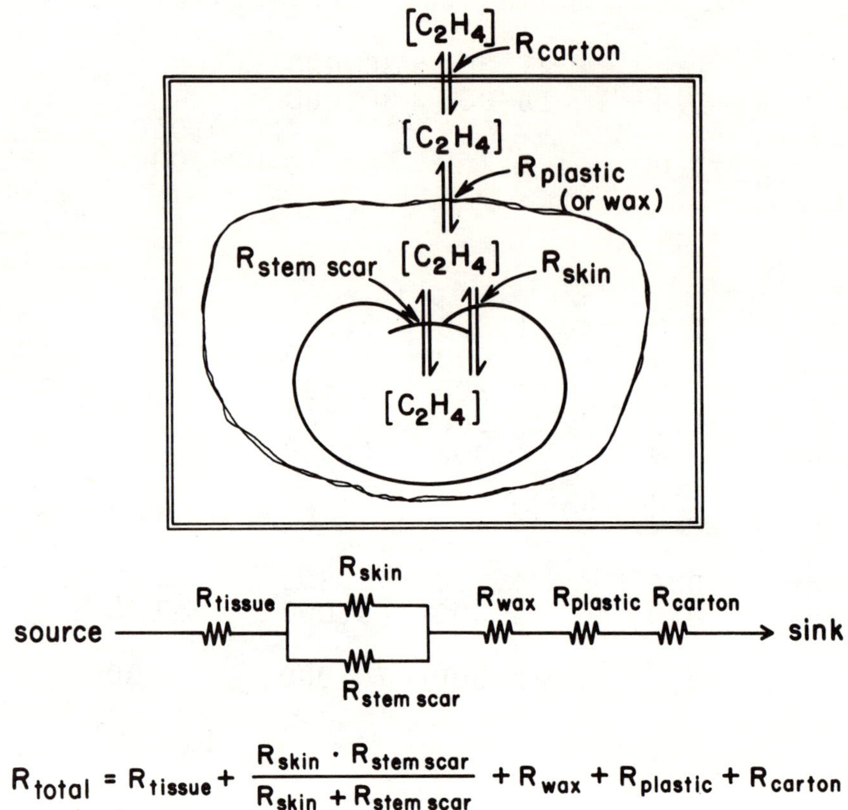

Figure 1. The resistance network approach applied to diffusion from bulky organs. The barriers to diffusion of ethylene from a tomato fruit are shown diagramatically and as a resistance network. The total resistance is calculated as for an electrical circuit.

CYANIDE-RESISTANT RESPIRATION IN POTATO TUBERS AS AFFECTED BY COLD STORAGE

Harry W. Janes, George Wulster, and Chaim Frenkel, Department of Horticulture and Forestry, Cook College, New Jersey Agricultural Experiment Station, Rutgers University, New Brunswick, New Jersey 08903.

ABSTRACT

Cyanide-resistant respiratory capacity increases in potato tubers from 10% to over 90% of the total respiratory rate following 12 days at 2°C. The development of cyanide resistance was seen to be enhanced by high O_2 tensions. Also, cyanide-resistant respiration which had developed in the cold could be maintained by holding the tubers at 22°C in 100% O_2. Similar effects were not seen at lower O_2 concentrations. This development of resistance to cyanide was accompanied by a rise in the CO_2 evolution of the tuber which was similarly enhanced by high and inhibited by low O_2 tensions.

The involvement of the alternative electron transport pathway was confirmed.

The data are discussed from the standpoint of enhanced metabolic flux through glycolytic and tricarboxylic acid cycle pathways and the maintenance of a homeostatic redox state of the tissue.

INTRODUCTION

Low temperature storage of potato tubers at 0-5°C is used to arrest degradative processes such as rotting, sprouting, and moisture loss (9), but is accompanied with an increase in sucrose and reducing sugar content of the tuber (4). These high sugar levels are undesirable since they are lost to discoloration and off-flavors in fried potato products, e.g. chips. Thus, low temperature storage of potatoes has some serious limitations and has led to research into the mechanisms and pathway of sweetening (1, 15, 16), aimed at understanding and regulating the metabolism of sugar accumulation.

Some of the research approaches include the effect of cold storage on glycolytic intermediates and adenine nucleotide levels. The early work conducted by Barker (1) on ethylene effect on tuber metabolism provided an interesting correlation to similar work by Solomos and Laties (19) and Rychter et al (18). Both Barker (1) and Solomos and Laties (19) note an increase in fructose diphosphate, dihydroxyacetone phosphate, and ATP levels, while measuring reduced levels of 3-phosphoglyceric acid and phosphoenolpyruvate in cold of C_2H_4 treated tubers, respectively. Solomos and Laties (19) report also an increase in sucrose and reducing sugar content following C_2H_4 treatment similar to the changes reported for low temperature treated tubers (9). Rychter et al (18) showed that ethanol treatment of potato tubers will cause changes in glycolytic and tricarboxylic acid cycle (TCA) intermediates similar to those induced by C_2H_4. These authors also report that high O_2 tension can enhance the effects of both C_2H_4 and ethanol.

Apparently ethylene, ethanol, and related volatiles stimulate aerobic glycolysis and metabolite flux into the TCA cycle. The concomitent development of the alternative cyanide resistant electron transport pathway by these treatments (12, 17) has led to the speculation that the alternative pathway may serve as a sink for the excess reducing power generated through increased glycolytic and TCA cycle activity and therefore prevent a change in the redox state of the tissue (13). Since low temperature storage elicit metabolic changes similar to those of ethylene and other volatiles we speculate that the treatment may lead also to the development of alternative pathway activity, and in the present work we test this hypothesis.

MATERIALS AND METHODS

Locally grown potatoes (Solanum tuberosum L., var. 'Norchip') were stored at 7°C prior to use. The tubers were preconditioned at room temperature for 2 weeks prior to treatments. Treatment consisted of holding the tubers at 2°C in 20-liter containers and continually ventilating with a 20% O_2 and 80% N_2 mixture. The tubers were stored in the cold for 0, 3, 6, or 12 days. Following the cold storage the tubers were transferred to room temperature (22°C) and kept in 4-liter jars for up to 72 h. These jars were ventilated with various N_2 and O_2 mixtures to provide 2, 20, or 100% O_2. At various times tubers were removed for the preparation of tissue slices and mitochondrial extraction. Slices were prepared and their respiration monitored using a Warburg manometric technique as previously described (17). A washed mitochondrial preparation was made using 100 g of potato tissue, representing a composite sample of 3 tubers according to the procedure outlined by Bonner (3). The final mitochondrial preparations contained 5 mg protein per ml. This preparation was used for measuring respiration with a YSI Model 53 Oxygen Monitor.

RESULTS

Figure 1 shows that the CO_2 evolution of whole potato tubers increases at room temperature (22°C) following 12 days of storage at 2°C in 20% O_2. It is further shown that this respiratory upsurge is amplified when the chilling treated potatoes are kept at room temperature in 100% O_2. There is no effect of O_2 treatment on tubers not first subjected to a cold pretreatment. The data showing the effect of low temperature storage on respiration correlated well with that of Barker (1), Boe et al (2), and Craft (7). Rychter et al (18) and Chin and Frenkel (6) showed that the ethylene or the ethanol stimulated respiratory upsurge is further stimulated in the presence of high O_2 tensions and the present work shows that the oxygen effect can be extended to the chilling stimulated respiration. The data raises the question of whether the cold pretreatment causes the production of C_2H_4 by the tuber in response to the stress condition and consequently a respiratory response and further stimulation by oxygen. However, no increase in C_2H_4 production by tubers could be detected following the cold treatment (data not shown). This supports the work of Creech et al (8) who also could not find an increase in C_2H_4 production by tubers stored for 28 weeks at 0°C.

The development of cyanide resistant respiration (CRR) in potato tubers as influenced by cold storage and post-storage treatment with different oxygen concentrations is depicted in Figure 2. Tubers taken out of cold storage after 0, 3, 6, or 12 days were kept at room temperature in oxygen concentrations of 2, 20, or 100%. During the post-storage period, sluces were prepared periodically from these tubers and respiration measured in the presence or absence of 10^{-3} M KCN. The results suggest that CRR develops in the cold since the percent CRR increases from less than 10% of the total respiration at day 0, to more than 90% when measured immediately after 12 days at 2°C. The data also indicate the involvement of O_2 in CRR development. For example, in tubers stored in the cold for 6 days, 30% of the total O_2 uptake is resistant to cyanide. However, when the low temperature storage treatment is coupled with post-storage application of 100% O_2 at 22°C there was a subsequent development of between 80-100% cyanide resistance. By comparison oxygen concentrations of 2 or 20% stimulate no such increase. High oxygen concentrations appear also to stabilize the CRR. After 12 days of storage at 2°C and subsequent post-storage treatment of tubers with 2 or 20% O_2 resulted in a decline in the percent CRR. On the other hand, a post-storage treatment of tubers in 100% O_2 maintained resistance to cyanide in the slices for at least 72 h (data not shown). Moreover, high oxygen concentrations lead to the enhancement of the respiratory rate following the addition of cyanide to the slices, resulting in a percent CRR greater than 100%. This phenomenon is of unknown origin and has been shown to occur on several other occasions following C_2H_4 plus 100% O_2 treatment of whole tubers (10). The effect of the oxygen on the development of CRR is dramatized by plotting the values obtained in Figure 2 after 48 h in oxygen (Fig. 3).

Figure 4 shows the effects of the cold storage and the post-storage effect of oxygen concentrations on the CO_2 evolution of tubers. The results show that similar to the effect on CRR, low temperature storage 3, 6, and 12 days progressively stimulate the CO_2 evolution by whole tubers. Furthermore, post-storage applications of oxygen stimulate whereas oxygen depravation (2%) does not. This figure is a correllary to Figure 3 and indicates the close parallel between whole tuber respiration and CRR activity in the potato tissue. The association between respiration and CRR development has been reported previously for tubers treated with ethylene (17) or acetaldehyde (12). Whether this stimulation of respiration by cold treatment is indicative of CRR in other tissues awaits further testing. However, Solomos and Laties (20) have demonstrated that the cyanide resistant electron transport path must be present for ethylene to stimulate respiration and we may speculate that the same could be true of the cold stimulation of respiration.

In order to more directly demonstrate that the development of CRR infers also the involvement of the alternative electron transport pathway the appropriate measurements were performed, and are outlined in Figure 5. A washed mitochondrial preparation was obtained from tubers stored for 12 days at 0°C and subsequently placed for 24 h at 22°C in 100% O_2. The tracing of the O_2 uptake of these mitochondria was compared with those obtained from tubers held for 12 days at room temperature and kept for 24 h at 100% O_2. While the mitochondria from the room temperature treated potatoes exhibit respiratory control those from the cold storage tubers do not; ADP had no effect on the O_2 consumption in these mitochondria and the state 2 rate is very similar to the state 3 rate of the control organelles. This loss or, in any case, decrease in respiratory control has been noted also in mitochondria with developed alternative pathway from potatoes treated with ethylene, or other volatiles including acetaldehyde, ethanol, or acetic acid (12, 17). Figure 5 also demonstrates that in control mitochondria cyanide inhibits respiration completely while in mitochondria from treated tubers cyanide inhibits only 10% of the O_2 consumption. The further addition of SHAM results in almost complete respiratory inhibition, whereas, the addition of SHAM, prior to cyanide application, results in a 66% reduction in O_2 consumption further suggesting that capacity for an alternative electron path has developed in these mitochondria. It has recently been shown that SHAM can inhibit lipoxygenase activity (14) and that it is possible to mistake this enzyme for the alternative oxidase, however, this has been shown not to be the case in potato (11). These results, therefore, demonstrate that the resistance to cyanide exhibited by the slices from cold stored potatoes is due at least to some extent to the alternative electron transport pathway.

DISCUSSION

The present work clearly shows that low temperature storage of potato leads to the stimulation of respiratory metabolism including the development of alternative pathway activity. Following 12 days in the cold, slices taken from tubers exhibited over 90% resistance to cyanide. Moreover, oxygen appears also to influence the development of CRR capacity and respiration. For example, the CRR was fully developed in tubers after 6 days at 2^oC when the tubers were subsequently stored in 100% O_2. Similar low temperatures storage coupled with a post-storage application of oxygen of 2 or 20% had no such effect. This effect of oxygen on respiration has been noted previously (10, 17) and suggests the involvement of O_2 in the induction of the alternative pathway. The present data is insufficient to resolve the question whether the requirement of the oxygen is for the development of CRR or as a substrate. Previous results (11) suggest, however, that a free radical scavenger, e.g. propyl gallate, can prevent the development of the alternative pathway in aged potato slices, indicating a more direct effect of O_2 or its active species in the development of CRR. Further experimentation is needed to discern this effect from the role of O_2 as a respiratory substrate.

The close similarity between the effects of C_2H_4, ethanol, and cold on potato tuber respiratory metabolism is interesting. The treatments lead to changes in glycolytic intermediate levels suggesting an increased aerobic glycolysis. We speculate that the enhancement in glycolytic and TCA cycle can result in an increase in tissue reducing power such that the alternative path develops as a compensatory response to re-establish homeostatic conditions (13). The possibility exists that the aforementioned treatments result in cytoplasmic perturbations leading in each case to increases in glycolysis and alternative pathway activity at the same time and independent of each other. It is known that fresh potato slice respiration from untreated tubers is 3-5 times higher than whole tuber respiration. It is also known that this respiration is via cytochrome oxidase. Therefore, in potato the cytochrome oxidase pathway has a capacity which would seemingly allow an increased glycolytic and TCA cycle rate before redox changes would be eminent.

The mechanism by which low temperature treatment initiate respiratory metabolism, and the effect of oxygen on respiration is an open question. Cold storage of potato tubers can result in an increased membrane permeability (22) as well as enhanced phospholipid breakdown as the underlying causes (5). Waring and Laties (21) have demonstrated, however, that phospholipid resynthesis is necessary for the development of alternative pathway activity in aged potato slices. These and other questions remain regarding the metabolic events associated with the chilling stimulated respiratory activity in potatoes, or other tissues.

In the present work we draw attention, however, to the fact that the metabolic effect of chilling temperatures may be similar to classical effects by ethylene, or other volatiles, and may represent a general stress response in tissues.

LITERATURE CITED

1. Barker, J., 1968, Studies in the respiratory and carbohydrate metabolism of plant tissues. XXV. Changes in rate of CO_2 output and in content of various phosphate compounds in potatoes. New Phytol., 67:495-503.

2. Boe, A. A., Woodbury, G. W., Lee, T. S., 1974, Respiration studies on Russet Burbank potato tubers: Effects of storage temperature and chemical treatments. Am. Potato J., 51:355-360.

3. Bonner, W. D., 1967, A general method for the preparation of plant mitochondria. Methods Enzymol., 10:126-133.

4. Burton, W. G., 1965, The sugar balance in some British potato varieties during storage. I. Preliminary observations. Eur. Potato J., 8:80-91.

5. Cheng, F. C., Muneta, P., 1978, Lipid composition of potatoes as affected by storage and potassium fertilization. Am. Potato J., 55:441-448.

6. Chin, C., Frenkel, C., 1977, Upsurge in respiration and peroxide formation in potato tubers as influenced by ethylene, propylene, and cyanide. Plant Physiol., 59:515-518.

7. Craft, C., 1963, Respiration of potatoes as influenced by previous storage temperature. Am. Potato J., 40:289-298.

8. Creech, D. L., Workman, M., Harrison, M. D., 1973, The influence of storage factors on endogenous ethylene production by potato tubers. Am. Potato J., 50:145-150.

9. Harkett, P. J., 1971, The effect of oxygen concentration on the sugar content of potato tubers stored at low temperature. Potato Res. 14:305-311.

10. Janes, W. H., Weist, S. C., 1980, Comparison between aging of slices and ethylene treatment of whole white potato tubers. Plant Physiol. 66:171-174.

11. Janes, H. W., Weist, S. C., 1980, The effects of propyl gallate on cyanide resistance in potato tubers. Plant Physiol. 65:S-6, 188.

12. Janes, H. W., Rychter, A., Frenkel, C., 1981, Development of cyanide-resistant respiration in mitochondria from potato tubers treated with ethanol, acetaldehyde, and acetic acid. Planta 151:201-205.

13. Lambers, H., Smakman, G., Respiration of roots of flood-tolerant and flood-tolerant and flood-intolerant <u>Senecio</u> species: Affinity for oxygen and resistance to cyanide. Physiol. Plant. 42:163-166.

14. Parrish, D. J., Leopold, A. C., 1978, Confounding of alternate respiration by lipoxygenase activity. Plant Physiol. 62:470-472.

15. Pollock, C. J., Ap Rees, T., 1975, Activities of enzymes of sugar metabolism

in cold-stored tubers of Solanum tuberosum. Phytochemistry 14:613-617.

16. Pressey, R., 1970, Changes in sucrose synthetase and sucrose phosphate synthetase activities during storage of potatoes. Am. Potato J., 47:245-251.

17. Rychter, A., Janes, H. W. Frenkel, C., 1978. Cyanide-resistant respiration in freshly cut potato slices. Plant Physiol., 61:667-668.

18. Rychter, A., Janes, H. W. Chin, C., Frenkel, C., 1979, Effect of ethanol, acetaldehyde, acetic acid, and ethylene on changes in respiration and respiratory metabolites in potato tubers. Plant Physiol. 64:108-111.

19. Solomos, T., Laties, G. G., 1975, The mechanism of ethylene and cyanide action in triggering the rise in respiration in potato tubers. Plant Physiol. 55:73-78.

20. Solomos, T., Laties, G. G., 1976, Effects of cyanide and ethylene on the respiration of cyanide-sensitive and cyanide-resistant plant tissues. Plant Physiol. 58:47-50.

21. Waring, A. J., Laties, G. G., 1977, Inhibition of the development of induced respiration and cyanide-insensitive respiration in potato tuber slices by cerulenin and dimethylaminoethanol. Plant Physiol. 60:11-16.

22. Workman, M., Kerschner, E., Harrison, M., 1976, The effect of storage factors on membrane permeability and sugar content of potatoes and decay by Erwinia carotovora var. Atroseptica and Fusarium roseum var. Sambucinum. Am. Potato J. 53: 191-204.

ACKNOWLEDGEMENTS

Paper of the Journal Series, New Jersey Agricultural Experiment Station, Cook College, Rutgers University, New Brunswick, New Jersey.

This work was performed as part of NJAES Project 12144, supported by the New Jersey Agricultural Experiment Station and Hatch Act Funds.

LEGEND FOR FIGURES

Figure 1. Whole potato tuber respiration as influenced by storage at $2^{o}C$ and subsequent post-storage at $22^{o}C$ in the presence of 20% (□) or 100% (○) oxygen. Solid symbols indicate tubers not subjected to cold storage.

Figure 2. The percent cyanide resistant respiration (CRR) of slices prepared from tubers stored for 0, 3, 6, or 12 days at $2^{o}C$ and post-storage at $22^{o}C$ in 2% (▲), 20% (●), or 100% (■) oxygen.

Figure 3. The effect of cold storage at $2^{o}C$ for 0, 3, 6, and 12 days on the development of CRR when the tubers are placed at $22^{o}C$ for 48 h in the O_2 concentrations noted in the figure.

Figure 4. Similar to Figure 3, except the CO_2 evolution of intact tubers is given.

Figure 5. Polarographic tracing of the O_2 uptake by mitochondria from tubers held at room temperature or at $2^{o}C$ for 12 days and post-storage application, at $22^{o}C$, of 100% O_2.

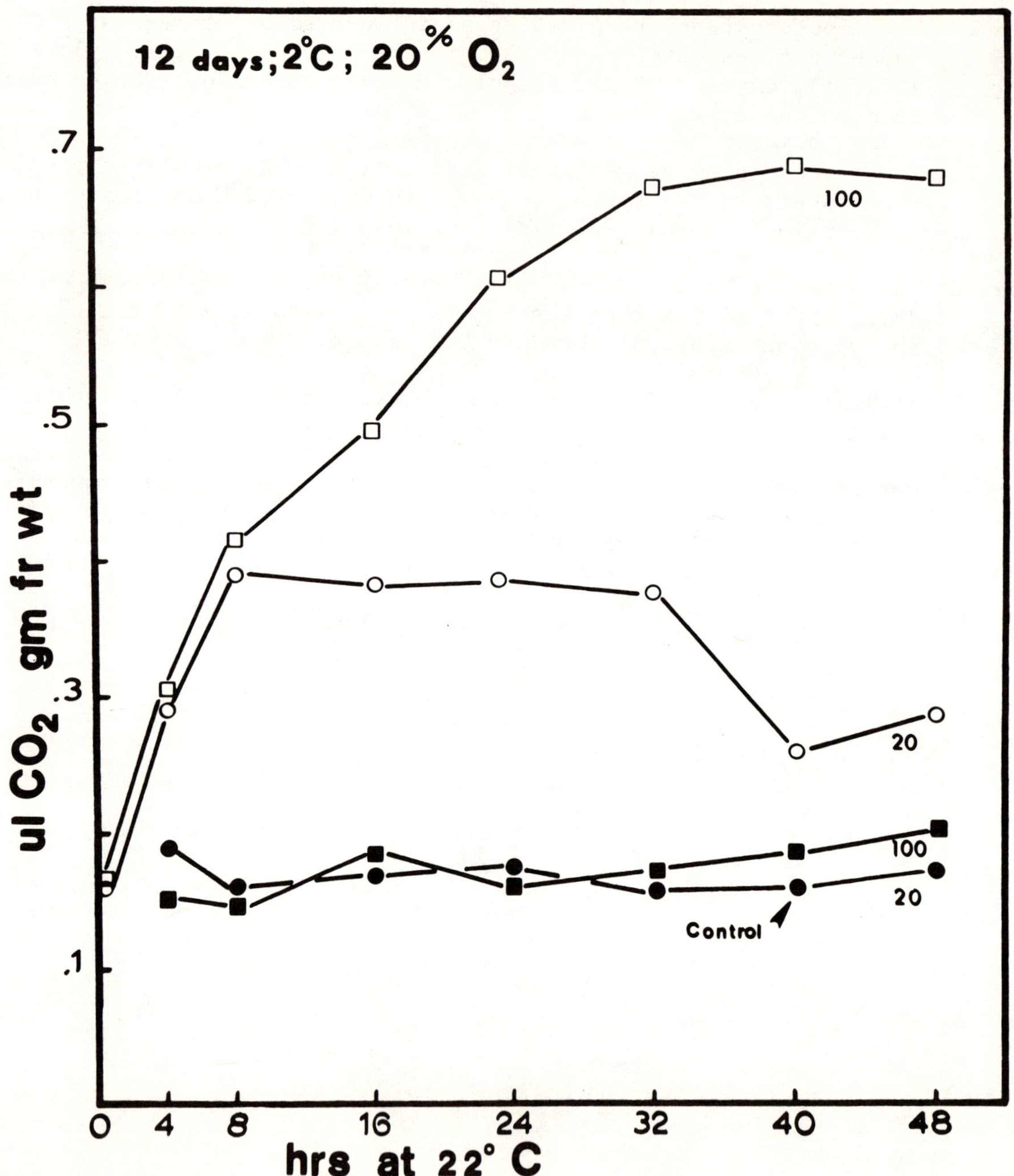
12 days; 2°C; 20% O2
.7
.5
.3
.1
ul CO2 gm fr wt
100
20
100
20
Control
0
4
8
16
24
32
40
48
hrs at 22° C

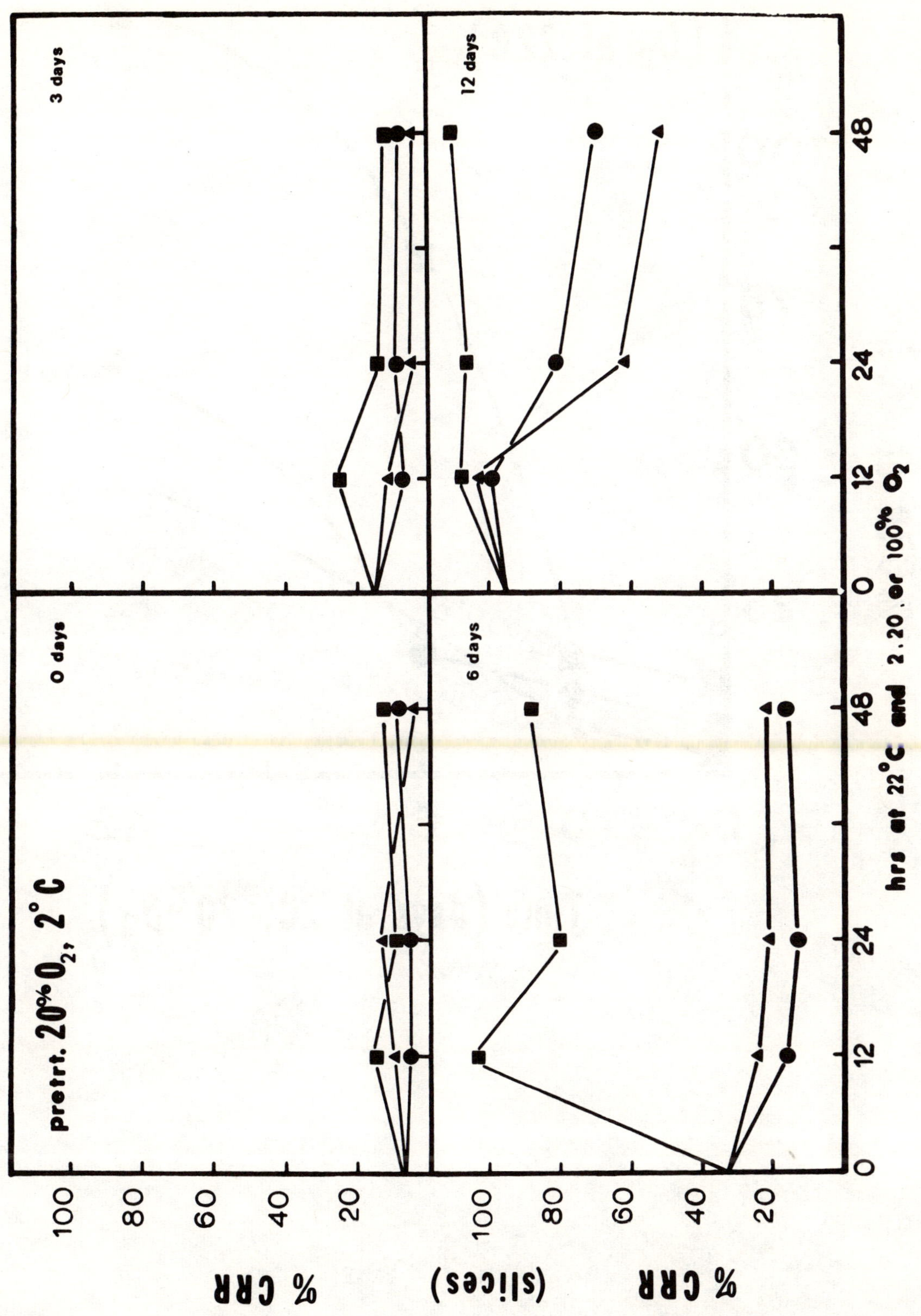

Pretrt. 20% O2, 2°C
0 days
3 days
6 days
12 days
% CRR
(slices)
% CRR
100
80
60
40
20
0
12
24
48
hrs at 22°C and 2,20, or 100% O2

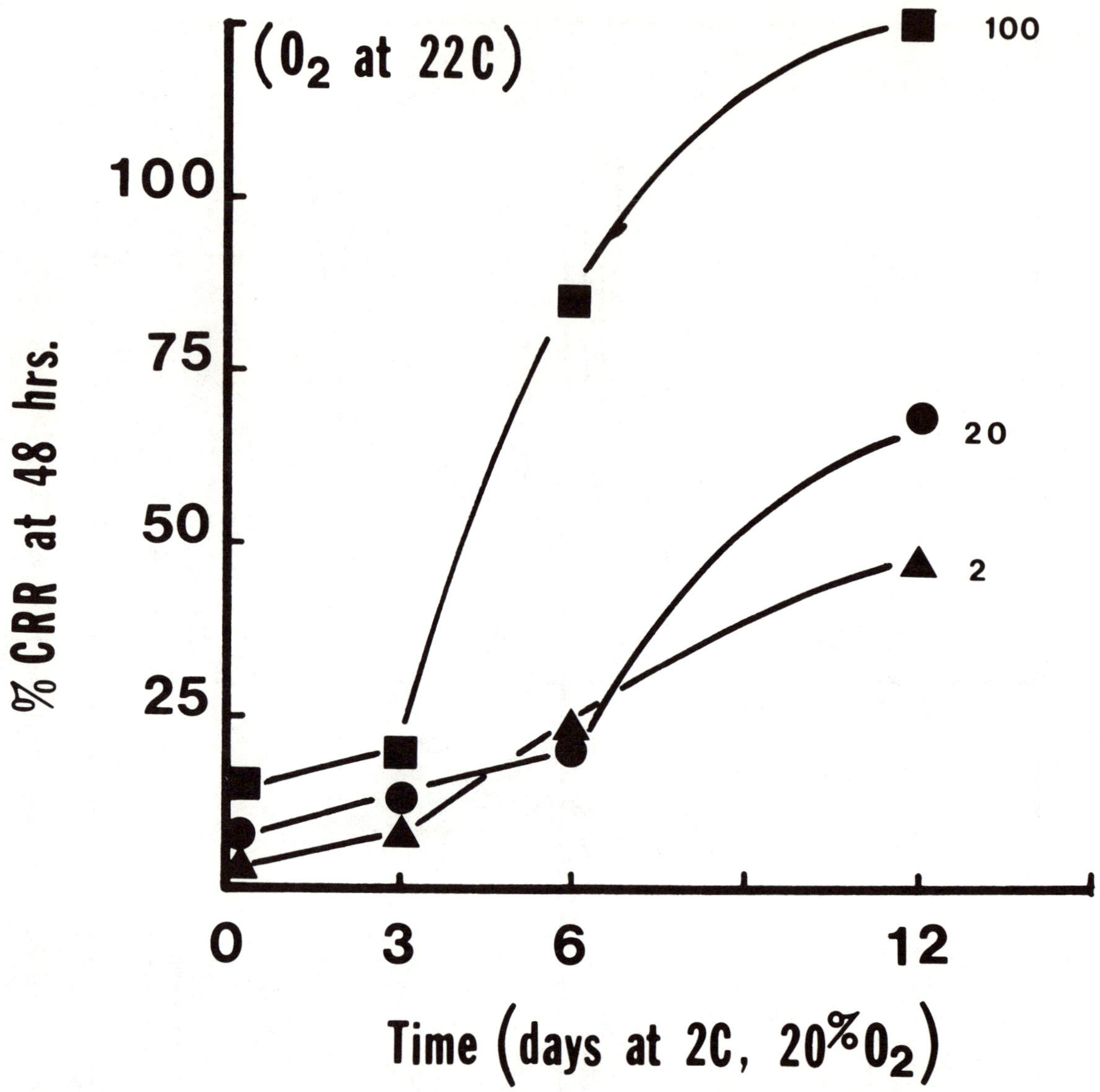

(O2 at 22C)
100
75
50
25
% CRR at 48 hrs.
100
20
2
0
3
6
12
Time (days at 2C, 20% O2)

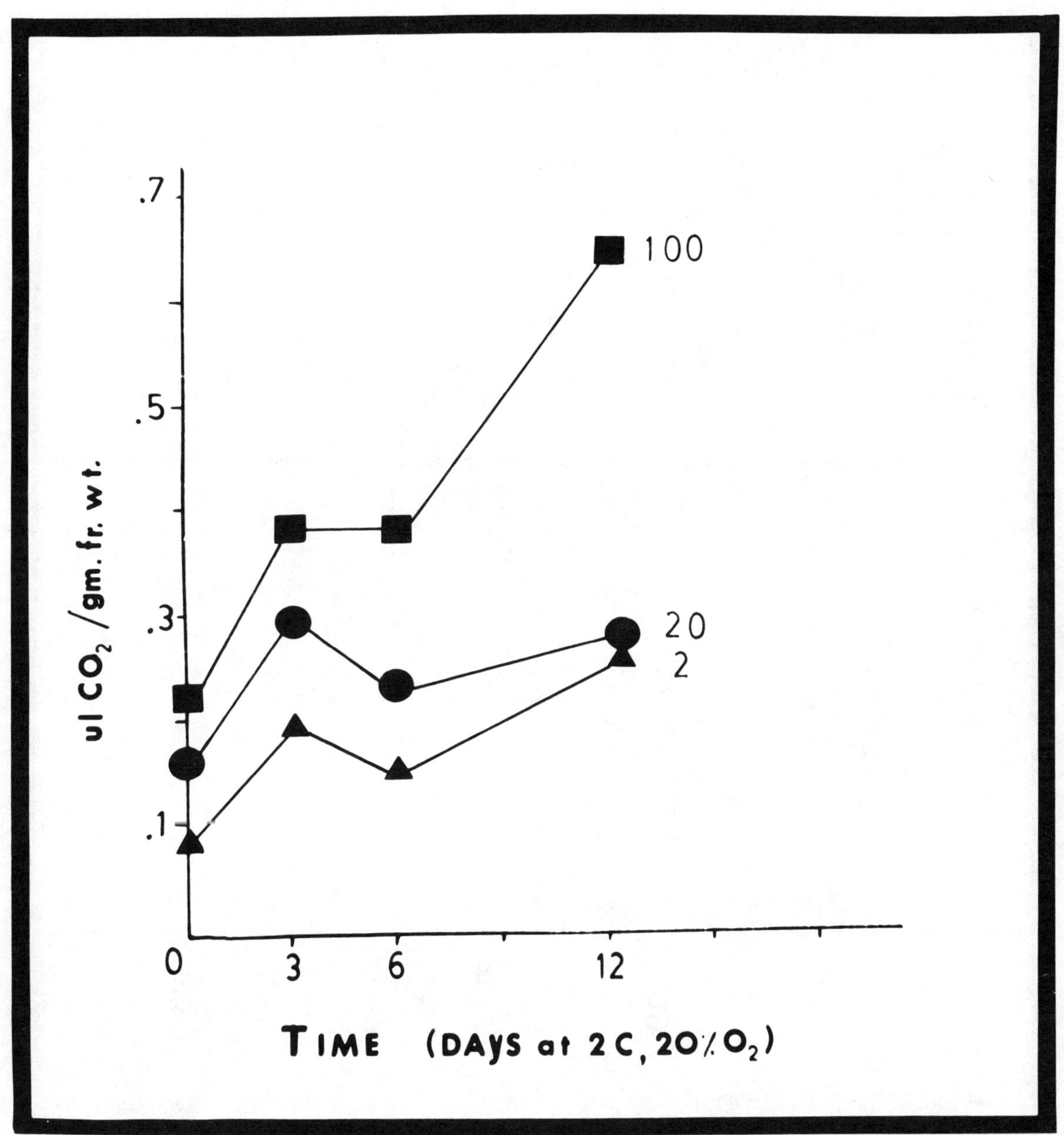

.7
.5
.3
.1
ul CO_2 /gm. fr. wt.
100
20
2
0
3
6
12
TIME (DAYS at 2 C, 20% O_2)

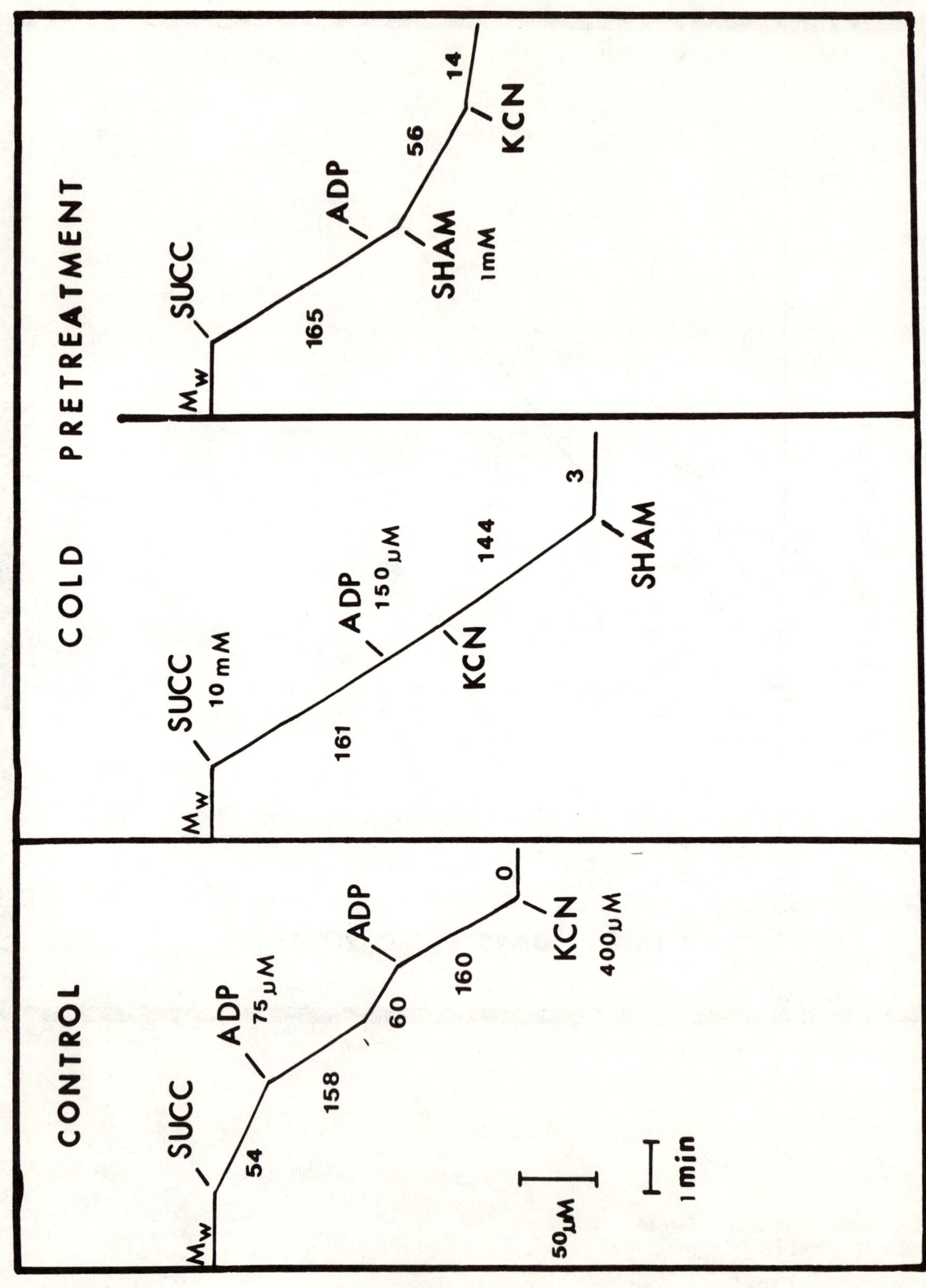
CONTROL
COLD
PRETREATMENT
Mw
SUCC
54
ADP
75μM
158
60
ADP
160
KCN
400μM
0
50μM
1 min
Mw
SUCC
10mM
161
ADP
150μM
KCN
144
SHAM
3
Mw
SUCC
165
ADP
SHAM
1mM
56
KCN
14

CONTRIBUTIONS OF MACRONUTRIENT MINERAL CONCENTRATIONS IN FRUIT TO LOSSES OF MASSACHUSETTS APPLES AFTER STORAGE

William J. Bramlage, Sarah A. Weis, and Mack Drake
Department of Plant and Soil Sciences
University of Massachusetts
Amherst, MA 01003 USA
(Presented by W.J. Bramlage)

Extensive studies in England quantified the influences of N, P, K, Ca, and Mg on fruit quality and the development of physiological disorders and rot after storage of apples and pears (8). Susceptibility to bitter pit, senescent breakdown, and *Gloeosporium* rot were negatively correlated with fruit Ca; susceptibility to bitter pit and rot were positively correlated with K/Ca and Mg/Ca ratios; fruit firmness was positively correlated with fruit P; low temperature breakdown was negatively correlated with fruitlet P. These studies led to recommended ranges of mineral levels for optimum storage of Cox's Orange Pippin apples, and to commercial prediction of storage life of these fruit from preharvest mineral analyses.

We have undertaken similar studies to determine if Massachusetts McIntosh apple storage life can also be predicted from mineral analyses. We report here the results of 2 years of surveys of commercial orchards to assess the quantitative influences of N, P, K, Ca, and Mg on characteristics of Massachusetts McIntosh apples after storage, and of an attempt to predict storage losses from mineral analyses.

In 1979-80, 34 commercial Massachusetts orchard blocks were sampled. We attempted to sample as diverse a group of blocks as possible, including different rootstocks, tree ages, McIntosh strains, and histories of problems. In some instances 2 or 3 different blocks in a given orchard were sampled since they represented different situations.

Approximately 2 weeks before commercial harvest began, 20 representative trees within each block were tagged and 1 fruit from each tree was taken for whole-fruit mineral analysis essentially according to Holland *et al.* (3). At the beginning of commercial harvest 2 bushels of fruit were taken from these same trees; 1 bushel was stored in 0°C air for 5 months, and the other was stored in 3.3°C CA storage (3% O_2, 5% CO_2) for 9 months.

Thus, all samples were harvested at about the same maturity and were stored in the same atmospheric conditions. Firmness was determined at harvest and 1 day after removal from storage, and disorders were counted after 1 week at room temperature (70-80°F) following storage. The 1980-81 survey included 25 orchards, in each of which both a block of trees on seedling rootstock and a block of trees on Malling 7 rootstock were sampled. Sampling, analysis, and storage were as in the previous year.

Perring's (7) summary of 20 years of analyses of Cox's Orange Pippin fruit mineral composition provides a comparison for the mineral levels of our commercial orchard samples (Table 1). Whereas our Mg levels were essentially identical to those of the English samples and our Ca levels covered about the same range, our P levels were somewhat lower and our K and especially our N levels were much lower. Furthermore, our Ca, N, K, and P levels were higher in 1980-81 than in 1979-80, probably due at least in part to somewhat smaller fruit the second year.

To assess relationships of these mineral levels to fruit characteristics after storage, correlation coefficients were calculated between firmness, senescent breakdown, rot (primarily due to Penicillium spp.) and superficial scald, and the whole-fruit concentrations of N, P, K, Ca, and Mg at harvest (Table 2). Bitter pit occurred too seldom to be considered.

The strongest relationships were the negative correlations between Ca and breakdown and Ca & rot.These were expected since we (1) and others (2) have consistently found that low fruit Ca levels are related to greater incidences of disorders after storage. There was also a strong negative relationship between Ca and scald after air storage, as we have sometimes seen previously (1).

High N and K concentrations can adversely affect fruit quality (2). Among our samples, the relationships of N to fruit quality were weak except for the negative correlation with fruit firmness at harvest in 1979 (Table 2). Relationships of K levels to fruit quality were also weak except for the positive correlation with scald after storage in 1979-80. These generally weak relationships probably reflect the relatively low fruit N and K levels in the samples (Table 1), which our commercial recommendations have been striving to achieve in recent years.

P concentrations in fruit were not correlated with any fruit characteristic in either year (Table 2), which is consistent with other research we have been conducting which has failed to associate fruit P levels with postharvest quality of McIntosh. This contrasts with studies in England (1, 4) and Australia (5), where low P has been associated with low temperature breakdown and soft fruit.

Mg was related to fruit firmness (Table 2) but since the relationship was negative in 1979-80 but positive in 1980-81, these correlations are probably either secondary or capricious. Mg concentrations in fruit appear to be very stable (Table 1).

Fruit firmness after storage is an extremely important quality factor for apples, especially an inherently soft cultivar such as McIntosh. In our samples there was no consistent relationship of mineral concentrations to fruit firmness. Greater firmness after storage has sometimes been found after Ca levels have been raised (6) but apparently the endogenous range of Ca levels does not normally affect McIntosh firmness.

Quantitative contributions of variations among these 5 minerals to variance in quality among the samples were determined by stepwise multiple regression (Table 3). By far the greatest effect was that of Ca on breakdown. N was the only other element significantly contributing to breakdown. Relationships of elements to occurrence of rot tended to weakly shadow the breakdown relationships. Both K and N concentrations contributed to scald incidence, and neither P nor Mg contributed significantly to any problem except for the strange Mg correlations with firmness.

It is interesting to note that in all cases, less total variance was accounted for by differences in mineral concentrations when CA fruit were examined than when air-stored fruit were examined. Thus, it appears that CA conditions may partially override the effects of adverse mineral concentrations in McIntosh apples.

Contributions of the 5 mineral elements to occurrences of disorders after storage are defined by the regression equations (Table 4). If these relationships are relatively constant from year to year, the regression equations can then be used as predictive indices for incidences of disorders. Since the strongest relationships were between minerals and senescent breakdown, we tested the 1979-80 regression equations as predictors of breakdown in the 1980-81 samples. In this prediction, we assumed that (1) the objective was to identify "high risk" samples before storage, not to predict the percent incidence of breakdown in each sample; and that (2) a "high risk" sample was one with the potential to develop breakdown in 10% or more of the fruit during a week at room temperature after storage. This value was chosen because it appears from previous data to be a natural break in arrays of commercial samples. Samples predicted to develop less than 10% breakdown were considered to possess "acceptable risk" for long-term storage.

The accuracy of these equations in predicting degree of risk is shown in Table 5. For air-stored fruit the equations were very effective predictors. Using the equation for Ca concentration alone, the prediction was correct for 48 of the 50 samples. When other elements were included the equations were less accurate, which is not surprising given the very weak contributions of N, P, K, and Mg to variance among these 1980-81 samples (Table 3), and the higher Ca concentrations in 1980-81 (Table 1).

For CA-stored apples the equations were not acceptable predictors of breakdown. The poor performance of the Ca-alone equation may have reflected the weaker relationship of Ca to CA breakdown in 1979-80 than in 1980-81 (Table 3), perhaps indicating the need for additional data to produce the appropriate equations. On the other hand, it might indicate that CA conditions override nutritional effects to an extent that accurate predictions of post-CA storage performance will not be possible.

In conclusion, these data suggest that among Massachusetts McIntosh orchards, variations in macronutrient concentrations in fruit are not contributing greatly to postharvest storage problems except for the adverse effects of low fruit Ca. This may reflect the efforts of these apple growers to maintain appropriate mineral levels in their crops and should not generate complacency. When samples were high in N and K, adverse effects usually occurred. Concentrations of most elements can change quickly from season to season, and other cultivars or other growing areas might possess stronger contributions of N and K to poststorage losses.

The successful prediction of susceptibility to senescent breakdown following air storage is encouraging to us, and we shall continue to test and attempt to refine the predictive indices for both air-stored and CA-stored apples.

Literature Cited

1. Bramlage, W.J., M. Drake, and J.H. Baker. 1979. Changes in calcium level in apple cortex tissue shortly before harvest and during post-harvest storage. Commun. Soil Science and Plant Anal. 10: 417-426.

2. Bramlage, W.J., M. Drake, and W.J. Lord. 1980. The influence of mineral nutrition on the quality and storage performance of pome fruits grown in North America. In: Mineral Nutrition of Fruit Trees. pp 29-39. Butterworths, Inc. London.

3. Holland, D.A., M.A. Perring, R. Rowe, and D.J. Fricker. 1975. Discrepancies in the chemical composition of apple fruits as analyzed by different laboratories. J. Hort. Sci. 50: 301-310.

4. Johnson, D.S. and N. Yogaratnam. 1978. The effects of phosphorus sprays on mineral composition and storage quality of Cox's Orange Pippin apples. J. Hort. Sci. 53: 171-178.

5. Letham, D.S. 1969. Influence of fertilizer treatment on apple fruit composition and physiology. 2. Influence on respiration rate and contents of nitrogen, phosphorus, and titratable acidity. Aust. J. Agric. Res. 20: 1073-1085.

6. Mason, J.L. 1979. Increasing calcium content of calcium-sensitive tissues. Commun. Soil Science and Plant Anal. 10: 349-371.

7. Perring, M.A. 1976. Fruit composition and the prediction of storage disorders of apples. Rep. E. Malling Res. Sta. for 1975: 169-170.

8. Sharples, R.O. 1980. The influence of orchard nutrition on the storage quality of apples and pears grown in the United Kingdom. In: Mineral Nutrition of Fruit Trees. pp. 17-28. Butterworths, Inc. London.

Table 1. Whole fruit mineral content (mg/100 g f. wt.) of English Cox's Orange Pippin apples and of Massachusetts McIntosh apples.

	Cox's Orange Pippin (20 year summary[z])		McIntosh			
			1979 - 1980		1980 - 1981	
	Mean	Range	Mean	Range	Mean	Range
Ca	5.0	3 - 8	4.4	2.8 - 6.6	5.5	4.0 - 8.7
N	65	40 - 100	24	12 - 35	37	21 - 48
K	135	100 - 200	95	66 - 116	106	82 - 131
P	13	8 - 20	9.7	6.8 - 11.7	10.7	6.4 - 13.6
Mg	5.2	4 - 7	5.4	4.5 - 6.2	5.3	4.6 - 6.4

[z]From Perring (7).

Table 2. Correlation coefficients for relationships between whole-fruit mineral analyses prior to harvest and characteristics of McIntosh apples. 1979-80. Values in parentheses are coefficients in 1980-81 that exceeded 0.29.

Characteristics and storage conditions[z]	Element				
	Ca	N	K	P	Mg
Breakdown,					
Air stg	-.58(-.53)	+.28	+.28(+.32)	-.11	+.02
CA	-.34(-.48)	+.33	+.21	-.26	+.06
Rot,					
Air stg	-.41	+.29	+.14(-.30)	-.16	-.31
CA	-.34(-.32)	+.25	+.03	-.03	-.28
Scald,					
Air stg	-.38	+.33(+.33)	+.45	-.12	+.22
CA	-.20	+.30	+.33	-.07	+.24
Firmness,					
Harvest	+.05	-.45	+.04(+.35)	+.14	-.37(+.32)
Air stg	-.06	-.19	+.17	.00	-.38(+.36)
CA	+.11	+.11	+.03	+.29	+.01

[z]0°C air storage for 5 months, or 3.3°C CA storage (3% O_2, 5% CO_2) for 9 months.

Table 3. Percent of total variance within McIntosh apple characteristics after storage that could be accounted for by differences in concentrations of macronutrients in fruit at harvest, as determined by stepwise multiple regression.

Parameter		Element					
		Ca	N	K	P	Mg	Sum
Bkdn, Air,	'79	33***	5	1	5	0	44**
	'80	29***	4	6	2	0	41***
Bkdn, CA,	'79	12*	12*	4	9	1	38*
	'80	23***	6	0	1	1	31*
Rot, Air,	'79	17*	8	2	4	8	39*
	'80	6	6	9*	1	0	22NS
Rot, CA,	'79	12*	6	1	1	6	26NS
	'80	10*	1	0	1	0	12NS
Scald, Air,	'79	1	11	20**	14	2	48**
	'80	6	11*	4	0	0	21NS
Scald, CA,	'79	0	9	11	7	3	30NS
	'80	5	1	0	0	4	10NS
Firmness, Air,	'79	1	2	5	0	14*	22NS
	'80	2	0	1	2	13*	18NS
Firmness, CA,	'79	2	1	0	8	0	11NS

Significance at 0.5 level (*), .01 level (**), and .001 level (***).

Table 4. Regression equations describing the relationships between whole-fruit analyses for macronutrients (in ppm, fresh wt basis) and occurrences of senescent breakdown in McIntosh apples after 5 months of air storage at 0°C or 9 months of CA storage at 3.3°C plus 1 week at room temperature. 1979-80.

Air storage:

% breakdown = 49 - .92 (Ca)
= 38 - .88 (Ca) + .039 (N)
= 62 - .92 (Ca) + .047 (N) -.24 (P)
= 53 - .85 (Ca) + .049 (N) -.27 (P) + .0085 (K)
= 48 - .87 (Ca) + .048 (N) -.27 (P) + .0075 (K) + .12 (Mg)

CA storage:

% breakdown = 30 - .46 (Ca)
= 57 - .52 (Ca) - .26 (P)
= 47 - .47 (Ca) - .32 (P) + 0.58 (N)
= 22 - .29 (Ca) - .39 (P) + .063 (N) + .024 (K)
= 18 - .30 (Ca) - .39 (P) + .062 (N) + .023 (K) + .12 (Mg)

Table 5. Accuracy of predictions of senescent breakdown after air storage or CA storage of 1980-81 McIntosh samples, using 1979-80 regression equations.

Air storage		CA storage	
Elements included	% accuracy	Elements included	% accuracy
Ca	96	Ca	71
Ca + N	88	Ca + P	59
Ca + N + P	86	Ca + P + N	61
Ca + N + P + K	84	Ca + P + N + K	57
Ca + N + P + K + Mg	84	Ca + P + N + K + Mg	55

THE USE OF RAPID CA TO MAXIMIZE STORAGE LIFE OF APPLES

O. L. Lau
Research Department
B. C. Tree Fruits Limited
Kelowna, British Columbia

Early work with McIntosh (10, 11) and with Delicious and Golden Delicious (6) had failed to show consistent practical benefits from rapid establishment of storage atmosphere. Opinion prevailed that prompt cooling was a more important factor than rapid establishment of the CA conditions. Consequently, the conventional CA procedure in British Columbia involves an 8 - 10 day loading period to facilitate rapid and thorough cooling and atmospheres containing 2.5% O2 are obtained by fruit respiration within 20 - 30 days from the time of room sealing. Lime scrubbers are used to regulate the CO2 concentrations (2 - 5%). Unfortunately, this 'slow' CA practice has failed to maintain a desirable level of firmness in such apples as McIntosh and Golden Delicious since fruit firmness of 5 kg (11 pounds) or less is often encountered in February or March.

This report shows that softening and acid loss of apples in CA storage, particularly Golden Delicious and McIntosh, can be greatly reduced by a rapid CA procedure (rapid loading plus rapid reduction of O2 to 2.5% or less). This procedure is as effective as or better than the pre-storage high CO2 treatment (1) or the postharvest calcium infiltration treatment (7) and does not injure the fruit. Results are also presented on the influence of rapid CA on the performance of apples stored at very low levels of O2 (4, 8).

<u>Experimental Procedures.</u> Samples of Golden Delicious, McIntosh, Spartan and Red Delicious, harvested at commercial maturity, were obtained from commercial orchards or at receiving docks of CA plants. Metal cabinets and commercial CA rooms (25,000 bushel capacity) were used in laboratory experiments and commercial trials, respectively.

Unless otherwise specified, replicated fruit samples were stored in air at 0° C or in a standard CA condition (2.5% O2 and 1.5 - 2% CO2 at 0° C for Golden Delicious, Red Delicious and Spartan; 2.5% O2 and 5% CO2 at 1.7° C for McIntosh). <u>Rapid CA</u> refers to a short loading time (within 2 - 3 days) plus a short period of O2 reduction (to 2.5% O2 within 1 - 3 days of room closure). The latter is achieved by the use of atmosphere generator or nitrogen gas. <u>Conventional CA</u>, on the other hand, involves a longer loading time (8 -10 days in B.C. to allow for thorough cooling) plus a less rapid rate of O2 reduction (O2 reduced to 2.5% by fruit respiration within 20-25 days of room closure).

In experiments involving very low levels of O2 (less than 2.5%), the O2 concentrations in the storage atmospheres were reduced to the specified levels by flushing with nitrogen gas within about 3 days of sealing (Rapid CA) or by fruit respiration within about 27 days of sealing of the CA cabinets (Slow CA).

Flesh firmness was determined with a Magness-Taylor pressure tester (with an 11.1-mm tip) and titratable acidity by titration of juice to pH 8.1 with 0.1N NaOH, upon removal of fruits from storage.

Effect of Rapid Oxygen Reduction. The rapidity with which O2 was reduced to 2.5% in the storage atmosphere clearly influenced the extent to which Golden Delicious apple softened during CA storage. The fruit from the intermediate CA (O2 reduced to 2.5% in 6 days by step-wise nitrogen flushings) was firmer than fruit from the conventional CA (with a 20-day O2 reduction period) but less firm than the rapid CA fruit (Table 1). The fruit from 1-day pull-down was almost 2 kg firmer than that stored in cold storage while that from 20-day pull-down was only about 0.5 kg firmer than the cold-stored fruit. Twenty-day O2 pull-down was undoubtedly less effective than 1-day pull-down in maintaining fruit firmness. Juice acidity differences were less dramatic but the rapid CA and intermediate CA fruits retained more acidity than the conventional CA fruit. Regular cold storage resulted in fruit clearly inferior to that from all CA treatments with respect to firmness and acidity.

Table 1. Effect of a rapid oxygen reduction procedure and other CA and non-CA options for maintaining flesh firmness and juice acidity of Golden Delicious apple.

Storage Type[y]	Days in Air 0° C	Days to 2.5% O2	April 10, 1979 Flesh Firmness (kg)	April 10, 1979 Juice Acidity (mg malate/100 ml)
Rapid CA (N_2)	1	1	7.78	463
Intermediate CA (N_2)	1	6	7.51	451
Slow CA (Respiration)	1	20	6.22	415
Cold Storage	Continuous	–	5.82	270
			S.E.= 0.093	S.E.= 13.2

[y]All CA atmospheres were held at 2.5% O2 and 2% CO2 and all storages held at 0° C. Oxygen in the air-tight cabinet was reduced to 2.5% in 1 day by nitrogen flushing (Rapid CA), in 6 days by stepwise nitrogen flushing (Intermediate CA), or in 20 days by fruit respiration (Slow CA).

Influence of Loading Time. It has been common practice in British Columbia to fill CA rooms in about 9 days to match the heat load to the refrigeration capacity. However, fruit subjected to a prolonged loading time (in air at 0° C) before implementation of the rapid O2 reduction procedure was clearly less firm and lower in juice acidity at the conclusion of the storage period (Table 2). Fruits with a rapid CA procedure were about 1 kg firmer than those with a conventional CA procdeure (9-day loading plus 20-day O2 reduction). In relation to conventional CA, progressive delays of 4 days reduced the benefit of rapid CA by 21%, 9 days by 55%, and 12 days by 70%.

Influence of Fruit Maturity and Responses of Various Fruit Lots. The rapid CA procedure is beneficial to all fruit lot tested, including those of different harvest maturity (Fig. 1) and those from various orchards with a wide range of initial fruit firmness (Table 3). The firmest fruits going

Table 2. The effectiveness of rapid CA compared to more conventional methods of storing Golden Delicious apples and the influence of a postharvest holding period in air at 0° C before implementation of the rapid CA procedures.

Storage Type	Days in Air at 0° C	Days to 2.5% O2	May 30, 1979 Flesh Firmness (kg)	May 30, 1979 Juice Acidity (mg malate/100 ml)
Cold Storage	Continuous	–	4.87	349
Conventional CA	9	18	6.06	436
Rapid CA (RCA)	0	1	7.14	481
Delayed RCA	4	1	6.78	495
" "	6	1	6.44	463
" "	9	1	6.38	456
" "	12	1	6.45	467
			r = – 0.8781*	r = – 0.8779*

* Correlation withdays delay significant at P = 0.05.

Table 3. Response of various lots of Golden Delicious apple to Rapid CA (2-day loading plus O2 reduction), conventional CA (8-day loading plus 20-day O2 reduction), and air storage at 0° C.

Grower No.	Initial Firmness (kg) Sept. 1978	kg Firmness: Rapid CA	kg Firmness: Conven. CA	kg Firmness: Cold Storage	kg Firmness Over: Conven. CA	kg Firmness Over: Cold Storage
		March 1979				
1	8.7	7.7	6.9	6.6	0.8	1.1
2	8.4	7.2	5.8	5.5	1.4	1.7
3	8.2	7.2	5.9	5.3	1.3	1.9
4	8.0	7.5	5.9	5.5	1.6	2.0
5	7.9	7.1	6.0	5.5	1.1	1.6
6	7.8	7.0	5.7	5.1	1.3	1.9
7	7.8	6.8	5.8	5.3	1.0	1.5
8	7.7	7.1	5.8	5.5	1.3	1.6
9	7.5	6.4	5.6	5.1	0.8	1.3
10	7.4	6.9	6.0	5.5	0.9	1.4
Average	7.9	7.1	5.9	5.5	1.2	1.6

into a given type of storage are always the firmest ones coming out (Fig. 1 and Table 3). In this regard, the rapid CA procedure and a fruit selection program should be implemented at the CA plant level to maximize the efficiency of CA storage.

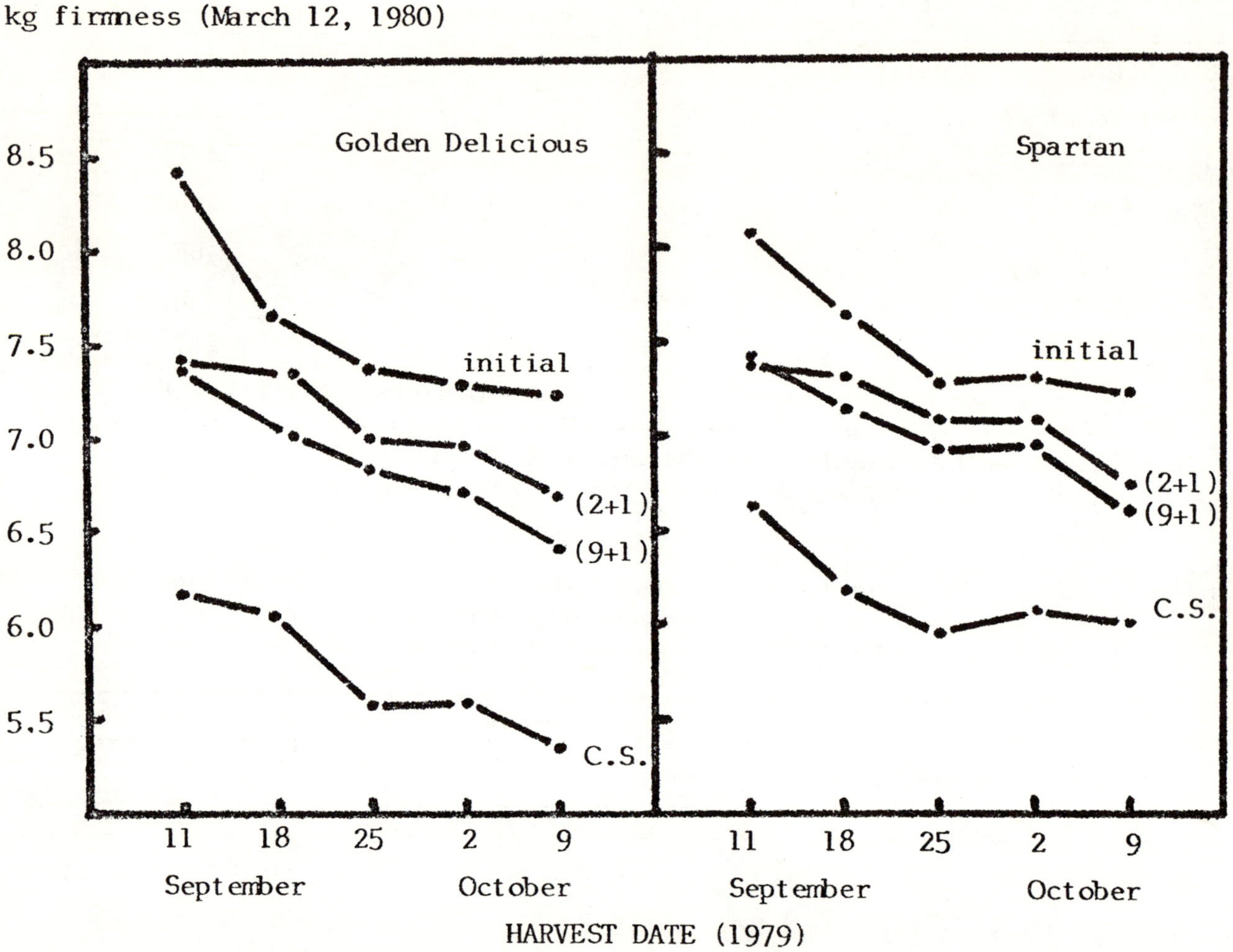

Fig. 1 Responses of Golden Delicious and Spartan apples harvested at different stages of maturity to cold storage (C.S.) or to rapid CA storage with different loading rates : 2-day loading plus 1-day O2 reduction (2+1) and 9-day loading plus 1-day O2 reduction (9+1).

Advantages of Rapid CA. The rapid CA procedure is an effective, simple and safe storage procedure to maximize storage life of apples in CA storage (Table 1, 2 and 3; Fig. 1). It is as effective as the pre-storage high CO_2 treatment (Table 4) and does not injure the fruit. Application of the high CO_2 treatment (1) in British Columbia has not been successful because of excessive CO_2 injury to the treated fruit (3, 5).

The rapid CA procedure is also better than a postharvest calcium infiltration treatment (7) in the maintenance of flesh firmness in Golden Delicious apple (Fig. 2). Non-treated fruits were much firmer in rapid CA than in slow CA or in air storage. They were also firmer than those infiltrated with calcium but kept in slow CA or in air storage. The calcium infiltration treatments retained more firmness than the check fruit held under air storage or slow CA but not under rapid CA (Fig. 2). Only the 8% calcium chloride treatment (which resulted in a flesh calcium level of 840 ppm) was significantly firmer than the other calcium-treated fuits or the check fruit stored with the rapid CA procedure but injury was generally very intensive with the calcium infiltration treatments (data not shown).

Table 4. Flesh firmness and juice acidity of Golden Delicious apples after storage at 0° C in air or in CA (2.5% O_2 and 2% CO_2) with different storage procedures.

Storage Treatment*	Flesh Firmness (kg)		Juice Acidity (mg malate/100ml)	
	Days in air at 0° C before storage treatment		Days in air at 0° C before storage treatment	
	No Delay	8-day Loading	No Delay	8-day Loading
Rapid CA (RCA)	6.9 a**	6.3 c	327 a**	282 bcd
17% CO_2 + RCA	7.0 a	6.5 b	308 ab	298 abc
Slow CA (Resp.)	6.1 d	5.9 e	276 cd	257 d
Continuous Cold Storage	5.2 f		163 e	

* The 2.5% O_2 in the storage atmosphere was established within about 2 days of sealing by introduction of nitrogen gas (Rapid CA) or within 20 days by fruit respiration (Slow CA). In the high CO_2 treatment, the O_2 level was reduced to about 5% within 6 hours by nitrogen flushing followed by introduction of CO_2 gas to bring the CO_2 concentrationn up to 17% and O_2 down to 4 - 5% at the conclusion of the first day. This atmosphere was maintained for the next 9 days before storage in rapid CA.

** Mean separation by Duncan's Multiple Range Test at 1% level.

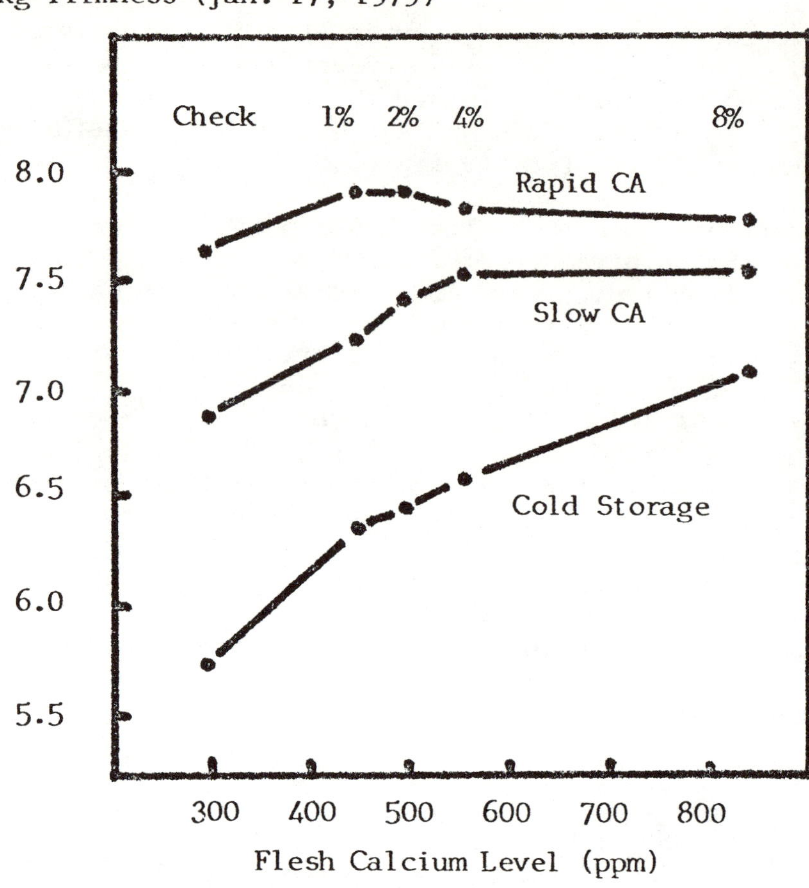

Fig. 2 Effect of a postharvest calcium chloride vacuum infiltration treatment and subsequent storage conditions on flesh firmness of Golden Delicious apple. Fruits were vacuum infiltrated (30 mm Hg for 5 minutes) with 1, 2, 4 or 8% calcium chloride solutions immediately after harvest, followed by rinsing with water and storage at 0° C in air or in CA (2.5% O2 and 2% CO2) with a rapid CA or slow CA procedure.

Commercial application of Rapid CA in British Columbia. Table 5 summarizes the response of B.C. Golden Delicious, McIntosh, Spartan and Red Delicious apples to the rapid CA procedure (rapid loading plus rapid reduction of O2 to 2.5%). Golden Delicious and McIntosh apples stored with a rapid CA procedure are at least 1 kg (2 pounds) firmer than those stored with a conventional CA procedure (long loading time and slow rate of O2 reduction). Slightly less benefit is, however, obtained with Spartan and Red Delicious apples. The latter can be stored successfully even in cold storage conditions. According to our experience, good storage results are obtained as long as the combined time of loading (preferably within 2-3 days) and O2 reduction (to 2.5% or less preferably within 2-3 days of sealing of CA room) does not exceed 4-6 days from harvest (Tables 1, 2, 3 & 5).

Table 5. Commercial application of the rapid CA procedure in British Columbia.

Variety	Crop Year	Storage Procedure	(Days loading + Days to 2.5% O2)		Firmness (kg)	kg benefit over	
						Conv.CA	Cold St.
Golden Del.	1978	Rapid CA (RCA)	(1 + 1)	May	6.9	1.0	1.7
		Conventional CA	(8 +20)		5.9		
		Cold Storage	-		5.2		
	1979	Moderate RCA	(4 + 4)	April	7.2	1.0	
		Conventional CA	(8 +20)		6.2		
McIntosh	1980a	Rapid CA	(1 + 1)	Feb.	6.6	0.9	1.1
		Conventional CA	(12 +19)		5.7		
		Cold Storage	-		5.5		
	1980b	Rapid CA	(4 + 4)	Jan.	6.1	1.0	
		Conventional CA	(7 +19)		5.1		
Spartan	1979	Rapid CA	(2 + 2)	April	6.6	0.4	1.1
		Conventional CA	(2 +21)		6.2		
		Cold Storage	-		5.5		
	1980	Moderate RCA	(4 + 3)	March	6.9	0.5	
		Conventional CA	(7 +20)		6.4		
Red Del.	1980a	Rapid CA	(1 + 1)	May	8.0	0.6	
		Conventional CA	(2 +23)		7.4		
	1980b	Moderate RCA	(4 + 3)	March	7.6	0.4	
		Conventional CA	(8 +21)		7.2		

Maximizing Storage Life of Apples by Rapid CA at Very-low Levels of Oxygen. Even though very-low O2 storage (less than 2.5% O2 in the storage atmosphere) is more effective than storage of apples at the conventional O2 levels of 2 - 3% (4, 8), the rapid CA procedure is still required to obtain the best storage results for Golden Delicious, Spartan, McIntosh and Red Delicious apples (Table 6). Golden Delicious and McIntosh apples stored at 1% O2 with a slow CA procedure were not any better than those stored at 2.5% O2 with a rapid CA procedure (Table 6).

Table 6. Softening rates of B.C. apples stored at 1.5% CO2 and various levels of O2 following a rapid CA (RCA) or slow CA (SCA) procedure.

Oxygen Conc.	Golden Del.		Spartan		Red Del.		McIntosh		
	RCA	SCA	RCA	SCA	RCA	SCA	RCA	SCA	C.S.
	% drop of initial flesh firmness after a 7-month storage period*								
1.0	2	8	2	5	7	5	5	25	-
1.5	4	9	3	6	7	8	-	-	-
2.0	7	14	8	12	9	10	-	-	-
2.5	7	15	11	17	8	15	15	28	-
21	-	-	-	-	-	-	-	-	36

* average readings of 1979/80 and 1980/81 experiments in a flow-through system (50 liters per hour).

Discussion. The results presented above clearly show that slow rates of loading and establishing low levels of O2 in the storage atmosphere are the primary causes of excessive softening of apples kept in CA conditions. It also suggests that the metabolic processess leading to fruit softening and acid loss throughout the storage period are induced in the field or shortly after harvest and continue to develop in air storage at 0° C and unless checked by the establishment of low O2 (or perhaps high CO2) regime immediately after harvest, these processes could have developed to the point where subsequent atmospheric modifications would become less or no longer effective.

This reasoning may explain why Olsen and Schomer (6) failed to show post-storage differences in firmness of Delicious and Golden Delicious apples when the O2 level was reduced to 3% in 10, 20,or 40 days. Similarly, Couey and Olsen (1) found that if a pre-storage high CO2 treatment occurred after 10 days in cold storage, the benefit was reduced by 50% compared to a treatment applied immediately after harvest; Fidler and North (2) achieved maximum retention of firmness and color when CA conditions for Cox's Orange Pippin apples were attained as soon as possible and Sharples and Munoz (9) observed that delays exceeding 7 days in establishing the desired O2 level resulted in significantly poorer results.

Conclusion. Post-storage condition of Golden Delicious, McIntosh, and to a less extent Spartan apples can be markedly improved by reducing the time between harvest and attainment of the desired storage temperature and low O2 environment. Delays in delivering harvested fruits to the storage facilities (data not shown), in filling the CA room, and in establishing the CA atmospheres can all contribute substantially to poor CA storage results.

Literature Cited

1. Couey,H.M. and K.L. Olsen. 1975. Storage response of Golden Delicious apples after high carbon dioxide treatment. J. Amer. Soc. Hort. Sci. 100: 148 - 150.

2. Fidler, J.C. and C.J. North. 1967. The effect on apples of the rate of establishment of conditions of storage. Ann. Rep. Ditton Lab. 1966-67. Agricultural Research Council, London. pp. 10 - 12.

3. Lau,O.L.and N.E. Looney. 1978. Effects of a pre-storage high CO2 treatment on British Columbia and Washington State Golden Delicious apples. J. Amer. Soc. Hort. Sci. 103: 341 - 344.

4. Lidster, P.D., F.R. Forsyth and H.J. Lightfoot. 1980. Low oxygen and carbon dioxide atmospheres for storage of McIntosh apples. Can. J. Plant Sci. 60: 299 - 301.

5. Meheriuk, M. 1977. Treatment of Golden Delicious apples with CO2 prior to CA storage. Can. J. Plant Sci. 57: 467 - 471.

6. Olsen, K.L. and H.A. Schomer. 1964. Oxygen and carbon dioxide levels for controlled-atmosphere storage of Starking and Golden Delicious apples. U.S.D.A. Agr. Marketing Service. Market Res. Report. 653: 1 - 11.

7. Scott, K.J. and R.B.H. Wills. 1977. Vacuum infiltration of calcium chloride: a method for reducing bitter pit and senescence of apples during storage at ambient temperatures. HortScience 12: 71 - 72.

8. Sharples, R.O. 1977, 1978 and 1979. Fruit Storage. Rept. East Malling Res. Stn. for 1976, 1977, and 1978.

9. Sharples, R.O. and G.C. Munoz. 1974. The effects of delays in the period taken to cool and establish low oxygen conditions on the quality of stored Cox's Orange Pippin apples. J. Hort. Sci. 49: 277 - 286.

10. Smock, R.M. and G.D. Blanpied. 1963. Some effects of temperature and rate of oxygen reduction on the quality of controlled atmosphere stored McIntosh apples. Proc. Amer. Soc. Hort. Sci. 83: 135 - 138.

11. Smock, R.M. and G.D. Blanpied. 1965. Effect of modified technique in CA storage. Proc. Amer. Soc. Hort. Sci. 87: 73 - 77.

CO_2 TREATMENT OF GOLDEN DELICIOUS APPLES IN VIRGINIA AS A SUPPLEMENT TO CA STORAGE

G. E. Mattus
Horticulture Department
Virginia Polytechnic Institute and State University
Blacksburg, Virginia

Research in Washington State (3) led to commercial CO_2 treatment of Golden Delicious apples in 1974. The CO_2 technique, resulting in firmer Goldens (1), has extended the distribution period of Golden Delicious which now appear on the market in firmer condition in the summer months.

Three years of research tests, 1976-77, 1977-78 and 1978-79, were conducted in Virginia with CO_2 treatment of apples picked in grower orchards in northern Virginia (2). Harvested test lots were transported from Winchester to the VPI & SU storage facilities at Blacksburg and the apples were treated with CO_2 from gas cylinders. CO_2 treatment was usually begun 3 days after apple harvest. After treatment, the apples were returned to Winchester to be held in commercial CA storage. After storing there were delays, on occasion up to two weeks, before CA room sealing. In the spring when the CA rooms were opened for the last time, the apples were brought to Blacksburg for testing.

Golden Delicious were picked at weekly intervals ("early", normal and "late" harvests) in commercial grower orchards. A test was also conducted with CO_2 treatment of Rome and Stayman apples in the 1978-79 storage season and with Red Delicious in each of the 3 test storage years.

Tables 1-5 indicate results of findings in the 1976-79 tests. Although Golden Delicious fruit firmness after CO_2 + CA treatment only averaged 1.2 pounds higher firmness than CA holdings, the CO_2 treated fruit was crisper, more acid, and had a better taste. All firmnesses were statistically higher following CO_2 treatment + CA storage of Golden Delicious compared with CA storage for every picking date of each year. It is surprising that the greatest firmness benefit of CO_2 treatment came from the later harvested Goldens each year and that this fruit also tested as firm or firmer than the "early" or "normal" harvest date pickings. This may be explained in part by delay in sealing rooms for the earlier harvested apples.

CO_2 injury in Virginia area grown Golden Delicious apples was rare, averaging about 0.1%, unusually low compared to Washington State or the high levels in British Columbia tests.

CO_2 treatment results were not favorable with Rome or Stayman. Romes did not benefit from CO_2 and Stayman were actually softened by the treatment. Red Delicious sustained CO_2 injury of 5, 7 and 2% in the 3 years of testing while fruit firmness was significantly increased in only one year out of 3.

Because of the potential for extending CA storage life and length of marketing for CO_2 treated Golden Delicious apples, this treatment was begun commercially for the first time in the Eastern part of the U.S. in September 1980, in 2 storages in Virginia. Apples from grower orchards in Pennsylvania, West Virginia and Virginia were treated with CO_2 and then CA stored at the Winchester Cold Storage. Fruit test lots removed from storage in May 1981 were crisp and juicy, testing 11.5 pounds pressure after a week at room temperature while test lots of Golden Delicious in regular storage had become soft by December 1980. Test lots harvested in 7 grower orchard blocks and commercially treated with CO_2 + CA storage averaged 10.8 pounds pressure on May 27, 1981 after a week at room temperature (68°F).

The other commercial CO_2 treatment was conducted at the Crown Orchard storage at Batesville, in the Charlottesville, Virginia area on their Golden Delicious. These Goldens averaged 11.1 pounds pressure in late May 1981 after a week at 68°F.

Both of Virginia's commercial CO_2 applications were successful in the 1980-81 season and may promise a better response in future seasons with improved CO_2 treatment and CA storage. Golden Delicious apples from the Pennsylvania, Maryland, West Virginia and Virginia area are physiologically similar and the CO_2 treatment technique would be expected to be of commercial value for this area of the U.S.

Our tentative recommendations for CO_2 treatment for the Virginia area Golden Delicious apples for the 1981-82 season are:

1. Place dry Golden Delicious in a CA storage room, do not stack tightly and do not treat with a decay and scald inhibitor before storage unless the fruit is dried before storing. Fill and seal room in 2 to 3 days.
2. Cool for at least 2 days, then build up CO_2 to near 15% with O_2 about 5%. Treat for 5 days.

3. Open and flush out the storage. When O_2 increases to normal, 21%, and CO_2 is 0%, remove bins and treat the Goldens with fresh ethoxyquin at 2700 ppm and also use 2 decay inhibitors, captan at 1 lb 50 WP/100 gal. + 8 oz benomyl (Benlate)/100 gal. Do not use DPA on Golden Delicious. DPA at 1000 ppm on our Goldens causes skin "scalding" injury. Tilt bins to drain excess liquid. Cover bins with a polyethylene film bulk bin cover and restore. Wetting Golden Delicious before CO_2 treatment could result in serious CO_2 injury which has not occurred on our Goldens when they were CO_2 treated while dry.

4. Generate a CA atmosphere with O_2 between 2.5 and 3.0% and CO_2 between 1 and 2% with a fruit temperature of 30° to 31°F.

If results are as favorable as in the past 4 years of research and application we could have good, crisp and juicy Golden Delicious for the fresh market into the summer months.

---

1. Bartram, Dick, 1979. CO_2-treated Goldens command premium late season market price. The Goodfruit Grower, Nov. 1, 1979: 10-11.

2. Mattus, George E., 1980. Carbon dioxide treatment of apples. Abstract 110. HortScience 15(3): 59.

3. Olsen, Kenneth L. and Richard D. Bartram, 1978. Carbon dioxide treatment of Golden Delicious apples. USDA AAT-W-2, 6 pp.

Table 1. CO_2 treatment of Golden Delicious apples.

Test Year	Apples Tested	CO_2 Treatment: % CO_2	% O_2	No. Days	°F Temperature
1976	April 1977	12	8	14	31°F
1977	May 1978	13	10	10	31°F
1978	June 1979	15	8	8	31°F

Table 2. Increase in firmness of Golden Delicious apples treated with CO_2 + CA storage over CA storage only* After storage + 7 days at 68°F

	CO_2 + CA, pounds pressure firmness higher than CA held fruit				
	No. of	Harvested:			Season
Year	Orchards	Early	Normal	Late	Average
1976-77	4	1.2	1.5	1.9	1.5
1977-78	8	0.7	0.5	1.3	0.8
1978-79	8	-	0.7	1.9	1.3
'76-'79 Average	-	1.0	0.9	1.7	1.2

* All firmness increases significant, Mean separation by Duncan's Multiple Range Test, 1% level.

Table 3. CO_2 treatment of Red Delicious, Rome and Stayman

	Fruit firmness upon removal from CA storage						
	Red Delicious			Rome		Stayman	
	CA	CO_2 + CA	% CO_2 injured	CA	CO_2 + CA	CA	CO_2 + CA
1976-77	13.8	14.2	5	-	-	-	-
1977-78	12.2	13.5	2	-	-	-	-
1978-79	15.6	15.3	7	13.4	13.4	13.9	12.4

Table 4. Golden Delicious apple scald, 1977-78, 8 grower test lots, commercial CA*

Treatment	% of apples scalded	% of apples with severe scald
Regular storage	64	23
CA storage	37	15
CO_2 + CA storage	16	5

* Fruit not treated with a scald inhibitor. Golden Delicious scalding was negligible in the 1976-77 and 1978-79 storage seasons.

Table 5. Golden Delicious apple scald, 1980-81, 7 grower test lots in commercial storage.

% scald after storage to May 1981 + 7 days at 68°F.

Storage Treatment	% of Apples Scalded	% Severe Scald
Regular storage	60	30
CA storage	4	1
CO_2 + CA storage	10	3
CO_2 + CA storage, Ethoxyquin treated	1	0

Conditions Affecting Storage Behavior of West Virginia Golden Delicious Apples

Morris Ingle
West Virginia University
Morgantown, WV 26506

Golden Delicious ranks second in tonnage among the apple cultivars produced in the United States, accounting for 18% of total production in the Appalachian area, which includes West Virginia (8). As data in this report will show, softening in storage is excessive in many years. Softening and loss of titratable acidity by Washington State-grown Golden Delicious can be reduced by prestorage treatment with 12-15% CO_2 followed by storage in atmospheres containing 2.5% O_2 and more than 2% CO_2 (2,7). British Columbia-grown fruit was badly injured externally and internally whereas no or only negligible injury was reported from Washington (3,6). Mattus (5) has reported improved storage behavior by high CO_2-treated Golden Delicious from Virginia. Recently, storage of several apple cultivars can be reduced by lowering the O_2 concentration from the conventional 2.5-3.0% to 1-3% (1,4). This report summarizes responses of West Virginia Golden Delicious to several storage conditions.

In 1976 and 1977 Golden Delicious apples from the University Experiment Farm in the Shenandoah Valley were stored in flow-through systems with 3% O_2 and varying levels of CO_2, and in air. Table 1 shows that high levels of CO_2 retarded softening and loss of green color. The response was the same the following year except that about 2% of the fruit in the highest CO_2 concentration had internal injury reminiscent of the type reported from British Columbia (3). That was the only time injury could be associated with high CO_2 either in prestorage treatment or in CA storage. Table 2 shows results from a small static system with fruit from the same source. Although a combination of 3% O_2 and over 3% CO_2 was not established with consistency, the benefits of increasing the CO_2 concentration seems fairly obvious. Also, there was less softening at the lower O_2 concentration. A post-harvest calcium chloride dip had little effect on softening but did retard loss of chlorophyll slightly.

Table 3 shows that high CO_2 prestorage treatment of Golden Delicious from an orchard at a higher elevation in West Virginia can effectively retard softening and loss of chlorophyll in CA storage; however, titratable acidity (malic acid) levels were not increased.

Table 4 attempts to show some effects of orchard location and harvest date or maturity on softening of Golden Delicious in CA storage. There are two principle apple growing areas in West Virginia. The Shenandoah Valley orchards are on deep, silty soils at elevations around 50-60 m. In the adjacent mountains orchards are at elevations of 115-150 m and on soils that are cherty and generally thin. Night temperatures and relative humidities are lower in the mountains than in the valley. In 1976 fruit from the higher elevations were firmer at harvest, due at least in part to the smaller sizes. Although firmness declines pretty consistently over the experimental picking period, there was not much difference after CA storage, except for orchard C where firmness after storage declined erratically from the earliest to the latest pickings. This storage is an inverse function of firmness at harvest.

If these findings are confirmed, perhaps ripening of Golden Delicious in storage, as measured by softening and loss of green color, could be best slowed by prestorage holding in an atmosphere containing 12-15% CO_2 and storage in 1.5-2.0% O_2 and 5% CO_2. The implementation of this system depends on identification of factors regulating the internal injury that has been observed, however slight it has been so far. It also has to be learned how some of the other cultivars that are commonly stored with Golden Delicious, such as Red Delicious or Stayman respond to low O_2 and high CO_2 in storage.

---

1. Bourne, M. L. and D. H. Dewey. 1980. Very low oxygen for the CA storage of McIntosh apples. HortScience 15(3):92. Abstract.

2. Couey, H. M. and K. L. Olsen. 1975. Storage response of 'Golden Delicious' after high-carbon dioxide treatment. J. Amer. Soc. Hort. Sci. 100:148-150.

3. Lau, O. L. and N. E. Looney. 1978. Effects of prestorage high CO_2 treatment of British Columbia and Washington State 'Golden Delicious' apples. J. Amer. Soc. Hort. Sci. 103:341-344.

4. Lougheed, E. C., J. T. A. Proctor, and S. R. Miller. 1980. Low-O_2 storage of three apple cultivars. HortScience 15(3): 92. Abstract.

5. Mattus, G. E. 1980. Carbon dioxide treatment of apples. HortScience 15(3):39. Abstract.

6. Meheriuk, M. 1977. Treatment of Golden Delicious apples prior to storage. Canad. J. Plant Sci. 57:467-471.

7. Olsen, K. L. and M. E. Petterson. 1977. Response of Golden Delicious apples to carbon dioxide in controlled atmosphere storage. HortScience 12(4):19. Abstract.

8. West Virginia Crop and Livestock Reporting Service. Fruit. January 23, 1981.

TABLE 1. Effect of carbon dioxide on behavior of Golden Delicious apples for six months in a flow-through system. Average of 3-5 kg replications.

% CO_2	% O_2	Firmness After Storage, kg	Color After Storage[x]
0	21	5.73	2.86
1	3	5.52	2.62
3	3	6.19	2.55
5	3	7.53	2.50

[x]1 = Bright green; 2 = Turning or silver green; 3 = bright yellow.

TABLE 2. Effect of atmospheric composition on behavior of Golden Delicious apples for 8 months in a static system. Average of 5-12 apple replications.

% O_2	% CO_2	Firmness After Storage, kg		Chlorophyll, mg cm^{-2}		Rots, %	
		+$CaCl_2$[1]	-$CaCl_2$	+$CaCl_2$	-$CaCl_2$	+$CaCl_2$	-$CaCl_2$
21	0	5.45	5.68	0.40	0.32	9	14
2.7 ± 1.4	0	6.05	6.18	0.57	0.53	9	15
2.2 ± 0.8	1.0 ± 0.6	6.64	6.86	0.77	0.74	1	2
2.9 ± 0.9	1.8 ± 1.8	6.50	--	0.64	--	9	--
4.4 ± 1.8	2.3 ± 1.8	6.41	6.55	0.72	0.62	4	2

[1] 4% v/v.

TABLE 3. Effect of preharvest CO_2 treatment on behavior of Golden Delicious apples for six months.

	Firmness after storage, kg	Chlorophyll mg cm^{-2}	Acid (mg malate 100 ml^{-1})
High CO_2, CA[x]	6.00	0.59	223
Control CA	5.56	0.46	264

[x]Fruit harvested 9-27-77. Cooled 20 hrs at 0°. Treated 13 days with 15% CO_2, 5% O_2. CA closed 10-12-77 (1% CO_2, 3% O_2). 20 fruit from 5 1-bushel treatments.

TABLE 4. Effect of location and harvest date on behavior of Golden Delicious apples for 6 months in CA storage.

Harvest Date	Days After Full Bloom	Firmness at Harvest, kg	Firmness After Storage, kg[y]	Firmness Change, kg	Color[x] After Storage
		A. Elevation 53 m[z]			
9/16/76	151	8.50	5.82	2.68	1.88
9/21/76	156	7.45	5.27	2.18	2.10
9/27/76	162	7.05	5.68	1.36	2.50
9/30/76	169	7.23	5.73	1.50	2.58
		B. Elevation 117 m[a]			
9/13/76	148	9.68	5.82	3.86	2.00
9/16/76	151	8.54	5.64	2.91	1.75
9/21/76	156	7.77	5.95	1.82	2.25
		C. Elevation 122 m[b]			
9/13/76	148	9.86	7.09	2.77	1.95
9/16/76	151	9.86	6.04	3.82	2.08
9/21/76	156	8.23	6.59	1.64	1.88
9/27/76	162	8.09	5.73	2.36	2.04
10/4/76	169	7.68	6.23	1.45	2.34

[x]1 = Bright green; 2 = Turning or silver green; 3 = bright yellow.
[y]Storage closed 10/15/76. 3% O_2, less than 1% CO_2.
[z]6 - 1 tree replications.
[a]4 - 1 tree replications.
[b]12 - 1 tree replications.

PROGRESS ON CONTROLLED ATMOSPHERE STORAGE AND INTERMITTENT WARMING OF PEACHES AND NECTARINES

C. Y. Wang and R. E. Anderson
Horticultural Crops Quality Laboratory
U.S. Department of Agriculture, Beltsville, Maryland

Introduction

The storage life of peaches and nectarines can be greatly extended by a controlled atmosphere (CA) of 1% O_2 and 5% CO_2 at 0°C (1,2,3). Fruit stored under these CA conditions were found to have 2 to 3 times longer storage life than that of fruit stored in air at 0°C. The inclusion of 5% CO_2 appears to be essential for extending storage life for both peaches and nectarines (2,20,22). Oxygen concentration below 1% may cause fruit to develop a fermented taste (3). However, peaches and nectarines are subject to low temperature injury even in CA conditions when they are held too long at 0°C. The injured fruit often lose the capacity to ripen and develop internal breakdown and woolliness at ripening temperatures. Intensities of bitterness, mustiness, and fermented attributes increase with increasing length of CA storage (31).

Intermittent warming has been reported to be effective in reducing or delaying the development of internal breakdown and woolliness and to further extend the storage life of peaches and nectarines (4,6). Intermittent warming apparently prevents woolliness by maintaining the capacity to produce adequate levels of pectolytic enzymes during ripening (9). The combination of CA and intermittent warming also tends to maintain higher acidity and lower respiration rate in the fruit (5).

The purpose of the present studies was to evaluate the effect of maturity, delayed CA storage, and multiple decay control treatments on the response of fruit to CA and intermittent warming, and to determine the effect of CA and intermittent warming on fruit quality, sugar content (fructose, glucose, and sucrose), and fatty acid composition of polar lipids in the fruit.

Experimental

'Redskin' and 'Rio Oso Gem' peaches and 'Regal Grand' nectarines were harvested from commercial orchards in New Jersey, Maryland, and Pennsylvania. They were brought to Beltsville Agricultural Research Center on the day of harvest and most lots of fruit were placed under test conditions by the following day. Two maturities were harvested one week apart. Fruit firmness was measured on pared surfaces of opposite sides of the fruit with a Magness-Taylor pressure tester with a 0.79-cm plunger. Firmness averaged 6.8 kg and 5.5 kg for the 2 maturities at harvest. In the delayed CA storage treatment, fruit were placed in air at 0°C for 2 weeks than transferred to CA.

To reduce the possibility of decay (particularly Brown rot *Monilinia fructicola* (Wint.) Honey), all fruit were dipped in a suspension of 100 ppm benomyl [methyl-1-(butylcarbamoyl)-2-benzimidazole carbamate] at 46°C for 2.5 min prior to storage. This treatment was found to be effective in controlling decay in peaches and nectarines during storage at 0°C and during subsequent ripening at 18.3°C without injuring the fruit (28).

Fruit samples were placed in stainless steel chambers (218 liters) or steel drums (114 liters) and held in either continuous-flow CA or air at 0°C. The desired CA (1% O_2 + 5% CO_2) was established and monitored as described previously (4).

For the intermittent warming treatment, fruit were removed from CA and placed in a second cold room in air. The refrigeration equipment in this room was then turned off and the temperature of the room was brought up to 18.3°C in 12 hours with a heater. The gradual warming minimized moisture condensation which previously was thought to contribute to skin discoloration of the fruit. After a total of 2 days, the CA was reestablished at 0°C. Fruit were warmed at 4-week intervals. Fruit samples in the multiple decay control treatment were dipped in a suspension of 100 ppm benomyl at 46°C for 2.5 min during each warming treatment. Titratable acidity, respiration rate, and internal appearance of the fruit were measured and evaluated as described previously (3,4).

For fatty acid analysis, flesh tissue was dipped in liquid nitrogen and then lypholized. The dried samples were stored under nitrogen at -15°C until analyzed. Polar lipids were extracted and quantitatively analyzed (26). The esterified fatty acids were analyzed using a Hewlett Packard 5711A gas liquid chromatograph equipped with a flame ionization detector and a 1.8-m stainless steel column packed with 10% SP-216-PS on 100/120-mesh Supelcoport. Temperature of the oven was 160°C; temperatures of the injector and detector were 200°C; and the flow rate of the carrier gas was 40 ml/min. A known amount of _n_-heptadecanoic acid was included in all samples as an internal standard, and methyl heptadecanoate was used as an external standard. Identification of individual fatty acid methyl esters was made by using known standards.

For sugar analysis, tissue was cut into small sections and immediately homogenized with a Polytron homogenizer in 80% ethanol. The resulting slurry was centrifuged at 500 g for 15 min. The residue was reextracted and washed twice by centrifugation and resuspension in 80% ethanol. The supernatants were combined and an aliquot of the extract was concentrated to dryness _in vacuo_ in derivatizing vials. Derivatization of the sugars was performed according to the procedures described by Li and Schuhmann (15). One ul of the derivatized samples was injected for gas chromatographic separation and quantification. Dual stainless steel columns (1.8 m x 1/3 cm) packed with 3% OV-17 on 80/100 mesh Supelcoport were used for separation of sugars. Temperatures of the chromatograph were injector 200°C, detector 300°C, and column 150°-300°C programmed at 8°C/min. Flow rates (ml/min) were H 40, N 30, and air 300. Separated sugars were compared with derivatized sugar standards for qualitative and quantitative determinations. A known amount of β-phenyl-D-glucopyranoside was included in all samples as an internal standard.

Response of fruit to CA and intermittent warming as affected by maturity, delayed CA storage, and multiple decay-control treatment

The internal quality of peaches and nectarines after 9 weeks of storage in air was very poor, whereas fruit in CA with intermittent warming treatment were in excellent condition (Table 1). The internal appearance rating of fruit placed in CA storage after a 2-week delay was lower after 15-week storage than that of fruit placed in CA without delay. The multiple decay-control treatments did not significantly reduce the percentage of decay, but after 20 weeks in storage, the CA fruit treated with benomyl at each warming ripened with a significantly better appearance than CA fruit not treated with benomyl at each warming. It appears that either benomyl, the hot-water treatment or a combination of the two may have some physiological effect on reducing the development of internal breakdown. Benomyl has also been shown to reduce chilling injury in grapefruit (25,30).

Respiration rate was highest in air-stored fruit and was lowest in fruit placed in CA immediately following harvest. The high rate of respiration tends to be associated with low quality and seems to be indicative of chilling injury in the fruit. CA fruit treated with benomyl at each warming retained the highest acidity. Fruit given delayed CA treatment were significantly lower in acidity than fruit placed in CA without delay. The less mature fruit had more acid and a lower incidence of decay than the more mature fruit, but the two maturities showed no difference in the respiration rates or internal appearance ratings throughout 20 weeks of storage (Table 2).

Changes of fatty acid composition as influenced by CA and intermittent warming

The degree of unsaturation of fatty acids in polar lipids of peaches was expressed as ratios of (A) C18:1 (oleic acid) + C18:2 (linoleic acid) + C18:3 (linolenic acid) to C18:0 (stearic acid), (B) C18:1 + C18:2 + C18:3 to C18:0 + C16:0 (palmitic acid), and (C) C18:2 to C18:3. There were significant differences between air samples and samples of CA plus intermittent warming in all three methods of measurement (Table 3). It is apparent that the difference in A ratio is greater than that in B ratio. Since all fatty acids were increased by CA plus intermittent warming treatment, this means that CA plus intermittent warming affect 18-carbon fatty acids more than 16-carbon fatty acid. The greatest difference exists in ratio C. This indicates that C18:3 may be the fatty acid affected the most by CA plus intermittent warming treatment, since both C18:3 and C18:2 were increased by this treatment.

Our data support the hypothesis that a higher degree of unsaturation of fatty acids is related to higher resistance to chilling injury (16). It is not clear how CA plus intermittent warming results in greater amounts of unsaturated fatty acids. It is possible that the shifting of fruit from cold to warm and then from warm to cold induces a rapid readjustment of metabolism that also includes increased synthesis of the unsaturated fatty acids. The higher CO_2 level and lower O_2 concentration of the CA conditions could also retard oxidative degradation of unsaturated fatty acids. Whether the higher degree of unsaturation was induced by CA, the warming treatment, or a combination of the two warrants further study.

Rapid changes in membrane lipids with altered temperatures have been reported in soybean roots (24). In these tissues, palmitic and stearic acids increased and oleic, linoleic, and linolenic acids decrease in both plasmalemma and mitochondrial membrane fractions as temperature was increased, and the reverse trend occurred as temperature was decreased. Elongation and desaturation of fatty acids have also been found in cell suspension cultures of *Catharanthus roseus* G. Don and *Glycine max* L. Merr. by changing temperatures (17).

Changes of sugars as influenced by CA and intermittent warming

Three major sugars in peaches were identified as fructose, glucose, and sucrose (Table 4). Although levels of fructose and glucose were lower in fruit in CA plus intermittent warming treatment than in air, a much higher amount of sucrose was found in CA plus intermittent warming treated fruit. Total sugar level was also higher in the fruit from CA plus intermittent warming than in fruit from air after 9 weeks' storage at 0°C. Whether sugar level is related to chilling susceptibility is not clear. It has been reported that both glucose and fructose levels decreased in tomato fruit ripened after 14 and 21 days of low-temperature storage (8) and that reducing sugars might play a significant role in promoting chilling resistance of grapefruit peel (23).

Conclusion

The combination of CA (1% O_2 + 5% CO_2) with intermittent warming reduced internal breakdown, extended storage life, and maintained good quality of peaches and nectarines. Treatment with benomyl in hot water at harvest and at each warming further improved the internal appearance. Cultivars such as 'Redskin', 'Rio Oso Gem' and 'Regal Grand' were kept for 20 weeks and still maintained good quality with these techniques. Fruit from CA plus intermittent warming treatment retained higher acidity, higher sugar contents, lower respiration rates and higher degree of unsaturation of fatty acids than those from air storage. After prolonged storage, fruit stored in CA immediately following harvest had significantly better quality than did fruit placed in CA after a 2-week delay. Less mature fruit had higher acidity than more mature fruit throughout the storage period but the two maturities showed no difference in respiration rates or internal appearance.

The use of intermittent warming to reduce chilling injury and extend storage life also has been successfully demonstrated in other fruits and vegetables such as apples (14), citrus (7,10), cranberries (11), cucumbers (29), okra (13), plums (27), potatoes (12), sweet peppers (29), and tomatoes (18). There is also evidence that the chilling-damaged ultrastructure of organelles in plant cells can be restored by rewarming (19, 21). However, the mechanism of how intermittent warming reduces chilling injury of fruits and vegetables is not understood. Does the warming treatment allow the tissues to metabolize toxic substances which are accumulated during chilling, or does warming allow tissues to restore materials which are depleted during chilling? What are the effects of alternating temperatures on physiological and biochemical reactions in the tissue? These are some of the questions that need to be answered in future investigations. With the increasing length of storage, good control of decay following extended storage is also a problem that remains to be solved.

Literature Cited

1. Anderson, R. E. and R. E. Hardenburg. 1977. Results and recommendations on controlled and modified atmosphere storage and transport of stone fruits. Hort. Rept. No. 28: 235-241. Proc. 2nd Natl. CA Res. Conf., Mich. State Univ.

2. Anderson, R. E., C. S. Parsons, and W. L. Smith, Jr. 1969. Controlled atmosphere storage of peaches and nectarines. Hort. Rept. No. 9: 66-68. Proc. 1st Natl. CA Res. Conf., Mich. State Univ.
3. Anderson, R. E., C. S. Parsons, and W. L. Smith, Jr. 1969. Controlled atmosphere storage of eastern-grown peaches and nectarines. U.S. Dept. of Agr., Mktg. Res. Rept. 836, 19 p.
4. Anderson, R. E. and R. W. Penney. 1975. Intermittent warming of peaches and nectarines stored in a controlled atmosphere or air. J. Amer. Soc. Hort. Sci. 100: 151-153.
5. Anderson, R. E., R. W. Penney, and W. L. Smith, Jr. 1977. Combining CA storage with intermittent warming extends the storage life of peaches and nectarines. Hort. Rept. No. 28: 149-155. Proc. 2nd. Natl. CA Res. Conf., Mich. State Univ.
6. Ben-Arie, R., S. Lavee, and S. Guelfat-Reich. 1970. Control of woolly breakdown of 'Elberta' peaches in cold storage, by intermittent exposure to room temperature. J. Amer. Soc. Hort. Sci. 95: 801-803.
7. Brooks, C. and L. P. McColloch. 1936. Some storage diseases of grapefruit. J. Agr. Res. 52: 319-351.
8. Buescher, R. W. 1975. Organic acid and sugar levels in tomato pericarp as influenced by storage at low temperature. HortScience 10: 158-159.
9. Buescher, R. W. and R. J. Furmanski. 1978. Role of pectinesterase and polygalacturonase in the formation of woolliness in peaches. J. Food Sci. 43: 264-266.
10. Davis, P. L. and R. C. Hofmann. 1973. Reduction of chilling injury of citrus fruits in cold storage by intermittent warming. J. Food Sci. 38: 871-873.
11. Hruschka, H. W. 1970. Physiological breakdown in cranberries--Inhibition by intermittent warming during cold storage. Plant Dis. Rept. 54: 219-222.
12. Hruschka, H. W., W. L. Smith, and J. E. Baker. 1969. Reducing chilling of potatoes by intermittent warming. Amer. Potato J. 46: 38-53.
13. Ilker, Y. and L. L. Morris. 1975. Alleviation of chilling injury of okra. HortScience 10: 324.
14. Kidd, F. and C. West. 1935. The cause and control of superficial scald of apples. Gt. Brit. Dept. Sci. and Ind. Res. Food Invest. Rept. 1934: 111-117.
15. Li, B. W. and P. J. Schuhmann. 1980. Gas-liquid chromatographic analysis of sugars in ready-to-eat breakfast cereals. J. Food Sci. 45: 138-141.
16. Lyons, J. M., T. A. Wheaton, and H. K. Pratt. 1964. Relationship between the physical nature of mitochondrial membranes and chilling sensitivity in plants. Plant Physiol. 39: 262-268.
17. MacCarthy, J. J. and P. K. Stumpf. 1980. The effect of different temperatures on fatty-acid synthesis and polyunsaturation in cell suspension cultures. Planta 147: 389-395.
18. Marcellin, P. and M. Baccaunaud. 1979. Effect of a gradual cooling and an intermittent warming on the cold storage life of tomatoes. Internatl. Congr. of Ref. XV: 6 - 7.
19. Moline, H. E. 1976. Ultrastructural changes associated with chilling of tomato fruit. Phytopath. 66: 617-624.
20. Munoz-Delgado, L., J. Caro, J. L. de La Plaza, and J. Espinosa. 1975. Treatment and controlled atmosphere storage of peaches of 'Confrentes' variety. 14th Internatl. Congress Refrig., Moscow, Bul. Internatl. Inst. Refrig. Vol. LV: 798.

21. Niki, T., S. Yoshida, and A. Sakai. 1979. Studies on chilling injury in plant cells. II. Ultrastructural changes in cells rewarmed at 26°C after chilling treatment. Plant & Cell Physiol. 20: 899-908.

22. Olsen, K. L. and H. A. Schomer. 1975. Influence of controlled-atmospheres on the quality and condition of stored nectarines. HortScience 10: 582-583.

23. Purvis, A. C., K. Kawada, and W. Grierson. 1979. Relationship between midseason resistance to chilling injury and reducing sugar level in grapefruit peel. HortScience 14: 227-229.

24. Rivera, C. M. and D. Penner. 1978. Rapid changes in soybean root membrane lipids with altered temperature. Phytochem. 17: 1269-1272.

25. Schiffmann-Nadel, M., E. Chalutz, J. Waks, and M. Dagan. 1975. Reduction of chilling injury in grapefruit by thiabendazole and benomyl during long-term storage. J. Amer. Soc. Hort. Sci. 100: 270-272.

26. St. John, J. B. and M. N. Christiansen. 1976. Inhibition of linolenic acid synthesis and modification of chilling resistance in cotton seedlings. Plant Physiol. 57: 257-259.

27. Smith, W. H. 1947. Control of low-temperature injury in the Victoria plum. Nature 159: 541-542.

28. Smith, W. L., Jr. and R. E. Anderson. 1975. Decay control of peaches and nectarines during and after controlled atmosphere and air storage. J. Amer. Soc. Hort. Sci. 100: 84-86.

29. Wang. C.Y. and J.E. Baker. 1979. Effects of two free radical scavengers and intermittent warming on chilling injury and polar lipid composition of cucumber and sweet pepper fruits. Plant & Cell Physiol. 20: 243-251.

30. Wardowski, W.F., L.G. Albrigo, W. Grierson, C.R. Barmore, and T.A. Wheaton. 1975. Chilling injury and decay of grapefruit as affected by thiabendazole, benomyl, and CO_2. Hort Sci. 10: 381-383.

31. Watada. A.E., R.E. Anderson, and B.B. Aulenbach. 1979. Sensory, compositional and volatile attributes of controlled atmosphere stored peaches. J. Amer. Soc. Hort. Sci. 104: 626-629.

Table 1. Effect of various storage treatments on internal appearance, respiration rate, titratable acidity, and decay of peaches and nectarines.[z]

Storage treatment	Internal appearance 9 wk	15 wk	20 wk	% Decay	Respiration rate (mg CO2/Kg-hr)	Titratable acidity (mg/100 g FW)
CA + IW	86 ab	80 b	68 c	11.3 a	41 c	678 b
CA + IW + BEN	94 a	88 ab	83 b	8.9 a	36 c	709 a
Delayed CA + IW	80 b	68 c	--	--	47 b	613 c
Air	21 d	--	--	--	77 a	--

CA = CONTROLLED ATMOSPHERE. IW = INTERMITTANT WARMING. BEN = BENOLATE TREATMENT

[z] Internal appearance rating: 100 = excellent (no internal breakdown)
20 = severe internal breakdown.

Data of 9, 15 & 20 weeks were combined for decay, respiration rate, and acidity.

Mean separation within each category by Duncan's multiple range test, 5% level.

Table 2. Influence of maturity on acidity, respiration rate, internal appearance, and decay of peaches and nectarines.[z]

Maturity	Titratable acidity (mg/100 g FW)	Respiration rate (mg CO_2/Kg-hr)	Internal appearance	% Decay
1	727 a	50 a	85 a	7.4 b
2	607 b	50 a	81 a	17.9 a

[z] Fruit firmness of maturity 1 - 6.8 Kg & maturity 2 - 5.5 Kg at harvest.

Mean separation within column by Duncan's multiple range test, 5% level.

Table 3. Degree of unsaturation of fatty acids in polar lipids of peaches.[z]

Treatment	A	B	C
Air	9.89 b	1.03 b	1.77 a
CA + IW	23.97 a	1.92 a	0.46 b

CA = CONTROLLED ATMOSPHERE, IW = INTERMITTANT WARMING, BEN = BENOLATE TREATMENT

[z] A = Ratio of C18:1 + C18:2 + C18:3 / C18:0.

B = Ratio of C18:1 + C18:2 + C18:3 / C18:0 + C16:0.

C = Ratio of C18:2 / C18:3.

Mean separation within column by Duncan's multiple range test, 5% level.

Table 4. Sugar levels in peaches as influenced by controlled atmosphere storage and intermittent warming.[z]

Treatment	Fructose	Glucose	Sucrose	Total
Air	31.5 a	22.6 a	34.2 b	88.3 b
CA + IW	24.2 b	17.4 b	53.8 a	95.4 a

CA = CONTROLLED ATMOSPHERE, IW = INTERMITTANT WARMING

[z] 9-week storage at 0°C.

Mean separation within column by Duncan's multiple range test, 5% level.

Sugar level expressed as mg/g of fresh weight.

RESPONSES OF HORTICULTURAL COMMODITIES TO MODIFIED ATMOSPHERES: AN OVERVIEW OF PUBLISHED LITERATURE AND FUTURE RESEARCH NEEDS

Adel A. Kader
Department of Pomology
University of California
Davis, CA 95616

By the end of 1980, the number of publications dealing with modified atmospheres reached 3,574 (1, 2, 3, 4). These included 3,301 references which were related to horticultural commodities. Publications on fruits and nuts represented 69.2% while those dealing with vegetable and ornamentals were 26.4% and 4.4%, respectively, of the total number. Among fruits and nuts, apples received the most attention with 46% of the total number of references on fruits and nuts. Pears were a distant second with 14% of the fruit publications. The other fruits among the top 10 were banana, peach and nectarine, strawberry, orange, grape, cherry, plum and prune, and avocado (Table 1).

Tomato and potato ranked #1 and 2 among vegetable crops, with about 16% each of the total number of references dealing with vegetables. Lettuce references represented 11%. Carrots, beans, cabbage, cauliflower, asparagus, dry onions, and peas were among the top 10 vegetables as to the number of publications (Table 2).

As shown in Table 3, carnations and roses received more attention than other cut flowers. An increasing interest in the effects of modified atmospheres on nursery stock is apparent from the number of references cited in Supplement No. 3 (1977-1980) relative to the previous lists (Table 3).

Most of the 3,301 publications on horticultural crops deal with the influence of modified and controlled atmospheres on visual quality and storage-life of the various commodities. Studies have also included effects on respiration and ethylene production rates, responses to ethylene, ripening, and physiological disorders. More recently, some attention has been given to composition and flavor quality of a few commodities as affected by modified atmospheres. Generally, research on the biophysical and biochemical basis of atmospheric modification has been limited.

Some of the areas of research which merits further investigation include the following:

1. Effects of modified atmospheres (MA) on flavor and nutritional quality of those commodities which respond favorably to MA in terms of appearance quality and storage-life.

2. Physiological and biochemical changes associated with short prestorage treatments with elevated CO_2.

3. Effects of ethylene and other volatiles under MA conditions on physiological disorders and quality attributes.

4. Reasons (anatomical, biochemical, etc.) for genotypic differences in tolerance to elevated CO_2 and/or reduced O_2.

5. Influence of CO addition to MA on physiological and biochemical changes in climacteric and non-climacteric fruits.

6. Definition of optimum MA ± CO combinations for those commodities which have previously received little or no attention, such as many of the tropical and subtropical fruits and chilling-sensitive ornamental crops.

7. Potential for using MA, alone or in combination with other treatments, on insect control in harvested horticultural commodities. Such treatments may include temporary exposure to anaerobic conditions, use of very high levels of CO or CO_2 for short durations, etc.

8. Continue efforts to develop MA & CA technology for increased use during transit, storage and marketing of those commodities which benefit from atmospheric modifications. Such efforts should include:

    a. Improved systems for generating, maintaining and monitoring desired MA & CA conditions.

    b. Better effective methods for removal of ethylene and other volatiles when needed.

    c. Improved safety procedures for CO use.

    d. Innovations which will facilitate the use of MA during transit.

References

1. Morris, L. L., L. L. Claypool, and D. P. Murr. 1971. Modified atmospheres. An indexed reference list through 1969, with emphasis on horticultural commodities. University of California, Division of Agricultural Sciences, 115 pp. (2326 titles).

2. Murr, D. P., A. A. Kader, and L. L. Morris. 1974. Modified atmospheres. An indexed reference list with emphasis on horticultural commodities, Supplement #1 (January 1, 197o to April 30, 1974). Vegetable Crops Series 168, University of California, Davis, 39 pp. (395 titles).

3. Kader, A. A. and L. L. Morris. 1977. Modified atmospheres. An indexed reference list with emphasis on horticultural commodities, Supplement #2 (May 1, 1974 to February 28, 1977). Vegetable Crops Series 187, University of California, Davis, 28 pp. (386 titles).

4. Kader, A. A. and L. L. Morris. 1981. Modified atmospheres. An indexed reference list with emphasis on horticultural commodities, Supplement #3 (March 1, 1977 to December 31, 1980). Vegetable Crops Series 213, University of California, Davis, 36 pp. (467 titles).

Table 1. Number of publications related tc effects of modified atmospheres on fruit and nut crops.

Commodity	Original Bibliography	Supplement No. 1	Supplement No. 2	Supplement No. 3	Total	Rank
Apple	714	98	103	137	1052	1
Apricot	19	5	3	3	30	13
Avocado	19	7	9	3	38	10
Banana	76	7	10	12	105	3
Berries						
Blackberry	4	0	0	0	4	26
Blueberry	4	1	0	2	7	23
Cranberry	8	4	2	0	14	21
Currant	5	2	2	1	10	22
Dewberry	3	0	0	0	3	29
Gooseberry	3	1	0	0	4	27
Raspberry	19	1	1	0	21	17
Strawberry	55	13	4	13	85	5
Cherimoya	0	0	0	1	1	35
Cherry	37	7	1	8	53	8
Citrus						
Grapefruit	16	7	6	6	35	12
Lemon	24	2	3	9	38	11
Lime	7	3	2	5	17	19
Orange	56	7	5	8	76	6
Others	14	10	3	2	29	14
Custard apple	0	0	0	1	1	36
Date	2	0	0	0	2	31
Fig	2	0	0	0	2	32
Grape	45	8	3	12	68	7
Guava	4	0	0	0	4	28
Kiwifruit	0	0	0	3	3	30
Mango	18	2	3	6	29	15
Nuts	23	1	0	3	27	16
Papaya	8	3	2	4	17	20
Passion fruit	5	0	0	0	5	25
Peach & nectarine	67	8	7	12	94	4
Pear	240	28	26	36	328	2
Persimmon	11	1	3	4	19	18
Pineapple	6	1	0	0	7	24
Plantain	0	0	1	1	2	33
Plum & prune	45	2	0	4	51	9
Raisin	1	0	0	1	2	34
Sapota	0	0	0	1	1	37
Total	1560	229	199	296	2284	

Table 2. Number of publications related to effects of modified atmospheres on vegetable crops.

Commodity	Original Bibliography	Supplement No. 1	Supplement No. 2	Supplement No. 3	Total	Rank
Artichoke	8	2	2	1	13	16
Asparagus	22	0	2	4	28	8
Beans	24	2	0	6	32	5
Beet, table	12	0	0	0	12	20
Broccoli	16	0	4	1	21	12
Brussels sprouts	7	0	0	2	9	24
Cabbage	15	3	3	10	31	6
Carrots	35	1	16	8	60	4
Cauliflower	12	4	5	8	29	7
Celeriac	0	0	1	4	5	27
Celery	8	0	3	1	12	21
Chicory	0	0	1	1	2	34
Chinese cabbage	0	0	0	2	2	35
Corn, sweet	9	1	0	1	11	22
Cucumber	9	2	5	5	21	13
Eggplant	2	1	0	0	3	32
Endive	0	2	2	0	4	30
Garlic	2	0	0	0	2	36
Kale	2	0	0	0	2	37
Kohlrabi	1	0	1	0	2	38
Leek	2	0	0	6	8	25
Lettuce	30	28	24	15	97	3
Melons	8	1	0	4	13	17
Mushrooms	5	4	6	4	19	14
Okra	0	1	1	3	5	28
Onion, dry	16	2	1	8	27	9
Onion, green	0	0	3	0	3	33
Parsley	0	1	3	2	6	26
Parsnip	1	0	1	2	4	31
Peas	19	1	1	1	22	10
Pepper, green	8	2	6	6	22	11
Potato	118	7	9	6	140	2
Radish	4	4	1	2	11	23
Rutabaga	2	0	0	0	2	39
Spinach	12	0	2	4	18	15
Squash & pumpkin	4	0	0	1	5	29
Sweet potato	11	0	0	2	13	18
Tomato	64	23	22	32	141	1
Turnip	10	0	1	2	13	19
Water cress	1	0	1	0	2	40
Total	499	92	127	154	872	

Table 3. Number of publications related to effects of modified atmospheres on ornamental crops.

Commodity	Original Bibliography	Supplement No. 1	Supplement No. 2	Supplement No. 3	Total
Cut flowers					
Carnation	14	1	2	2	19
Rose	17	2	1	2	22
Others	13	2	3	1	19
General	16	4	2	3	25
Flowering bulbs	12	2	0	5	19
Nursery stock	21	2	2	16	41
Total	93	13	10	29	145

MODIFIED-ATMOSPHERE STORAGE OF KIWIFRUITS (*Actinidia chinensis*)

F. Gordon Mitchell, Mary Lu Arpaia and Gene Mayer
Pomology Department
University of California
Davis, CA 95616

The kiwifruit is a relatively new commercial fruit crop of increasing importance in world commerce. Originally from the Yangtze Valley in China, this subtropical fruit was first commercialized in New Zealand. Cultivars developed there were introduced into California, and have provided the basis for a rapidly developing industry. The California industry is almost exclusively based on the 'Hayward' cultivar.

The kiwifruit has a good storage potential if properly handled. Much of its commercial success has been based on an apparent 6 month storage life and ability to carry through extended export shipment. The fruit has been shown to be climacteric (3), and has a high starch content at harvest. Usually harvested at 6.5 to 7.5% soluble solids content, the fruit will often increase to 14 to 16% soluble solids content after ripening.

A major problem in the successful storage of kiwifruits is softening of the flesh which occurs even during 0°C storage in air. Patterns of change in flesh firmness of California grown 'Hayward' kiwifruits for the 1976 through 1980 fruit seasons are shown in Fig. 1. In all but one season the most rapid decline in flesh firmness occurred during the first month of storage, and by 3 months the flesh firmness had declined to near one kg (penetrometer reading with 8 mm tip).

This flesh softening is a major problem in the handling of kiwifruits for several reasons. Because of it, fruits cannot be stored in bulk and packed later. Soft fruit will be more subject to mechanical damage during subsequent handling and transport. If flesh softening is accompanied by other physiological changes, as expected, then the storage softening could be accompanied by senescent breakdown in storage and/or reduction in subsequent market life. Finally, ethylene production which occurs in small quantities in firm kiwifruits could be accelerated in softer fruits; and ethylene gas is associated with the kiwifruit storage softening problem (4).

Poly Bag Storage

One method of slowing flesh softening of kiwifruits during storage might be the use of sealed poly bags, in which an elevated carbon dioxide atmosphere might inhibit ethylene action and production. Another modification might be the inclusion of an ethylene oxidizer (such as $KMnO_4$) inside the poly bag to prevent any ethylene build up.

Both of these approaches were tested in the 1977 and 1978 seasons. In these tests the CO_2 content in sealed 1.5 mil polyethylene bags varied around 4 to 6% (with great variability among bags), and in vented bags remained at ambient. Ethylene concentration in sealed bags ranged around 50 ppb, and when $KMnO_4$ impregnated aluminum oxide pellets were included, it dropped to below 5 ppb. In the vented bags ethylene concentration varied around 25 ppb.

During the two seasons of study the use of sealed bags resulted in less flesh softening than with air storage, but the results were variable among bags, as were the CO_2 concentrations encountered. Flesh firmness of fruits in sealed bags dropped in 1977 to about 2.5 kg in two months; in 1978 to about 2.5 kg in 5 months. At the same times, flesh firmness of air stored fruits measured 0.9 kg and 1.0 kg, respectively. The inclusion of $KMnO_4$ had a slight but unclear effect in further reducing the rate of softening.

Carbon Dioxide Storage and Controlled-atmosphere Storage

During 1977 a test was established to evaluate the effect of 2.5, 5 and 10% CO_2 atmospheres on flesh softening, and to compare those treatments with air storage and with 2% O_2 + 5% CO_2 controlled-atmosphere storage. After 4 months storage (end of test) the results were:

Atmosphere	Firmness - kg
Air	0.9
2.5% CO_2	1.9
5% CO_2	2.8
10% CO_2	3.8
2% O_2 + 5% CO_2	4.8

Studies conducted here and elsewhere show that fruits held at 10% CO_2 develop injury symptoms, with New Zealand researchers suggesting 8% CO_2 as an upper limit of tolerance (2). The complete CA treatment provided the best protection from flesh softening, and the fruit ripened well following storage.

Controlled-atmosphere Manipulations

During 1978, a 6 month CA storage test was conducted comparing 1% O_2 + 5% CO_2, 2% O_2 + 5% CO_2, 2% O_2 + 10% CO_2 and an air storage control. At

6 months the air control reached 0.9 kg firmness, the 2% O_2 + 10% CO_2 - 2.4 kg, and both 5% CO_2 combinations (with 1 and 2% O_2) - about 5.4 kg firmness. Most of the firmness reduction associated with 10% CO_2 occurred during the last two months and are presumed to be related to CO_2 injury.

Both of the 5% CO_2 combinations responded similarly, and flesh firmness changes were small enough to avoid handling or deterioration problems during the 6 month storage season. With the increased danger of fruit injury at lower O_2 levels, especially under commercial CA conditions, it was decided to pursue further studies using 2% O_2 + 5% CO_2.

Seasonal Variations

Controlled-atmosphere studies, using 2% O_2 + 5% CO_2 have now been conducted over a 4 season period. While slight variations in the softening pattern have occurred during the 4 seasons, the pattern has been generally similar (Fig. 2). In each season fruits from the storage studies have ripened to good eating texture and quality without difficulty.

In developing a commercial treatment from this work it is essential that the effect of other environmental modifications (such as ethylene) be considered. Results of follow up studies of ethylene effects on kiwifruits in CA storage are reported elsewhere in this proceedings (1).

References

1. Arpaia, M. L., F. G. Mitchell, A. A. Kader and G. Mayer. 1981. The ethylene problem in modified atmosphere storage of kiwifruit. Proceedings of the National Controlled-Atmosphere Research Conference. Oregon State University, Corvallis, pp. 331-335.

2. Harman, Jane. 1981. Personal communication.

3. Pratt, H. K. and M. S. Reid. 1974. Chinese gooseberry: seasonal patterns in fruit growth and maturation, ripening, respiration and the role of ethylene. J. Sci. Fd. Agric. 25:747-757.

4. Reid, M. S. and S. Harris. 1977. Factors affecting the storage life of kiwifruits. The Orchardist of N.Z., April.

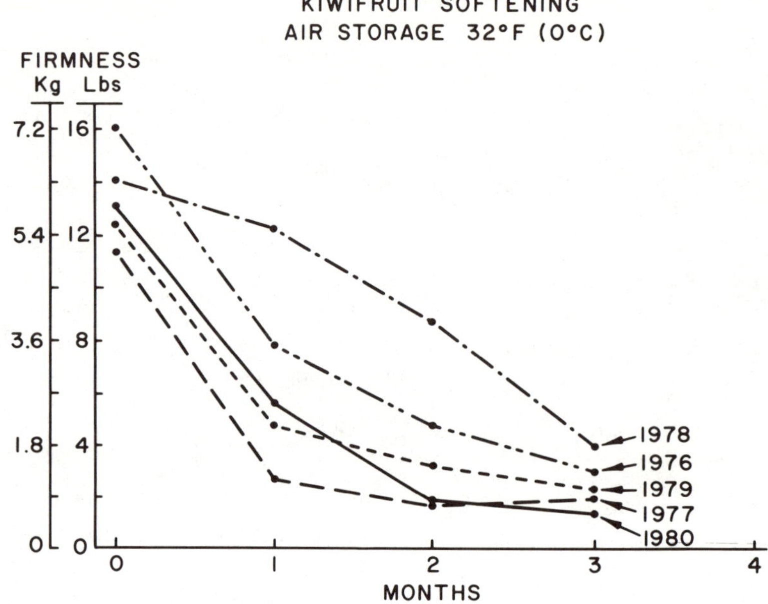

Fig. 1. Flesh softening patterns during storage in air at 32°F (0°C) of California grown 'Hayward' kiwifruits - 1976 through 1980 seasons.

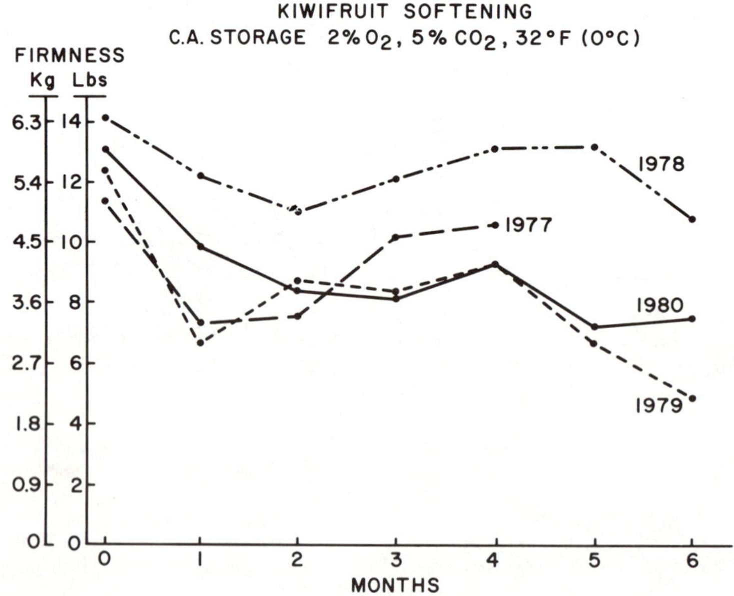

Fig. 2. Flesh softening patterns during controlled-atmosphere storage (2% O_2 + 5% CO_2) at 32°F (0°C) of California grown 'Hayward' kiwifruits - 1977 through 1980 seasons.

FLORAL CROP STORAGE

G.L. Staby, J.W. Kelly, and M.S. Cunningham
Department of Horticulture
The Ohio State University
Columbus, OH 43210

(Presented by J.W. Kelly)

Floral crop storage frequently refers only to the storage of one category of floral crops:cut flowers. In this paper, however, we will refer to the storage of potted flowering and foliage crops, cuttings, greens, bedding plants, and cut flowers as well as the storage of floral tissue cultures. Readers are encouraged to read reviews by Aarts (1957), Staby (1977), Carow (1978), Eisenberg (1977), and Halevy and Mayak (1979) for historical perspectives of some of the literature in this subject area.

Three general types of storage systems are presented for each crop category where data is available, namely normal refrigerated storage, modified atmosphere storage including controlled atmosphere, and low pressure or hypobaric storage.

As a final introductory comment, we highly recommend to our readers that they make use of our computerized literature retrival system at The Ohio State University. The main subject area is the postharvest care, handling, storage, and physiology of floral crops. We presently have about 5000 references stored, all of which have been translated into English if required. References can be extracted by keywords, authors, journal, and/or year published. All requests for such information should be made to the senior author.

Literature cited - general

Aarts, J.F. Th. 1957. De ontwikkeling en houdbaarheid van afgesneden bloemen. Meded. Dir. Tuinb. 20:690-702.

Carow, B. 1978. Frischhalten von schnittblumen. Ulmer-Fachbuch, Stuttgart, 144 pp.

Eisenberg, B.A. 1977. Low pressure and common cold storage of rooted and unrooted ornamental cuttings. M.S. thesis, The Ohio State University, Columbus, 69 pp.

Halevy, A.H. and S. Mayak. 1979. Senescence and postharvest physiology of cut flowers, part I. Hort. Rev. 1:204-236.

Staby, G.L. 1977. Controlled atmosphere and low pressure storage of floral crops-overview. Controlled atomspheres for the storage and transport of perishable agricultural commodities, Second National Controlled Atomsphere Research Conference. p60-70.

Cuttings

An extensive review by Eisenberg (1977) has been presented in which he summarizes cutting storage work prior to 1977. Of the 65 species noted, only 5 were floral crops (Altstadt and Holley, 1964; Anon., 1975; Laurie, et al. 1967; Judd, 1976; Pryor and Stewart, 1964; Mastalerz, 1957) with the remaining being woody species. All of these data were obtained using normal refrigerated storage facilities.

As more became known concerning low pressure storage, it seemed logical to expand the floral cutting research into this area. Data presented in the following 2 tables exemplify the results obtained.

Table 1. Visual evaluation of foliage 3 weeks after removal from normal refrigeration or low pressure storage of *Euphorbia pulcherrima* (poinsettia) cuttings.[z]

Cutting type	Storage time (weeks)			
	Normal[y]		Low pressure[x]	
	1	3	1	3
rooted	6.0	3.9	6.0	5.8
unrooted	1.0	1.0	4.6	1.0

[z]Rating: 1=dead; 6=excellent
[y]760mm Hg at 10°C.
[x]35mm Hg at 10°C.

Table 2. Visual evaluation of foliage 4 weeks after removal from normal refrigerated or low pressure storage of *Pelargonium* x *hortorum* (geranium) cuttings.[z]

Cutting type	Storage time (weeks)					
	Normal[y]			Low pressure[x]		
	2	4	6	2	4	6
rooted	5.4	1.7	1.0	6.0	5.8	5.2
unrooted	2.8	1.1	1.0	4.3	3.4	1.1

[z]Rating: 1=dead; 6=excellent
[y]760mm Hg at 5°C.
[x]35mm Hg at 5°C.

The data shows that low pressure storage is generally better than normal refrigerated storage but not sufficiently so as to warrent its use commercially. Pretreatment with a fungicide is a must, regardless of storage type. In many cases using poinsettia and geranium cuttings the plants were visually acceptable immediately after removal but deteriorated to an unacceptable condition with 12 to 24 hours (Eisenberg, 1978). In a more recent publication (Jensen and Rasmussen, 1979), low pressure storage had a positive effect on cutting storage only at warm temperatures (6-12°C) but not at lower temperatures (1-2°C). Therefore, low pressure storage may be more beneficial to cuttings that are chill sensitive.

At the commercial level, successful use of low pressure storage for chrysanthemum and carnation cuttings has been employeed for a number of years by Yoder Brothers, Fort Myers, Florida. They use this storage technique to help control inventory levels.

Literature cited - Cuttings

Altstadt, R.A. and W.D. Holley. 1964. Effects of storage on the performance of carnation cuttings. Colo. Flw. Grw. Assoc. Bull. 173:1-2.

Anon. 1975. Hypobaric storage and transportation of perishable commodities. Grumman Allied Industries, Inc., Garden City, NY. 23 pp.

Eisenberg, B.A. 1977. Low pressure and common cold storage of rooted and unrooted ornamental cuttings. M.S. thesis, The Ohio State Univ., Columbus, 69 pp.

Eisenberg, B.A., G.L. Staby and T.A. Fretz. 1978. Low pressure and refrigerated storage of rooted and unrooted ornamental cuttings. J. Amer. Soc. Hort. Sci. 103:732-737.

Jensen, H.E.K. and P.M. Rasmussen. 1979. Lavtryksopbevaring af stiklinger. I. Principper og indledende forsog. (Hypobaric storage of cuttings. I. Principles and preliminary experiments). Tidsskrift Planteavl. 82:623-632.

Judd, R.W. 1976. Storage and rooting of impatiens cuttings. Conn. Greenhouse Newsletter. 71:8-9.

Laurie, A., D.C. Kiplinger, and K.S. Nelson. 1967. Commercial flower forcing. McGraw-Hill, New York, 381-383 pp.

Mastalerz, J.W. 1957. Low temperature storage of foliage plant cuttings. Penn. Flw. Grw. Bull. 74:5-6.

Pryor, R.L. and R.M. Stewart. 1964. Storage of unrooted softwood azalea cuttings. Proc. Amer. Soc. Hort. Sci. 82:483-484.

Tissue Culture

The need to store plant germplasm has been well established. Tissue culture may be the ideal technique to fill this need since it requires a minimum of space, plant cells can be continuously sub-cultured and divided when needed, and the plants are aseptic and possibly disease and insect free. However, the tissue culture's need for frequent subculturing creates several problems including the possibility that genetic change may occur, the cultures may lose their morphogenetic ability, or the cultures may become contaminated. These problems could be prevented by a storage technique that delays the need to subculture.

The two major techniques that have been studied to delay or prevent tissue subculturing are cold storage and freeze preservation (Bajaj and Reinert, 1977 and Withers, 1978). Other forms of modified atmospheres for tissue culture storage have not been studied; therefore the purpose of our research under the direction of Mark Bridgen, was to evaluate the effects of low pressure (hypobaric) and low oxygen storage conditions on the growth of differentiated and undifferentiated plant tissue cultures.

Undifferentiated callus and differentiated shoots and stem sections of chrysanthemum tissue cultures were stored for up to 6 weeks under various oxygen partial pressures using low pressure and modified atmospheric storage techniques. Data presented in Table 3 summarizes some of the findings.

Table 3. Fresh weight gain and number of leaves of Chrysanthemum x morifolium 'Nob Hill' after 6 weeks of storage.

Treatment			
Total Pressure (mm Hg)	Oxygen partial pressure (mm Hg)	Fresh wt. gain (mg)	Leaves (no.)
760	152	383	7.6
760	54	317	6.6
300	61	325	7.5
760	28	152	5.8
150	30	233	6.6
760	8	55	1.7
70	8	50	2.3
SE		0.19	1.81

Success of such storage systems is measured by the maximum reduction in tissue culture growth with no loss in subsequent growth and development performance. With this in mind, the rate and amount of growth decreased when the oxygen partial pressure was below 50 mg Hg regardless if obtained by low pressure or modified atmospheric techniques. Also, no phenotypic growth differences were noted on plants after removal from storage.

Further information on this subject has been published in a thesis by Bridgen (1979) and in an article accepted for publication in Plant Science Letters which will be out in 1981.

Literature Cited - Tissue Culture

Bajaj, Y.P.S. and J. Reinert. 1977. Cryobiology of plant cell cultures and establishment of gene-banks. In: Applied and fundamental aspects of plant cell, tissue, and organ culture, J. Reinert and Y.P.S. Bajaj (eds.), Springer-Verlag, Berlin. 757-777 pp.

Bridgen, M.P. 1979. Low pressure and controlled atmosphere storage of plant tissue cultures. M.S. thesis, The Ohio State University, Columbus, OH 33 p.

Bridgen, M.P. and G.L. Staby. 1981. Low pressure and low oxygen storage of plant tissue cultures. (accepted for publication in Plant Science Letters).

Withers, L.A. 1978. Freeze-preservation of cultured cells and tissues. In: Frontiers of plant tissue culture, T. Thorpe (ed.), International Assoc. Plant Tissue Culture, Calgary, Alberta. 297-306 pp.

Bedding Plants

This crop category has suffered in that very few people have done much of any type of research with them. Plant genera like *Petunia*, *Pelargonium*, *Tagetes*, *Lycopersicon*, *Alyssum*, *Cucumis*, *Begonia*, and *Impatiens*, all considered to be bedding plants, often are classfied somewhere between floral and vegetable crops in regard to research interests. Also, as an industry *per se*, bedding plants have only been very strong for about 15 years. Hence, large gaps of knowledge exist in all areas of bedding plant production, marketing, and expecially in the postharvest area. What follows is a summary of the storage research results with bedding plants.

Ethylene has long been known to damage bedding plants (Fischer, 1950). More recently, seed geraniums plants were stored for 4 days in darkness and at various ethylene levels. Leaf yellowing increased as storage increased while ethylene had no effect. When dry stressed, synthesis of ethylene by geraniums decreased while leaf yellowing increased (Dean and Carlson, 1979).

Nelson *et al.* (1980) examined the keeping quality of marigolds and impatiens as affected by growing and storage temperatures. The following summarizes some of their results.

Table 4. Keeping quality (days) of marigold and impatiens bedding plants as affected by production night temperatures and postharvest storage temperatures.

Storage temperature (°C)	Night temperature (°C) 10	16	21
	-Marigolds-		
10	17	14	12
21	17	13	10
32	4	4	4
	-Impatiens-		
10	12	15	10
21	15	15	15
32	0-2	0-2	0-2

The above data suggests that the best production night temperatures for marigolds and impatiens are 10 and 16°C, respectively when one measures postharvest keeping quality. Also, the best postharvest storage temperature is from 10 to 21°C.

In 1979 and 1980, Gehring and Lewis published a series of articles related to extending the shelf life of bedding plants by altering container volume, the addition of a hydrogel to the growing medium, and by the use of antitranspirants. All three methods were geared towards increasing plant water availability after harvest. The only technique not effective was the use of antitranspirants. Data in the following table summarizes the advantages of using a hydrogel material in extending shelf life.

Table 5. Hydrogel (potassium propenoate propenamide copolymer) rates as it influences shelf life of 3 bedding plant species.

Hydrogel conc. (kg/m^3)	Shelf life (hours)^z Ageratum	Marigold	Zinnia
0	85	83	79
4	97	110	103
8	113	150	124
12	120	154	135
16	126	164	153

^{z}Until wilted.

Literature cited - bedding plants

Dean, S. and W.H. Carlson. 1979. Effect of ethylene on the shipment of seed geraniums (*Pelargonium* x *hortorum Bailey*). *HortScience* 14(3):444-445.

Fischer, C.W. 1950. Ethylene gas a problem in cut flower storage. *NY State Flower Growers Bul.* 61:1,4.

Fischer, C.W., Jr. 1951. *A study of some factors affecting the storage of cut flowers.* M.S. thesis. Cornell Univ., Ithaca, NY.

Gehring, J.M. 1979. *Shelf life extension of bedding plants.* M.S. thesis. Virginia Polytechnic Inst. 39 pp.

Gehring, J.M. and A.J. Lewis. 1979. Extending the shelf life of bedding plants. I. Container volume. *Flor. Rev.* 165(4278):17,60-61.

Gehring, J.M. and A.J. Lewis. 1980. Extending the shelf life of bedding plants. II. Hydrogel. *Flor. Rev.* 165(4282):30, 46.

Gehring, J.M. and A.J. Lewis. 1980. Extending the shelf life of bedding plants. III. Antitranspirants. *Flor. Rev.* 165(4286):31, 65.

Gehring, J.M. and A.J. Lewis. 1980. Effect of hydrogel on wilting and moisture stress of bedding plants. *J. Amer. Soc. Hort. Sci.* 105(4): 511-513.

Neff, M.S. and W.E. Loomis. 1935. Storage of french marigolds. *Proc. Amer. Soc. Hort. Sci.* 33:683-685.

Nelson, L.J., A.M. Armitage, and W.H. Carlson. 1980. Keeping quality of marigolds and impatiens as affected by night temperature and duration. *Flor. Rev.* 167(4318):28-29, 62, 74.

Potted Crops

It is estimated that less than 5% of the literature available on the post-harvest care and handling of floral crops deals with plants grown in containers or more commonly refered to as potted crops. This great void in information is present even though the sale of potted crops represents 53% of total floral sales. The following summary of potted crop postharvest storage research is divided into foliage and flowering crops.

Flowering potted crops. A glance at the references cited in this section reveal that only 15 are listed. These research findings can be summarized as follows:

Table 6. Comments regarding the storage of the three major flowering potted crop species.

Crop	Reference	Comments
Poinsettia	Hammer and Kirk (1977) Shanks (1975) Staby *et al.*(1978,80,81)	Store at 10°C for a maximum of 6 days with resulting petiole epinasty possible. Epinasty can be reduced if stored at 5°C but maximum time is 2 days or else a chilling damage will occur which is a blueing of the bracts. Low pressure storage did not prevent blueing.
Chrysanthemum	Harrison *et al.*(1972) Staby *et al.*(1978) Staby *et al.*(1981, Hort-Science, in press)	'Yellow Mandalay' plants in 10 cm pots stored for 10 days at 21°C in 4 mill polyethylene packages had a useful post-storage life of 32 days. Storage for 17 days resulted in a 21 day life. Plants in paper sleeves with no poly-ethylene can be stored for 5 days at any temperature between 3 and 24°C with equal post-storage life. Plants stored over 5 days do better if stored at 3°C verses higher temper-ature.
Easter Lily	Healy *et al.*(1979) Maxie and Hasek (1974) Staby and Erwin (1977)	Results are not conclusive. Maxie and Hasek say they can be stored for 4 weeks at 0 to 3°C. Staby and Erwin noted that 1 or 6°C were equal & that storage for 1 or 2 weeks reduced longevity by 25 and 35%, respectively. Healy *et al.*showed no advantages of spraying with N^6 benzyladenine or light in storage.

Literature cited - flowering potted crops

Anon. 1981 Flowering pot plant research described. *Flor. Rev.* 167(4341):63.

Fischer, C.W. 1950. Keeping cut flowers. New methods make possible successful long term storage. *Flor. Rev.* 107(2764):21-22.

Hammer, P.A. and T. Kirk. 1977. Poinsettias - droopy bracts. *Focus on Floriculture* 5(1):2-11.

Harrison, M.R., D. Durkin, and F. Hanlon. 1972. The effect of pressurized sealed film packaging on the keeping quality and shelf life of three Chrysanthemum cultivars. Penn. Flower Growers 258:6.

Healy, W.E., R.D. Heins, H.F. Wilkins. 1979. Short-term storage of Lilium longiflorum thunbergia in the 'puffy' flower bud storage of development. Flor. Rev. 163(4231):21.

Heins, R.D. and J.J. Hanan. 1980. Climatic influence on postharvest life of commercial potted plants. HortScience. 15(3):387.

Hurst, L.K. 1973. Potted chrysanthemums Floral Facts Nov.:1-4.

Marousky, F.J. and B.K. Harbaugh. 1979. Ethylene-induced floret sleepiness in Kalanchoe blossfeldiana Poelln. HortScience 14(4):505-507.

Maxie, E.C. and R.F. Hasek. 1974. Cold storage of 'Georgia' Easter lilies. Flower and Nursery Rpt., Agr. Ext. Univ. CA Jan:3-4.

Saltveit, M.E., D.M. Pharr, and R.A. Larson. 1979. Mechanical stress induces ethylene production and epinasty in poinsettia cultivars. HortScience 14(3):454.

Shanks, J.B. 1975. Poinsettias and their greenhouse culture. The Maryland Florist 197:31 pp.

Staby, G.L. and T.D. Erwin. 1977. The storage of Easter lilies. Flor. Rev. 161(4162):38.

Staby, G.L., J. Thompson, A. Kofranek. 1978. Precooling potted chrysanthemums. HortScience 13(3):346.

Staby, G.L., J.F. Thompson, A.M. Kofranek. 1978. Postharvest characteristics of poinsettias as influenced by handling and storage procedures. J. Amer. Soc. Hort. Sci. 103(6):712-715.

Staby, G.L., B.A. Eisenberg, J.W. Kelly, M.P. Bridgen, and M.S. Cunningham. 1980. Leaf petiole epinasty in poinsettias. HortScience 15(5):635-636.

Staby, G.L., J.F. Thompson, A.M. Kofranek and V.R. Walter. 1981. Cooling of potted chrysanthemums. HortScience (in press).

Staby, G.L., M.S. Cunningham, J.W. Kelly, P.S. Konjoian, and C.L. Holstead. 1981. Poinsettia epinastic response under low pressure storage and with prestorage sprays. HortScience 16:404.

Foliage potted crops. Until now, most of the work relating to the postharvest characteristics of foliage plants has been related to acclimatization practices at grower level. Recently, there has been a major push for postharvest storage research with foliage plants at the Apopka Research Station in Florida under the direction of Dr. Charles Conover.

The work done to date has shown that many species of foliage plants can be stored for long periods of time when sealed in polyethylene packages at room temperature (Harbaugh, et al. 1976, 1978). Also, it takes rather large amounts of ethylene to create problems with philodendrons (Marousky and Harbaugh, 1977, 1979). No data is yet available as to how long various species can be held at various temperatures and when, if ever, chilling damages occur.

Literature cited - foliage potted plants

Dickey, R.D. 1954. Location in box, insulation used influence cold injury to plants during shipping. Southern Florist & Nurseryman 67(39):87-90.

Harbaugh, B.K., G.J. Wilfret, A.W. Engelhard, and W.E. Waters. 1976. Evaluation of 40 ornamental plants for a mass marketing system utilizing sealed polyethylene packages. Proc. Fla. State Hort. Soc. 89:320-323.

Marousky, F.J. and B.K. Harbaugh. 1977. Influence of ethylene on *Philodendron oxycardium* Schott. HortScience 12(4):404.

Marousky, F.J. and B.K. Harbaugh. 1979. Interactions of ethylene, temperature, light, and CO_2 on leaf and stipule abscission and chlorosis in *Philodendron scandens* subsp. *oxycardium*. J. Amer. Soc. Hort. Sci. 104(6):876-880.

Florist Greens

The Dutch and Germans have lead the way in research regarding the storage of florist greens (Barendse, 1978, 1979; Behrens, 1967). Much of the work has centered around the storage of *Asparagus* species with the following variables tested: grower source; crop age; water quality; harvest stage; time of year; use of preservative; storage relative humidity and temperature; and recutting of stems. These factors had little influence on storage life and subsequent longevity if not held over 7 days. Maximum storage life is 10 days at 4°C and 90% relative humidity. Storage in water is better than stored dry, namely, wet storage at 20°C is better than dry storage at 4°C. The use of a vapor barrier to reduce water loss is a must. In general, each day shortens subsequent longevity by an equal time period.

Barendse (1979) classified florist green species into three categories depending on their storage performances as follows: Group I-can be stored longer than three weeks - *Polystichum* and *Cyrtomium*; Group II-storage for two to three weeks - *Asparagus meyerii*, *A. retrofractus*, *A. setaceus*, and *Pteris cretica*; Group III-ten to fourteen days - *Asparagus falcatus*, *A. virgatus*, and *Nephrolepis*.

Unpublished results from our laboratory have shown that very low to non-detectable levels of ethylene are produced from a wide number of florist greens used at Christmas time. These results suggest that these greens can be stored with ethylene sensitive flowers like carnations assuming that no diseases are present to any great extent which could enhance ethylene production and that proper storage temperatures are utilized.

Literature cited - florist greens

Barendse, L.V.J. 1978. Houdbaarheid snijgroen kan goed zijn. (Keeping quality of cut greens really can be good). Vakblad Bloemist. 33(48):25.

Barendse, L.V.J. 1979. Bewaring en verpakking van snijgroen. (Storage and packing of cut greens). Vakblad Bloemist. 34(45):39.

Barendse, L.V.J. 1979. Gemiddelde houdbaarheid siergroen goed. (On the average, vaselife of cut greens is good). Vakblad Bloemist. 34(1):33.

Behrens, W. 1967. Schnittblumenlagerung in kuhlraumen. (Storing cut flowers in refrigerated cells). Zierpelanzenbau 7(23):883-884.

Cut Flowers

The literature cited for this section is divided into normal refrigerated, modified atmosphere, and low pressure storage. Unlike the other crop categories, much information is available in this subject area. Each storage system will be discussed separately as it relates to cut flowers.

Modified atmosphere. Much of the research prior to 1977 has been summarized (Staby, 1977) with the general conclusion that such storage systems (low O_2, high N_2, and/or high CO_2) are not acceptable for the storage of cut flowers. While it is true that some species/cultivars can respond favorably (Asen and co-workers, 1963-64; Buneman and Dewey, 1956; Hannan, 1966-67; Hardenberg etal, 1967; Longley, 1933; Marousky, 1977; Parsons, 1964; Thorton, 1930; Uota, 1963, 69; Uota and Garazsi, 1967; and Van Stuivenberg, 1949-51) commercial implimentation is considered to be doubtful because the margin of safety is small before phytotoxicity occurs, costs are high, and there is not enough volume of any one cultivar to warrant its use.

Regardless of the commercial aspects, the following table summarizes some of the results only when using various forms of modified atmospheres.

Table 7. Modified atmosphere storage parameters of cut flowers.

Cut Flower	Storage Conditions	Storage time	Reference	Comments
Carnation	22^oC 0.2% C_2H_4O	1 day	Asen, 1963	C_2H_4O prevented sleepiness, with or without the addition of 1 ppm C_2H_4. Slight damage noted at 0.2% didn't occur at 0.05 or 0.1% levels.

Cut Flower	Storage Conditions	Storage time	Reference	Comments
Carnation	1-24°C 0.1% O_2 and 0% CO_2 up to 9% CO_2 and CO_2 - 10%	3 days- 9 weeks	Hanan, 1967	Many experiments: no gas combination was consistently better than the controls (containers with circulating air).
Carnation	2°C 0.5 or 1% O_2	4-5 weeks	Hardenburg 1967	This was the best treatment. Had less Botrytis than those stored in air. At 0°C storage over 2 wks. resulted in petal injury to pink and reds.
Carnation	5-20°C 125-500 ppb C_2H_4 5-20% CO_2 in air	1-2 days	Uota, 1969	10 and 20% CO_2 could counteract the effects of 205 and 500 ppb C_2H_4 respectively at 20°C.
Carnation	0°C 0.5-1% O_2	4-5 wks.	Uota & Garazsi 1967	Low O_2 reduced decay but did not greatly extend vase life. The addition of CO_2 was not beneficial.
Chrysanthemum	2°C 1.4-1.5% O_2 Balance N_2	6 weeks	Marousky, 1977	Leaves stored in CA were green on removal whereas those stored in air were chlorotic. There was some petal browning in CA. The deterioration after removal was more rapid than unstored.
Daffodil	4.5°C 100% N_2	up to 3 weeks	Asen, 1964	Vaselife of CA stored was equal to that of freshly cut flowers, and vaselife was better than normal atmosphere. CA was also effective at 21°C for 2.5 days at 0°C for 3 weeks.

Cut Flower	Storage Conditions	Storage time	Reference	Comments
Daffodil	4.5°C 100 N_2	3 weeks	Hardenburg, 1967	CA stored were equal to freshly cut and better than those stored in air.
Delphiniums	7°C 24% CO_2, 16% O_2	7 days	Van Stuiven-berg, 1951	Retards petal abscission
Gladiolus	0-45°C Average 4% CO_2 12% O_2	6-8 days	Hardenburg 1967	This was the best treatment combination. It helped retain green color and aided floret opening.
Gladiolus	2.5°C 1.4-1.5% O_2 Balance N_2	2-4 weeks	Marousky, 1977	Stored were not as good as freshly harvested, those stored 4 wks. were very poor.
Rose	5.5°C 0.25% C_2H_4O	20 hr.	Asen, 1963	Treatment delayed the rate of maturation without injury, also resulted in better color and longer petal resention.
Rose	0°C 0.5% O_2, 5% CO_2	2-3 wks.	Hardenburg, 1967	This was the best gas combination used, still blueing can develop and vaselife is less than fresh cut blooms. CA was not consistently better than air, in some cases low O_2 decreased blueing.
Rose	0-0.5°C 8-10% CO_2	2 wks.	Langley, 1933	Flowers in CA opened less rapidly during storage. Vaselife was longer than flowers in ordinary atmo-sphere.
Rose	0°C 99-99.5% N_2	3 wks.	Parsons, 1964	CA suppressed bud open-ing and maintained quality.

Cut Flower	Storage Conditions	Storage time	Reference	Comments
Rose	3.5 or 10°C	7 days	Thornton, 1930	CA controlled petal abscission, and blooms opened slowly. Storage time can be lengthened to 7 days (3 under normal conditions).
Rose	15°C 0.5-1% O_2 5 or 10% CO_2	40 hrs.	Uota, 1963	Supresses bud opening and maintains vase-life.
Rose	1°C 16 or 5% CO_2 2 or 16% O_2	22 days	Van Stuivenberg, 1951	Blooms stored in CA opened more slowly.
Snapdragon	2°C 1.4-1.5% O_2 Balance N_2	2-4 days	Marousky, 1977	CA was no different than control for each storage period when stored in water. When stored in preservative, in CA there were more viable florets, quality was better than the control, but there were aborted florets.
Tulip	-0.5°C 10% CO_2, 10% O_2	up to 8 weeks	Bunemann & Dewey, 1956	Storage 8 wks. in an open cooler in water at 0.5°C was satisfactory.

Literature cited - modified atmosphere storage of cut flowers

Akamine, E.K. 1973. Postharvest handling and storage of Hawaiian flowers. Proc. 2nd Annual Flower Growers Short Course, Coop. Ext. Ser. Univ. Hawaii. Mis. Pub. 105:12-17

Anon. 1957. More tests on keeping flowers in gasbags. The Grower 47:375.

Anon. 1958. No-ice film holds flowers in own humidity. Modern Packaging 31(5):103.

Anon. 1964. Added life for cut flowers. Agr. Res. 13:4.

Anon. 1966. Cut shipping loses, extend cut flowers life. Florist and Nursery Exchange 145(8):7, 16.

Anon. 1967. Longer life for cut flowers. Agr. Res. 16:4-5.

Anon. 1970. Progress in flower packaging helps entire industry. Florist and Nursery Exchange. 153(7):4-6.

Anon. 1970. Controlled atmosphere storage. The Refrigeration Res. Foundation 70-2:1-2.

Asen, S. and M. Lieberman. 1963. Ethylene oxide found to combat deterioration of carnations. Flor. Rev. 132(3431):21, 43.

Asen, S. and M. Lieberman. 1963. Ethylene Oxide-a possible break through for increasing longevity of cut flowers. Flower News Jan:1-3.

Asen, S. and M. Lieberman. 1963. Ethylene oxide experimentation aimed at cut flower longevity. Flor. Rev. 131(3398):1, 4.

Asen, S. and M. Lieberman. 1964. A tonic for cut flowers. Agr. Res. 12:14.

Asen, S., N. Stuart and O.S. Parsons. 1964. Added life for cut flowers. Agr. Res. 13:4.

Asen, S., C.S. Parsons and N.W. Stuart. 1964. Controlled atmosphere for storing flowers. The Exchange. 141(24):30, 45, 54.

Asen, S., C.S. Parsons and N.W. Stuart. 1964. Experiments aim at prolonging narcissus display life. Flor. Rev. July 11:25, 69, 71-72.

Ben-Yehoshua, S., B. Juven, M. Fruchter, and A.H. Halevy. 1966. Effect of ethylene oxide on opening and longevity of cut rose flowers. Proc. Amer. Soc. Hort. Sci. 89:677-682.

Bunemann, G. and D.H. Dewey. 1956. Cold storage of cut tulips with and without water. Quarterly Bul,, Mich. Agr. Exp. Sta. 38:580-587.

Burg, S.P. 1968. Ethylene plant senescence and abscission. Plant Physiol. 43:1503-1511.

Carrier, L.E. 1952. Cut flower storage. Calif. State Floral Assoc. Bul. Setp:7-9.

Dewey, D.H., R.C. Herner and D.R. Dilley, (ed). 1969. Controlled Atmospheres for the Storage and Transport of Horticultural Crops. Proc. Natl. controlled atmosphere res. conference, Mich. State Univ. Number 9:155 pp.

Dewey, D.H., Ed. 1977. Controlled Atmospheres for the Storage and Transport of Perishable Agricultural Commodities. Second National Controlled Atmosphere Research Conference, Michigan State Univ. Number 28:301 pp.

Dilley, D.R. and W.J. Carpenter. 1975. Principles and application of hypobaric storage of cut flowers. Acta Hort. 41:249-268.

Eaves, C.A. and F.R. Forsyth. 1970. Longer flower life through ethylene removal. Commerical Grower 3896:307-314.

Engle, S. 1976. Staby stresses better product quality. The Packer 83:9B.

Fischer, C.W., Jr. 1951. A study of some factors affecting the storage of cut flowers. M.S. thesis. Cornell Univ., Ithaca, NY.

Hague, A., W. Bryant, A. Laurie. 1947. Packaging of cut flowers. Proc. Amer. Soc. Hort. Sci. 49:427-432.

Halevy, A.H., S. Mayak, M. Fruchter, and S. Ben-Yehoshua. 1965-1966. Studies on postharvest physiology of cut flowers. Res. Rpt., Horticulture, Hebrew Univ. of Jerusalem. 1965-1966:563.

Hanan, J.J. 1966. Controlled atmosphere (CA) storage of carnations. Colo. Flower Growers Assoc. Bul. 193:1-6.

Hanan, J.J. 1967. Experiments with controlled atmosphere storage of carnations. Proc. Amer. Soc. Hort. Sci. 90:370-376.

Hardenburg, R.E. 1964. Developments on postharvest use of controlled or modified atmospheres for quality retention of horticutlure crops. Post-Harvest Physiol. Confer. Phila. PA. Oct.:1-16.

Hardenburg, R.E., M. Uota and C.S. Parsons. 1967. Refrigeration and modified atmospheres for improved keeping quality of cut flowers. Mkt. Quality Res. Div. U.S.D.A. 4.02:1-9.

Hardenburg, R.E., M. Uota and C.S. Parsons. 1969. Refrigeration and modified atmospheres for improving keeping quality of cut flowers. Proc. 12th Int. Cong. Refrig. (Madrid) 3:339-347.

Koon, G.D. 1949. Studies of the fundamental principles underlying the prepackaging of cut flowers. M.S. thesis, Ohio State University. 154 pp.

Kosugi, K., M. Yokoi, and Y. Motoki. 1974. Studies on the storage of cut flowers. Pro. Japan Soc. Hort. Sci. 1974:402-403. (with English abstr.)

Laurie, A. and W. Bryant. 1948. Packaging cut flowers proves successful. Ohio Farm and Home Res. Jan-Feb:11-16.

Lecrenier, A. and R. Linden. 1967. Application du froid aux fleurs et cultures florales. (Application of low temperatures to flowers and flower cultivation). XIII Inter. Congr. Refrig. 3:323-238. Madrid.

Longley, L.E. 1933. Some effects of storage of flowers in various gases at low temperature on their keeping quality. Proc. Amer. Soc. Hort. Sci. 30:607-609.

Magie, R.O. 1961. Controlling gladiolus botrytis bud rot with ozone gas. Proc. Fla. Hort. Sco. 73:373-375.

Marousky, F.J. 1977. Controlled atmosphere and low pressure storage of gladiolus, chrysanthemums and snapdragons. Controlled atmos. for storage and transp. of perishable commodities. Second Nat. Contr. Atm. Res. Conf., Mich. St. Univ. 28:116-121.

Marousky, F.J. 1977. Commodity requirements and recommendations for flower and nursery stocks. Contr. atm. for the storage and transp. of perishable agr.

commodities. Second Nat. Conf. Atm. Res. Conf., Mich. St. Univ. 28: 287-291.

Mastalerz, J.W. 1952. Blueing of 'Better Times' roses. N.Y. State Flower Grower Bul. 87:2-3.

Mayak, S. 1968. Cooling during transport of gladiolus. (in Hebrew). Ann. Rpt. Hebrew Univ., Ornamental Hort. 25 p.

Mayak, S. and D.R. Dilley. 1976. Regualtion of senescence in carnation (Dianthus caryophyllus). Plant Physiol. 58:663-665.

Molenaar, W.H. 1977. Kwaliteitsbehoud van snijbloemen. (Maintaining quality of cut flowers.) Vakblad Bloemist. 32(38):60-63.

Neff, M.S. 1942. Prolonging the life of flowers in cool storage. Flor. Exch. and Hort. Trade World. 99(8):10.

Nichols, R. 1971. Refrigeration and storage of cut flowers. Refrig. and Air Cond. 74(880):36-39.

Nichols, R. 1977. A descriptive model of the senescence of the carnation (Dianthus caryophyllus) inflorescence. Acta Hort. 71:227-232.

Nowak, J. and R.M. Rudnicki. 1979. Long term storage of cut flowers. Acta Hort. 91:123-133.

Parsons, C.S. 1964. Storage in nitrogen:effect on fruits, vegetables and flowers. Ioma Broadcaster Feb 4:1-7.

Parsons, C.S., S. Asen and N.W. Stuart. 1967. Controlled-atmosphere storage of daffodil flowers. Proc. Amer. Soc. Hort. Sci. 90:506-514.

Parvin, P.E. 1965. Studies of the effects of postharvest handling practices on the keeping quality and marketability of 'Better Times' roses Ph.D. dissertation, Michigan State University. 117 pp.

Paulin, A. 1976. L'Amelioration du traitement des fleurs coupees. La conservation frigorifique des fleurs coupees. (Improving the treatment of cut flowers. Cold storage). Fleuriste de France 106(Suppl):7-14.

Pope, T.E. and D.C. Kiplinger. 1959. Respiration rates of 'Better Times' roses as affected by modified atmosphere and low temperature storage. Ph.D. disseration Ohio State University 15 pp.

Post, K. 1955. Cut flowers. Florists crop production and marketing Orange Judd Pub. Co. Inc., NY chapter 10, 220-250 pp.

Pratella, G.C., G. Tonini, and R. Tesi. 1967. La conservazione dei garofani in atmosfera controllata. Rio Orto Floro Fruttie. 51:3-12.

Rogers, M.N. 1962. Sell flowers that last. Flor. Rev. 131(3378):13; later parts through No. 3385.

Rothenberger, B. 1963. Symposium in Missouri: Focus on floral longevity. Flor. Rev. 131:23-24, 75-78.

Rudolphij, J.W., W. Verbeek, and F.H. Fockens. 1977. Measuring heat production of respiring produce under normal and CA-storage conditions with an adiabatic calorimeter. Lebensm.-Wiss. U.-Tech. 10(3):153-158.

Rule, D.E. 1977. Development of selected plant species in low pressure and controlled atmospheres. M.S. thesis, Ohio State Univ. 107 pp.

Serini, G. 1960. La conservazione della frutte degli ortaggi dei fiori in atmosfera confinata a bassa percentuale di ossigeno-notra 2. Rivista Della Orto Florofrutticoltura 44:474-483.

Sherwood, C.H. 1957. Effect of a plastic coating on cut flowers and greens. Ph.D. disseration, Michigan State University 50 p.

Siegelman, H.W. and V.T. Stoutmeyer. 1949. Recent developments in cut flower storage and shipment. Sci. Monthly 69:126-127

Siegelman, H.W. 1951. The respiratory metabolism of flowers. Ph.D. thesis, University of Calif., Los Angeles. 161 pp.

Smith, W.H. 1965. Vase-fresh flowers. Great Britain Min. Agr. Fisheries and Food Agriculture, London 72:190-193.

Smith, W.H. 1965. "Gas Packs" for flowers. Report, Ditton and Covent Garden Lab. 1964-65:29-30.

Smith, W.H. 1965. Ethylene and the storage of flowers. Report, Ditton and Covent Garden Lab. 1964-65:30-31.

Smith, W.H. 1967. Flower storage and marketing. VI. A manual of carnation production. Bulletin, Ministry Agr, Fish, and Food, London 151: 152-155.

Souter, D. 1976. Le transport des fleurs coupees. L'amelioration du traitement des fleurs coupees. Fleuriste de France 106(suppl.):15-22.

Staby, G.L. 1977. Controlled atmosphere and low pressure storage of floral crops-overview. Controlled atmospheres for the storage and transport of perishable agricultural commodities, Second National Controlled Atomsphere Research Conference. 28:60-70.

Staby, G.L. 1978. Effects of modified storage atmospheres on floral crops. Univ. CA Cooperative Extension Ser. Bul. 40:8-9.

Staby, G.L., M.S. Cunningham, B.A. Eisenberg, M.P. Bridgen, and J.W. Kelly. 1979. Low pressure storage of carnations and roses. HortScience 14(3):446

Stuart, N.W., C.S. Parsons, and C.J. Gould. 1970. The influence of controlled atmospheres during cool storage on the subsequent flowering of easter lilies and bulbous iris. HortScience 5:356

Sytsema, W. 1968. Bewaring en houdbaarheid van snijbloemen. (Storage and keeping quality of cut flowers). Jarr. Proefst. Bloem. Aalsmeer 1968: 111-137.

Tesi, R. 1968. La Conservazione dei fiori recisi. Infa Orotoflorofruttic 9:143-146.

Thornton, N.C. 1930. Carbon dioxide storage of fruits, vegetables, and Flowers. Industrial and Engineering Chemistry 22:1186-1189.

Thornton, N.C. 1934. Carbon dioxide storage VII. Changes in flower color as evidence of effectiveness of carbon dioxide in reducing the acidity of plant tissue. Contrib. Boyce Thompson Inst. 6:403-405.

Tincker, M.A.H. 1942. The care of cut flowers. Royal Hort. Soc. 67:373-395

Tinga, J.H. 1956. The effect of modified atmosphere storage at low temperature and treatments after storage which affect the keeping quality of cut flowers. Cornell Univ. Diss. Abstr. 16:623.

Tinga, J.H. 1956. The effect of modified atmosphere storage at low temperature and treatments after low temperature storage which affect the keeping quality of cut flowers Ph.D. Thesis, Cornell University. 189pp.

Tonini, G. 1968. Conservazione e distribuzione dei fiori recisi. Frutticoltura 30:177-182.

Tonini, G. and R. Tesi. 1969. La conservazion dei gladioli in atmosfera controllata. Riv. Dell Ortoflorofrutt. 4:356-366.

Tonini, G. and R. Tesi. 1969. Risultati di una prova semicommerciale di conservazione in atmosfera controlleta dei garofani resisi. Riv. Ortofloro Fruttic 53:520-526.

Ulrich, R. 1960. Traitment des fleurs coupees et des bulbes. (Treatment of cut flowers and bulbs.) Internat'l. Inst. Refrig. Bul. Sup. 3:341-350.

Uota, M. 1963. Summary of current research on shipping and storing cut roses. Ann. Meeting. Rpt., Roses, Inc. Bul. 1963.

Uota, M. 1963. Controlled atmosphere research in storage and transportation. Roses Inc. Bul. Oct.:1.

Uota, M. 1965. Current research on the storage and shipping of roses. Roses Inc. Bul. Oct.:1-2.

Uota, M. and M. Garazsi. 1967. Quality and display life of carnation blooms after storage in controlled atmosphere. U.S.D.A. Marketing Research Rpt. 796:9pp.

Uota, M. 1969. CO_2 prevents "Sleepy" carnations. Agric. Res. 18(3):12.

Uota, M. 1969. Commodity requirements and recommendations for flowers and nursery stock. Controlled atmosphere for the storage and transport of horticultural crops. Natl. Controlled Atmosphere Research Conference, Mich. State Univ. 9:109-112.

Van Stuivenberg, J.H. 1951. Gas storage of cut flowers. Proc. 8th Intl.

Cong. Refrig. 1951: 425-429.

Van Stuivenberg, J.H.M. 1946. Gebruik van afwijkend gasmilieu. Tuinbouw 6:5-8.

Van Stuivenberg, J.H.M. 1949. De mogelijkheid van verlenging van de levensduur van snijbloemen. (The possibilities of increasing the durability of cut flowers). Meded. Dir. Tuinb. 12:717-734.

Von Oppenfeld, H., R.S. Linddstrom, D.H. Dewey and J.W. Goff. 1955. Cold storage of field-grown cut tulips. Mich. Agr. Exp. Sta. Quar. Bul. 38:273-278.

Waters, W.E. 1966. The influence of post harvest handling techniques on vase-life of gladiolus flowers. Proc. Fla. State Hort. Soc. 79:452-456.

Waters, W.E. 1977. The influence of post-harvest handling techniques on the vase life of gladiolus flowers. Gladio Gram 25:6-10

Weinstein, L.H. and H.J. Laurencot, Jr. 1958. Senescence of Roses. II. Dark fixation of CO_2 by cut 'Better Times' Roses at different stages of senescence. Contrib. Boyce Thomspon Inst. 19:327-240.

Wilkins, H.F. 1965. Factors effecting carbon dioxide and ethylene gas production in flowers of the carnation (Dianthus caryophyllus, Linn.) cultivar 'Red Gayety'. Ph.D. thesis, University of Ill. 105 pp.

Low pressure. Much publicity has been given to this storage technique for cut flowers since the early work of Burg (1973), Carpenter and Dilley(1974, 1975), and Dilley and co-workers (1973-77). Other researchers contributing significantly to this area include: Bangerth (1973), Bredmose (1979-80); Dressler and Jamieson (1979), Lougheed et al. (1974, 1978) Marousky (1977), Nowak and Rudnicki (1979), Patterson (1975), and Staby et al. (1976-79).

In an attempt to summarize the key findings using the three major cut flower crops of roses, carnations, and chrysanthemums, we will use some of the data recently obtained in our laboratory.

In our studies we have used three low pressure systems of the following configurations: 40 liter milk cans; cylinder unit measuring 6.1 meters long by 2.4 meters in diameter; and a trailer-sized unit measuring 11.6 meters long by 2.1 meters wide by 1.9 meters high. Testing all three units showed that similar results could be obtained in each if the temperature, relative humidity, pressure, and air exchange rates were similar.

The data in Table 8 and other data not presented shows that:

1. The use of an anti-ethylene compound, silverthiosulfate, is just as important if not more important than low pressure storage with carnations.
2. The best storage system for roses is low pressure.
3. Chrysanthemums respond very well to low pressure.
4. All flowers except silverthiosulfate-treated carnations store better regardless of storage system if handled dry prior to storage.

5. Contrary to expectations, flowers relatively insensitive to ethylene (roses, chrysanthemums) do better under low pressure systems than ethylene sensitive crops like carnations.

6. A key to the success of low pressure systems is the control of temperature, relative humidity, air exchange, and pressure and not just the low pressure aspect.

7. Flowers entering any type of storage system must be properly harvested, precooled, and have a vapor barrier in the package (generally corrugated fiber) they are stored in.

In summary, sufficient research results, both in the laboratory and at commercial level, will be available by the end of 1981 to help determine whether or not low pressure storage will become a reality in the floricultural industry.

Table 8. Longevity (days) of carnations, roses, and chrysanthemums after storage under normal and low pressure storage systems for 6 weeks.[z]

Crop	Pre-storage treatment	Control	Storage system[y] normal	Low Pressure
'Scania 3C' Carnation	none	10.4	5.9	4.8
	silverthiosulfate[x]	20.8	12.4	16.0
'Forever Yours' Rose	wet[w]	10.4	2.7	4.9
	dry	10.0	0	6.0
'Yellow Albatross' Chrysanthemum	none	15.9	10.2	14.6

[z]Mums were stored for 3.5 weeks.
[y]Control was not stored. Normal refrigeration: 40 liter milk cans; 6.12 air exchanges/day; and 2-4°C. Low pressure: 0.67 air exchanges/day; 0-1°C; 10mm Hg.
[x]80 ppm Ag, 1:4 molar ration $AgNO_3$ to Sodium Thiosulfate treated for 1 hr.
[w]Dry: flowers were held dry before storage. Wet: flowers were placed into preservative solution for 2 hrs. prior to storage.

Literature cited - low pressure storage of cut flowers

Akamine, E.K. 1973. Postharvest handling and storage of Hawaiian flowers. Proc. 2nd Annual Flower Growers Short Course, Coop. Ext. Ser. Univ. Hawaii. Misc. Pub. 105:12-17.

Anon. 1975. Grumman Dormavac hypobaric storage and transportation of perishable commodities Grumman Allied Industries, Inc. Garden City, New York. 23 pp.

Anon. 1975. Proper handling of plant shipments. N.Y. State Flower Ind. Bul. 65:7.

Anon. 1976. Commercial use of hypobaric storage is now possible. Florist 10:80-83.

Anon. 1976. 90-day life at low pressure. The Grower 85(21):1087.

Anon. 1980. Transportation-storage system extends cut flower life. Link Magazine 3(9):5.

Anon. 1980. Saying it longer with flowers. Newsweek Aug.11:53.

Anon. 1981. Study rose storage. Ohio Report 65(6):93.

Bangerth, F. 1973. Zur wirkung eines reduzierten drucks auf physiologie, qualitat und lagerfahigkeit von obst, gemuse und schnittblumen. Die Gartenbauwissenschaft 38:479-508.

Blumenfeld, A. 1975. Ethylene and the annona flower. Plant Physiol. 55:265-269.

Boer, W.C. and O. Wiersma. 1974. Heeft vacuumkoeling ook voor snijbloemen perspectief? (Has vacuum cooling good prospects for cut flowers too?) Vakblad Bloemist. 29(48):16-17.

Bredmose, M. 1979. Influence of subatmospheric pressure on storage life and keeping quality of cut flowers of Belinda roses. Acta Agricult. Scand. 29(3):287-290.

Bredmose, N. 1980. Low pressure storage and cut rose keeping quality trials. Flower Trades Journal 2(10):27.

Burg, S.P. 1968. Ethylene plant senescence and abscission. Plant Physiol. 43:1503-1511.

Burg, S.P. 1973. Hypobaric storage of cut flowers. HortScience 8:202-205.

Burg, S.P. and W. Hentschel. 1974. Low pressure storage of metabolically active matter with open cycle refrigeration. U.S. Patent No. 3, 810, 508.

Butters, R.E. 1976. Promise of longer storage under low pressure confirmed in UK. The Grower 86(8):395.

Carpenter, W.J. and D.R. Dilley. 1974. Improving the decorative life span of cut flowers and ornamental plants. Report of Progress, M.S.U., (Soc. Amer. Florists) p 1-13.

Carpenter, W.J. and D.R. Dilley. 1975. Investigations to extend cut flower longevity. MSU Res. Rpt. 263:1-10.

Claypool, L.L., L.L. Morris, W.T. Pentzer and W.R. Barger. 1958. Air transportation of fruits, vegetables and cut flowers: temperature and humidity requirements and perishable nature. U.S.D.A., Agr. Mkt. Serv. 280:1-27.

Della-Vedowa, R. 1950. Troposphere protection of perishables - it's in the bag presented at the CA. Air Freight Clinic Held in Oakland. Aug. 19-20: 34 pp.

Della-Vedowa, R.P. and B.A. Rose. 1950. Altitude tests on flowers. Lockheed Air Cargo Progress Rpt. Slr - 942-27 pp.

Dewey, D.H., Ed. 1977. Controlled atmospheres for the storage and transport of perishable agricultural commodities. Second National Controlled Atmosphere Research Conference, Michigan State Univ. Number 28:301 pp.

Dilley, D.R. and W.J. Carpenter. 1973. Hypobaric storage of cut flowers a physiological interpretation. HortScience 8:273.

Dilley, D.R., W.J. Carpenter, and S.P. Burg. 1974. Principles and application of hypobaric storage of cut flowers. Mich. State Univ. Agr. Exp. Sta. Journ. 6644:1-18.

Dilley, D.R. and W.J. Carpenter. 1975. Principles and application of hypobaric storage of cut flowers. Acta Hort 41:249-268.

Dilley, D.R. 1976. Application of the hypobaric system for storage and transportation of perishable agricultural commodities. Intl. Conf. On Handling Perishable Agr. Commodities Sept 8-10:135-149.

Dilley, D.R. 1977. The Hypobaric concept for controlled atmospheric storage. Mich. State Univ. Hort. Rpt. 28:29-37.

Dressler, B. and W. Jamieson. 1979. Hypobaric storage of several cut flower varieties. Grumman Allied Ind. Publication Oct. 1979:17 pp.

Hamner, C.L., R.F. Carlson, and H.B. Tukey. 1945. Improvement in keeping quality of succulent plants and cut flowers by treatment under water in partial vacuum. Science 102:332-333.

Hasek, R.F. 1975. Hypobaric storage of cut flowers. Flower and Nursery Rpt. Agr. Ext. Univ. CA Sept.-Oct.:9-11

Hauge, A.G. 1946. A study to improve wholesale and retail methods of packaging cut flowers. M.S. thesis, Ohio State Univ. 101 pp.

Lougheed, E.C., E.W. Franklin, D.J. Papple, D.R. Pattie, H.K. Malinowski and A. Wenneker. 1974. A feasibility study of low pressure storage. Hort Science Dept. School of Engineering, Univ. Guelph, Canada 47pp.

Lougheed, E.C., D.P. Murr, and L. Berard. 1978. Low pressure storage for horticultural crops. HortScience 13:21-27.

Marousky, F.J. 1977. Controlled atmosphere and low pressure storage of gladiolus, chrysanthemums and snapdragons. Controlled atmos. for storage and transp. of perishable commodities. Second Nat. Contr. Atm. Res. Conf., Mich. St. Univ. 28:116-121.

Mayak, S. and D.R. Dilley. 1976. Regulation of senescence in carnation (Dianthus caryophyllus). Plant Physiol. 58:663-665.

Mermelstein, N.H. 1979. Hypobaric transport and storage of fresh meats and produce earns 1979 IFT food technology industrial achievement award. Food Technol. 33(7):32-35, 38-40.

Molemaar, W.H. 1979. Testprogramma stekmateriaal - lagedruck container (Grumman) - Koelcel met vochtigheids regeling. (Geerlofs humicold, filacel.) Sprenger Ins. Werk. Proj. 35(3):72 pp.

Nicholas, S.L. 1980. Dormavac test gets mixed reviews. Florist 13(10):52-55.

Nowak, J. and R.M. Rudnicki. 1979. Long term storage of cut flowers. Acta Hort. 91:123-133.

Parvin, P.E. 1965. Studies of the effects of postharvest handling practices on the keeping quality and marketability of 'Better Times' roses.Ph.D. dissertation, Michigan State University. 117 pp.

Patterson, D. 1974. The use of opening solutions and hypobaric storage to promote chrysanthemum longevity. Washline 2(4):7.

Patterson, D.S. 1975. The effects of preservative solutions in hypobaric storage on chrysanthemum flower longevity. M.S. thesis Wash. State Univ. 58 pp.

Patterson, D.S. 1975. Hypobaric storage: a means of extending chrysanthemum cut flower longevity. Washington Floral Assoc. Washline 3:4-5.

Pentzer, W.T. 1941. Cut flowers tested on stratosphere flight. Flor. Rev. 88(2263):23.

Rule, D.E. 1977. Development of selected plant species in low pressure and controlled atmospheres. M.S. thesis, Ohio State Univ. 107 pp.

Staby, G.L. 1976. Hypobaric storage - an overview. Comb. Proc. Intl. Plant Prop. Soc. 26:211-215.

Staby, G.L., B.A. Eisenberg, T.A. Fretz and T.D. Erwin. 1976. Low pressure storage of bud-harvested Chrysanthemum morifolium Ramat, and rooted and unrooted Pelargonium hortorium, Bailey. HortScience 11:296. (abstr.)

Staby, G.L. 1977. Controlled atmosphere and low pressure storage of floral crops-overview. Controlled atmospheres for the storage and transport of perishable agricultural commodities, Second National Controlled Atmosphere Research Conference. 28:60-70.

Staby, G.L. 1978. Effects of modified storage atmospheres on floral crops. Univ. CA Cooperative Extension Ser. Bul. 40:8-9.

Staby, G.L., M.S. Cunningham, B.A. Eisenberg, M.P. Bridgen, and J.W. Kelly. 1979. Low pressure storage of carnations and roses. HortScience 14(3):446.

Van Nieuwenhuizen, I.G.H., C.M. Iedema, H.J. Van Laar and W.H. Molenaar. 1976. Het Geisoleerd transport van vacuumgekoelde bloemen in een 20' container. Sprenger Instit. Rapport 1937:53 pp.

Normal refrigeration. Numerous reports have consistently shown that high humidity (90% plus) and low temperatures (-1 to 1°C) are the correct levels for the storage of many cut flower types including roses, carnations, and chrysanthemums (see references cited by Halevy and Mayak, 1980-81; Staby, 1977; Nowak and Rudnicki, 1979; and references since 1976 cited in this publication under normal refrigeration). Yet, even with all this knowledge, the floral industry frequently completely disregards relative humidity settings and believes that 3 to 6°C is cold enough. As mentioned previously in this article, we believe that a great deal of the success of low pressure storage is the accurate control of temperature and relative humidity. Hence, significant improvements at commercial level in cut flower storage could be obtained by closely monitoring temperature and relative humidity levels.

In an attempt to promote a better understanding of this and other postharvest care and handling knowledge at commercial level, we at The Ohio State Univ. in cooperation with the Society of American Florists, established an educational program called the Chain of Life. Through this program and under the trademarked logo shown below, all types of postharvest information is distributed to the trade and ultimate consumer. Also, firms in the trade can become qualified members of the Chain of Life program by demonstrating that they do have proper storage temperatures and relative humidities, use floral preservatives, have high quality water, and reduce ethylene - induced problems by keeping their work areas clean.

Since the subject at hand relates specifically to normal refrigeration, the specifics relating to the Chain of Life program include that the maximum temperature be 3°C and the minimum relative humidity be 80%. By achieving these goals, industry members have reduced their cooler temperatures on average by 3 to 5°C and increased the relative humidity by about 15%. In return, shrinkage levels have been reduced by 20 to 40% at each marketing level.

Literature cited - normal refrigeration of cut flowers (since 1976)

Barendse, L.V.J. 1977. Slappe Nekken bij rozen te voorkomen. (Bentnecks in roses can be prevented.) Vakblad Bloemist. 32(20):23.

Barendse, L.V.J. 1978. Houdbaarheid snijgroen kam goed zijn. (Keeping quality of cut greens really can be good). Vakblad Bloemist. 33(48):25.

Barendse, L.V.J. 1979. Bewaring en verpakking van snijgroen. (Storage and packing of cut greens). Vakblad Bloemist. 34(45):39.

Barendse, L.V.J. 1979. Gemiddelde houdbaarheid siergroen goed. (On the average, vase-life of cut greens is good). Vakblad Bloemist. 34(1):33.

Boer, W.C. 1976. Invoed van de verpakking op het kwaliteitsverloop bij snijbloemen. (Packing of cut flowers). Sprenger Inst. Ann. Rep. 1976: 120-121.

Boer, W.C. 1976. Bewaarbaarheid van snijbloemen. (Tenability of cut flowers.) Sprenger Inst. Ann. Rep. 1976:119-120.

Boer, W.C. 1977. Bewaarbarheid van snijbloemen. (Keeping quality of cut flowers.) Sprenger Inst. Ann. Rep. 1977:104.

Boer, W.C. 1977. Distributie-aspecten bij snijbloemen. (Handling aspects of cut flowers.) Sprenger Inst. Ann. Rep. 1977:109-111.

Boer, W.C. 1978. Distributie-aspecten bij snijbloemen. (Handling aspects of cut flowers). Sprenger Inst. Ann. Rep. 1978:114:116.

Boer, W.C. 1978. Bewaarbaarheid van snijbloemen. (Storage life of cut flowers). Sprenger Inst. Ann. Rep. 1978:108-110.

Farnham, D.S. and T.G. Byrne. 1980. Marguerite daisies - postharvest handling methods. A review and update. Flower and Nursery Rpt. Summer 1980:5-7.

Goszczynska, D. and J. Nowak. 1979. The effect of growth regulators on quality and vase-life of dry-stored carnations. Acta Hort. 91:143-146.

Halevy, A.H. and P. Hannaford G. 1976. Conferencia Sobre Empaque, Almacenamiento y Cuidado de las Flores (Resume of conference on packing, storage and care of flowers) Bogata, Columbia, July 22, 1976.

Halevy, A.H., A.M. Kofranek, and S.T. Besemer. 1978. Post-harvest handling methods for bird-of-paradise flowers. (Strelitzia reginae, AIT). J. Amer. Soc. Hort. Sci. 103(2):165-169.

Hasegawa, A., M. Manabe, M. Goi, and Y. Ihara. 1977. Studies on the keeping quality of cut flowers. II. On the floral preservative "Kagawa Solution" and its practical use for cut carnation. Hort. Abstracts 47(3):No. 2758.

Jorgensen, G.H. and H. Brag. 1977. Knoppoppning och lagring av snittkrysanthemum. Lantbrukshogskolans Meddelsnden A266:11 pp.

Kaminaka, M.S., C.N. Thai, and J.F. Thompson. 1979. Handling of cut flowers at the wholesale level. Am. Soc. Ag. Eng. Win. Mtg. 1979:8.

Kofranek, A.M., E. Evans, J. Kubota, and D.S. Farnham. 1979. Chemical pretreatment of China aster to increase flower longevity. Progress report. Flower and Nursery Rpt. Spring 1979:1-3.

Marousky, F.J. 1979. Influence of conditioning treatments on floret opening of cut gladiolus spikes. Gladiograms 34:7.

Mayak, S. and E. Accati-Garibaldi. 1979. The effect of micro-organisms on susceptibility to freezing-damage in petals of cut rose flowers. Scientia Hort. 11:75-81.

Mayak, S. and A. Kofranek. 1976. Altering the sensitivity of carnation flowers (Dianthus caryophyllus L.) to ethylene. J. Amer. Soc. Hort. Sci. 101:503-506.

Molenaar, W.H. 1977. Kwaliteitsbehoud van snijbloemen. (Maintaining quality of cut flowers.) Vakblad Bloemist. 32(38):60-63.

Natarella, N.J. and S.J. Kays. 1979. Effect of storage temperature on cut flowers of Ranunculus asiaticus L. HortScience 14(4):504-505.

Nichols, R. and L.W. Wallis. 1976. Benefits and hazards with cool storage of narcissus. The Grower 85:848, 850, 852.

Nowak, J. and R.M. Rudnicki. 1979. Long term storage of cut flowers. Act Hort. 91:123-133.

Paulin, A. 1976. Vues nouvelles sur la conservation des roses coupees. (New views on maintaining keeping quality of cut roses). Amis Des Roses 325:10-15.

Paulin, A. 1976. L'Amerlidration du traitement des fleurs coupees. La conservation frigorifique des fleurs coupees. (Improving the treatment of cut flowers. Cold storage). Fleuriste de France 106(suppl):7-14.

Paulin, A. 1977. Le transport des fleurs coupees. Rev. Gen. Froid 1:55-60.

Rij, R.E., J.F. Thompson, and D.S. Farnham. 1979. Handling, precooling, and temperature management of cut flower crops for truck transportation. U.S.D.A. Adv. Agr. Tech. 5:26 pp.

Shuffler, H.J. 1979. Care and handling of cut flowers. I. What price profits. Link Magazine 2(9):6-9.

Souter, D. 1976. Le transport des fleurs coupees. L'Amerlioration du traitement des fleurs coupees. Fleuriste de France 106(suppl.):15-22.

Souter, D., J.M. Bureau, and A. Paulin. 1977. Vues nouvelles sur la conservation et le transport des oeillets. (Recent views on carnation conservation and transport.) Acta Hort. 71:265-272.

Sytsema, W. 1976. Nacherntebehandlung von schnittblumen. (Post-harvest treatment of cut flowers.) Gartenwelt 76(20):409-410.

Waters, W.E. 1977. The influence of post-harvest handling techniques on the vase life of gladiolus flowers. Gladio Gram 25:6-10

Woltz, S.S., W.E. Waters. 1976. Effects of light and temperature on keeping quality of cut flowers. Gladio Grams 23:2-5.

Van Doesburg, J. and L.V.J. Barendse. 1976. Onderzoek ter verbetering houdbaarheid van roos. (Research for improving the keeping quality of rose). Vakblad Bloemist. 31(15):18-19, 21.

Van Marsbergen, M. 1977. Oorzaken zwarte bloemknoppen in 'Ilona' opgespoord. (Causes of black buds in 'Ilona' have been found). Bloemist. 32(46):34-35.

EFFECTS AND FATE OF CARBON MONOXIDE IN SHREDDED LETTUCE

Galen D. Peiser, M. Concepcion C. Lizada and Shang Fa Yang
Department of Vegetable Crops
University of California
Davis, CA. 95616

Recently much interest has been directed towards the use of carbon monoxide for prolonging the storage life of fruits and vegetables as well as meats (Kader et al., 1978; Woodruff, 1977; Wolfe, 1980). During transit or temporary storage of vegetables and fruits, CO has been used as a supplement in modified atmosphere to inhibit discoloration of cut surfaces in lettuce and to reduce decay development in several fruits and vegetables. Under appropriate conditions CO has little or no adverse effects upon fruits and vegetables based on visual and sensory evaluations. Absorption and metabolism of CO has been examined in algae (Chappelle, 1962) and higher plants (Krall and Tolbert, 1957; Chappelle and Krall, 1961; Bidwell and Fraser, 1972; Kortschak and Nickell, 1973; Bidwell and Bebee, 1974). Krall and Tolbert (1957) observed that ^{14}CO was mainly incorporated into serine in barley leaves in the light, but dark fixation was too small to determine labeled products. In algae, $^{14}CO_2$ was the primary product of ^{14}CO metabolism in both light and dark although nonvolatile products from dark fixation were not examined (Chappelle, 1962). Exposing bean leaves to ^{14}CO, Bidwell and Fraser (1972) concluded that in the light most of the ^{14}CO was reduced and incorporated into serine and this was subsequently converted to sucrose; in the dark, over 90% of the ^{14}CO was converted to $^{14}CO_2$ although the labeling pattern of the acid-stable products suggested that primarily ^{14}CO and not $^{14}CO_2$ was being incorporated.

Since CO is presently being used commercially as a component of modified atmosphere during transit and storage of shredded lettuce, the present study was conducted to examine the effect of CO on compositional changes in shredded lettuce and to determine to what extent CO is incorporated in lettuce leaf discs.

Materials and Methods

Two different experimental conditions were used to study the effects of CO on crisphead lettuce: experiments conducted in flasks with controlled CO, designated "controlled CO", and experiments conducted in polyethylene bags designated "modified CO". Lettuce for the controlled CO experiments was obtained from a local market, cut into approximately 1 x 3 cm pieces and 50 g placed in 250 ml Erlenmeyer flasks. Control flasks contained air and treated flasks contained approximately 3.4% CO. Samples were held at 2.5 C and assays were run at 0, 5, 10, 15, and 20 days. All flasks were flushed with air when the CO_2 concentration reached 2.5% (in no flask did the $[CO_2]$ rise higher than 3.0% during the experiment) and the same level of CO was reintroduced into the treated flask. Flushing was necessary every 2 to 3 days. Shredded lettuce packaged in polyethylene bags obtained from Red Coach was used in the modified CO experiments. Control bags contained air only and treated bags contained elevated O_2 (ca. 30%) and CO (ca. 7%). The lettuce was held at 2.5 C and samples were taken at 2, 9, 16, and 23 days after the lettuce was processed and bagged.

Assay for ascorbic acid, amino acids, reducing sugars and total sugars: The procedure of Terada et al. (1978) was followed for determination of ascorbic acid. Amino acid content was determined by the method of Yemm and Cocking (1954). Reducing sugars and total sugars were determined by the procedures of Somogyi (1952) and Dubois et al. (1956), respectively.

Reducing sugars and total sugars were determined by the procedures of Somogyi (1952) and Dubois et al. (1956), respectively.

Respiration rate: Respiration rate was determined periodically be measuring the CO_2 concentration accumulated in the flasks or bags with a gas chromatograph unit.

CO determination: Gas samples from flasks or bags were periodically taken and their CO concentration was determined with a gas chromatograph unit.

^{14}CO generation: ^{14}CO was prepared from sodium [^{14}C]formate with concentrated H_2SO_4.

^{14}CO measurement: ^{14}CO was measured by oxidizing it to $^{14}CO_2$ quantitatively with I_2O_5 (National Academy of Science, 1977). $^{14}CO_2$ was assayed by a liquid scintillation counter after absorption in ethanolamine.

Comparison of labeled products from ^{14}CO and $^{14}CO_2$: Two g of lettuce leaf discs (11 mm diameter), which were prepared with a cork borer, were placed into a 50 ml Erlenmeyer flask which was sealed with a serum stopper and wrapped with aluminum foil. Either ^{14}CO (ca. 2 x 10^7 cpm, final concentration 84 µl/l) or $^{14}CO_2$ (ca. 6 x 10^6 cpm, final concentration 53 µl/l) was injected into the flasks which were incubated for 48 h at 2.5 C with continuous shaking. At the end of incubations, gas samples were taken for ^{14}CO and $^{14}CO_2$ determination. Then leaf discs were ground in 80% (v/v) ethanol in 0.4 M acetic acid, and the extracts were allowed to set at room temperature for 2 days. After concentration the extracts were fractionated into neutral, cationic and anionic fractions by ion exchange resins. The amino acids in the cationic fraction and the organic acids in the anionic fraction were further identified by paper chromatography and electrophoresis. Radioactivity was located using a radioscanner and the products were identified by co-migration with standards.

Dependence of the rate of CO oxidation to CO_2 on CO concentration: One g leaf discs with 0.5 ml water were placed in 25 ml Fernbach flasks which were wrapped with aluminum foil, stoppered and placed at 2.5 C. After temperature equilibration, various amounts of ^{14}CO were injected into the flasks. Thirty min later, a 3 ml gas sample was taken for $^{14}CO_2$ determination.

Results

Respiration rates for the controlled CO experiment are given in Figure 1. No apparent differences between the air and CO treatments were observed. The fluctuations in respiration rate were presumably linked in some way with flushing the flasks with air. Usually the respiration rate was lowest just prior to flushing and highest just after flushing. This may have resulted from suppression of the respiration rate by the buildup of CO_2 or the manipulation of the flasks during flushing may have stimulated the respiration rate.

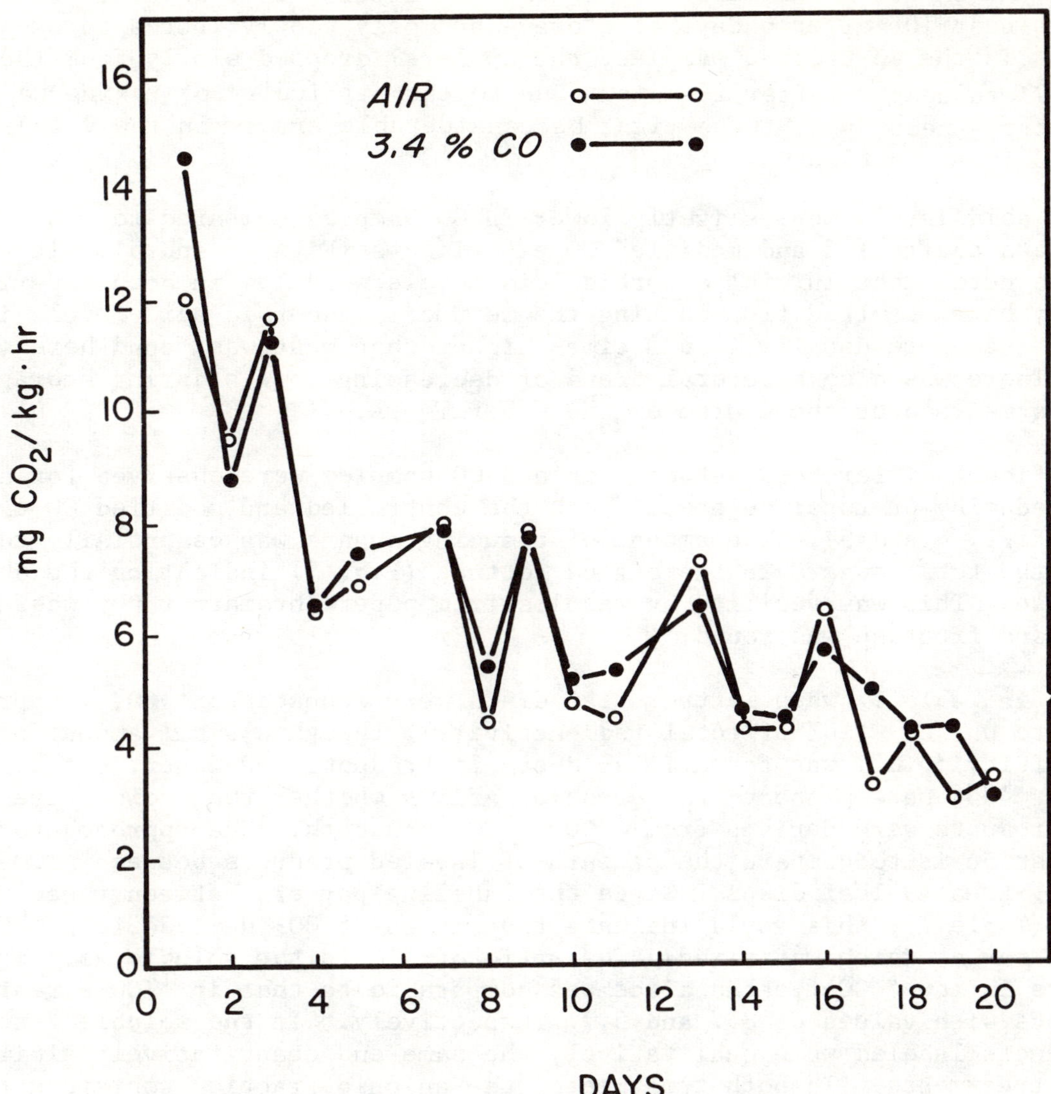

Figure 1. Respiration rates of air samples (○) and CO samples (●) of shredded lettuce during the course of the controlled CO experiment.

Changes in gas composition in the bags of shredded lettuce during the MA experiment are given in Figures 2 and 3. For both the CO-treated and air samples, the oxygen level decreased to 1 to 2% after 7 to 8 days and the CO_2 level was near 10% after 6 days of storage and only slowly increased subsequently. In the CO-treated samples, the CO level dropped slowly from the initial 7% to near 2% after 22 days. Due to the difficulty of taking gas samples from these bags, there might be considerable errors in the values presented in these figures.

Ascorbic acid levels were slightly lower in CO samples compared to those in air in both controlled and modified CO experiments (Figs. 4 and 5). It should be noted that initial ascorbic acid levels were low in both experiments, which may have resulted from cutting the lettuce. Ascorbic acid levels in uncut lettuce were usually 1 to 3 times higher than what was found here (Wu, 1978). There was also a general trend of decreasing levels during storage in both treatments over the course of the experiments.

No significant differences between air and CO samples were observed for amino acids, reducing or total sugars in both the controlled and modified CO experiments (Figs. 4 and 5). The amount of reducing sugars was essentially the same as the total sugars in the bagged lettuce (Fig. 5) indicating the absence of sucrose. This was verified by results from paper chromatography where only glucose and fructose was found.

As shown in Table I, when lettuce leaf discs were exposed to ^{14}CO, its primary metabolite was CO_2 (16% of total radioactivity), though a small amount of radioactivity (0.6%) was found in acid-stable products. However, since both ^{14}CO and $^{14}CO_2$ were present, the question arises whether the radioactive acid-stable products were derived from ^{14}CO, $^{14}CO_2$, or both. One approach to answer this question is to compare the pattern of labeled products between ^{14}CO-treated and $^{14}CO_2$-treated leaf discs. Since the labeling pattern between these two is similar (Table I), this would indicate that it was $^{14}CO_2$ derived from ^{14}CO and not ^{14}CO *per se* which was fixed. The ratio of ^{14}C in the soluble vs. insoluble fractions in the ^{14}CO-treated discs was comparable to that in $^{14}CO_2$-treated leaf discs with values of 8.7 and 5.7, respectively. In the soluble fraction, the products labeled were qualitatively the same and quantitatively similar in both treatments. In both treatments, the anionic fraction contained the most radioactivity, with malate being the predominate labeled product. In another experiment (data not shown) the cationic fraction was slightly higher in radioactivity than the anionic fraction, but again the distribution of label in the neutral, anionic and cationic fractions was comparable in the ^{14}CO and $^{14}CO_2$-treated discs and the labeled products in the soluble fraction were identical in the two treatments.

For most experiments the leaf discs were incubated at 2.5 C since this is a good commerical storage temperature for lettuce. However, a comparison was made between 2.5 and 20 C with respect to the distribution of label in the neutral, anionic and cationic fractions. No major differences between the two temperatures within each treatment or between the ^{14}CO and $^{14}CO_2$ treatments were observed (data not shown). As expected, more ^{14}CO was oxidized to $^{14}CO_2$ at the higher temperature (about 55% at 20 C compared with about 17% at 2.5 C), but less ^{14}C was incorporated in both treatments at 20 C, presumably due to the greater dilution of $^{14}CO_2$ by respiratory CO_2.

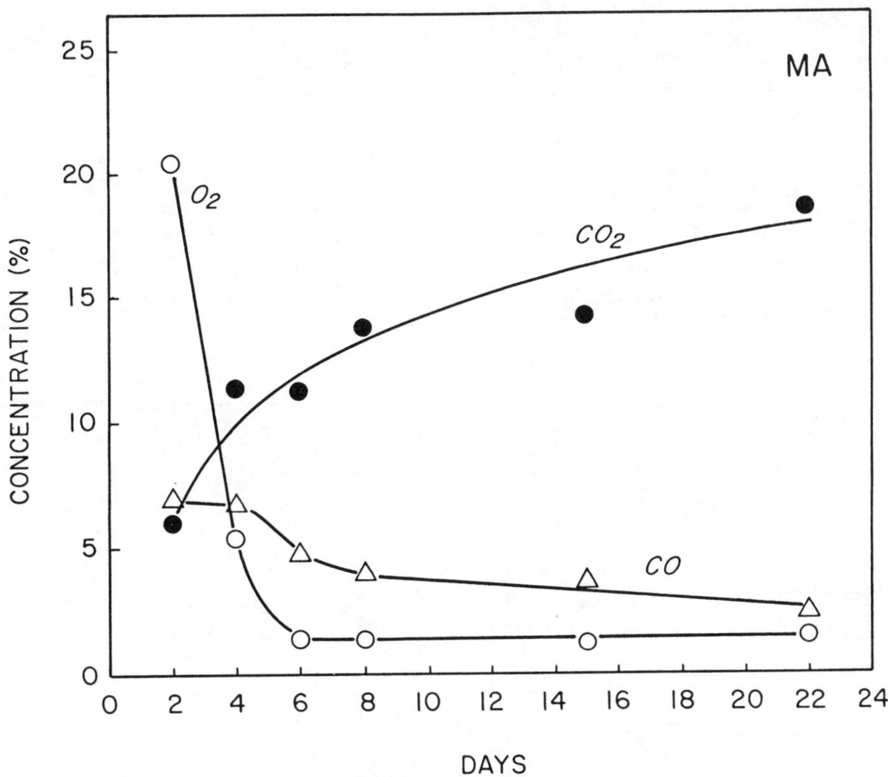

Figure 2. Changes in concentration of CO_2 (●), O_2 (○) and CO (△) in the CO-treated samples of polyethylene-bagged lettuce during the course of the modified CO experiment.

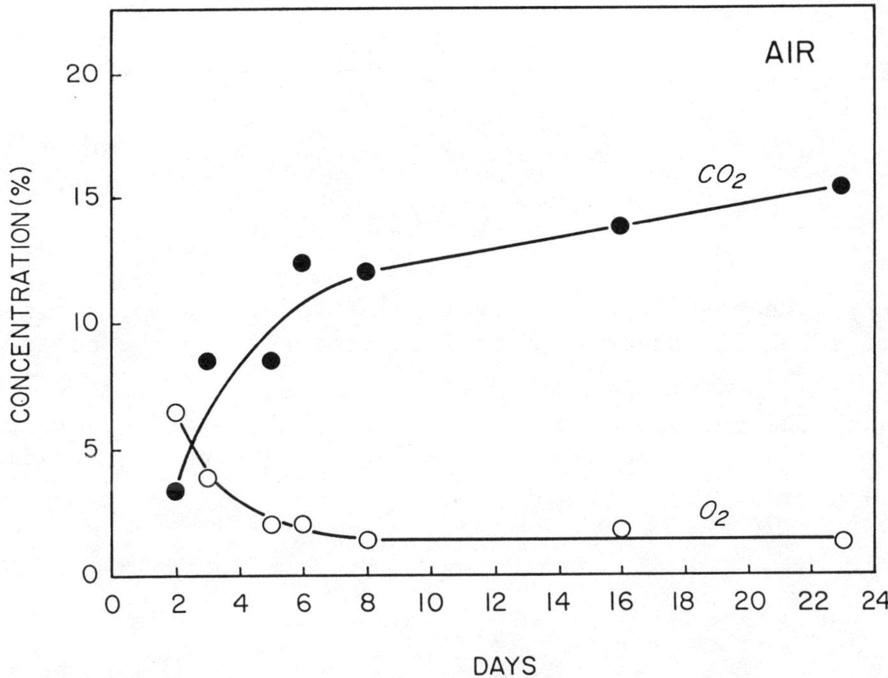

Figure 3. Changes in concentration of CO_2 (●) and O_2 (○) in the air samples of polyethylene-bagged lettuce during the course of experiment.

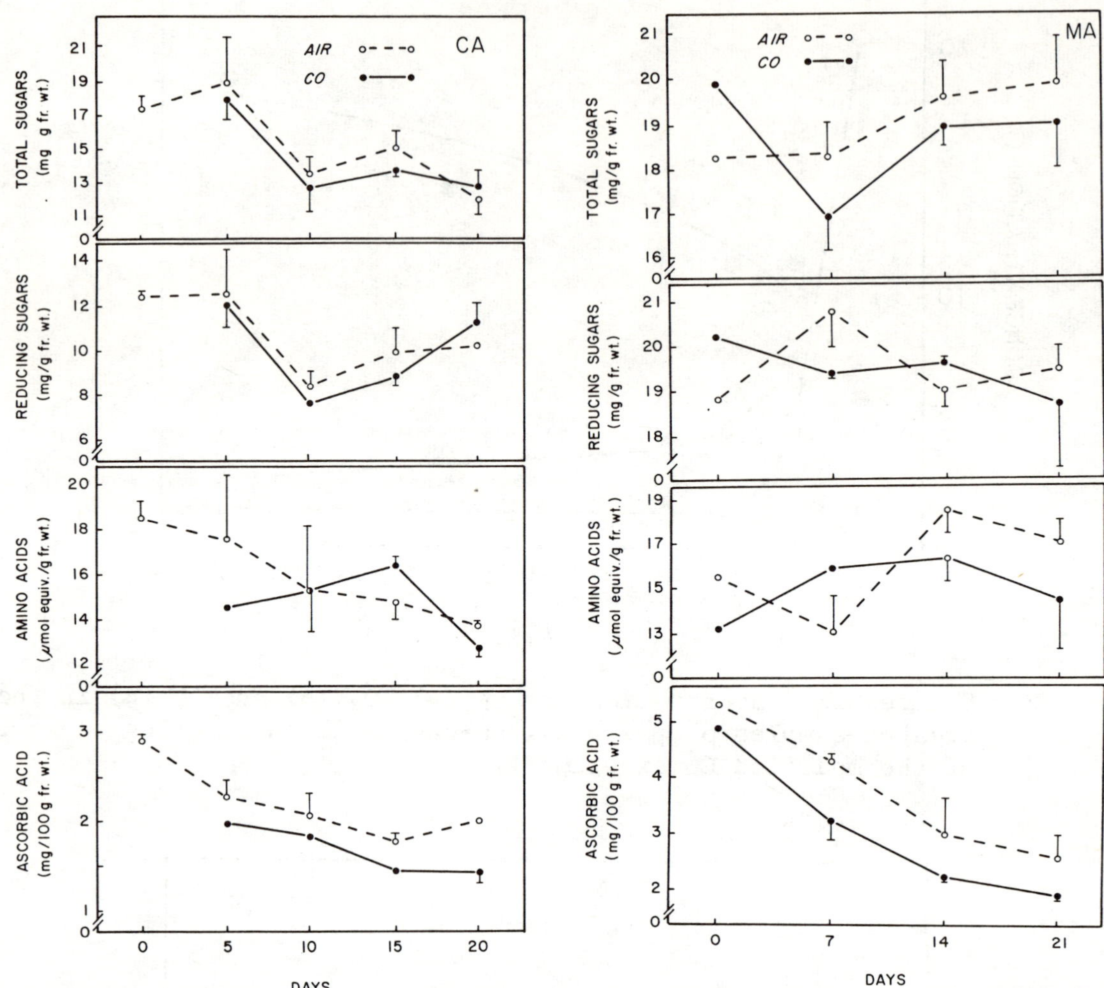

Figure 4. (left) Changes in concentrations of total sugars, reducing sugars, amino acids and ascorbic acid in lettuce during the controlled CO experiment.

Figure 5. (right) Changes in concentrations of total sugars, reducing sugars, amino acids and ascorbic acid in lettuce during the modified CO experiment.

Table I. Distribution of ^{14}C Into Products From Leaf Discs Exposed to ^{14}CO or $^{14}CO_2$.

Two g of leaf discs were incubated with either ^{14}CO (20×10^6 cpm, final conc. 84 μl/l) or $^{14}CO_2$ (6×10^6 cpm, final conc. 53 μl/l) for 48 h at 2.5 C.

Labeled Products	^{14}CO-treatment	$^{14}CO_2$-treatment
	% of ^{14}C supplied	
CO_2	16.7	-
Soluble fraction	0.61	6.38
Insoluble fraction	0.07	1.12
	% of soluble radioactivity	
Neutral fraction	3	5
Cation fraction	40	46
Aspartate	17	10
Glutamate	7	7
Asparagine	6	14
Glutamine	5	10
Serine	2	2
Alanine	1	2
Unidentified Cation	2	1
Anion fraction	57	49
Malate	40	28
Citrate	17	21

In barley (Krall and Tolbert, 1957) and bean leaves (Bidwell and Fraser, 1972; Bidwell and Bebee, 1974) serine in addition to sucrose was heavily labeled after exposure to ^{14}CO in the light. Bidwell and Bebee (1974) concluded that in bean leaves, sucrose was labeled by ^{14}CO either via serine, which does not involve the intermediacy of CO_2, or via the oxidation of CO to CO_2, which in turn was incorporated into sucrose via 3-phosphoglyceric acid. However, in corn leaves, ^{14}CO was incorporated exclusively via CO_2 and 3-phosphoglyceric acid into sucrose (Bidwell and Bebee, 1974). This was attributed to differences between C_3 and C_4 plants. In barley (Krall and Tolbert, 1957) and bean leaves (Bidwell and Bebee, 1974) it was suggested that ^{14}CO incorporation into serine occured via the tetrahydrofolate system.

In the dark, Bidwell and Fraser (1972) found that $^{14}CO_2$ was the main product of ^{14}CO metabolism in bean leaves as we observed in lettuce. However, contrary to our results, they found marked differences in labeled products when ^{14}CO and $^{14}CO_2$ were administered. They reported that with ^{14}CO, 50% of the soluble radioactivity was recovered in sucrose, but no label was recovered in sucrose with $^{14}CO_2$; only ^{14}CO labeled serine and only $^{14}CO_2$ labeled malate and phosphoglycerate. These results suggest that ^{14}CO was incorporated into serine as had been observed in the light. One main difference between our work and that of Bidwell and Fraser was the duration of exposure and temperature; 48 h at 2.5 C in our experiments and 20 min at room temperature in theirs. However the labeling pattern we observed from both ^{14}CO and $^{14}CO_2$ treated leaf discs was very similar to that found by Benson and Calvin (1950) in barley leaves exposed to $^{14}CO_2$ for 50 min in the dark. They observed that malate was the primary labeled product and the other labeled amino acids found were the same as we found in lettuce leaf discs. The direct incorporation of CO without the involvement of CO_2 did not appear to be a significant pathway since the radioactivity in serine from CO was low and essentially the same as that recovered from $^{14}CO_2$-treated leaf discs.

Since $^{14}CO_2$ was observed to be the major product of ^{14}CO metabolism (Table I) and the production of $^{14}CO_2$ from ^{14}CO appears to be an important step leading to labeled acid-stable products, we examined the dependence of CO oxidation on CO concentrations by leaf discs at 2.5 C. It was estimated that the system had an apparent Km of approximately 5 µl/l and a Vmax of 9.2 $nl \cdot g^{-1} \cdot h^{-1}$ (Fig 6). In the experiment of Table I, in which the leaf discs were incubated for 48 h at 2.5 C with 84 µl/l of CO, the oxidation rate of CO was 7.3 $nl \cdot g^{-1} \cdot h^{-1}$. This value is in agreement with the Vmax determined in Fig. 6.

The oxidation of CO to CO_2 appears to be a widespread phenomenon of living organisims including bacteria (Hegeman, 1980; Meyer and Schlegel, 1979), animals (Fenn, 1970), algae (Chappelle, 1962) and higher plants (Bidwell and Bebee, 1974; Bidwell and Frazer, 1972). In algae, Chappelle (1962) observed that this reaction required molecular oxygen, was effectively inhibited by cyanide, azide and hydroxylamine, and low light intensity increased the oxidation rate several fold over that of the dark rate.

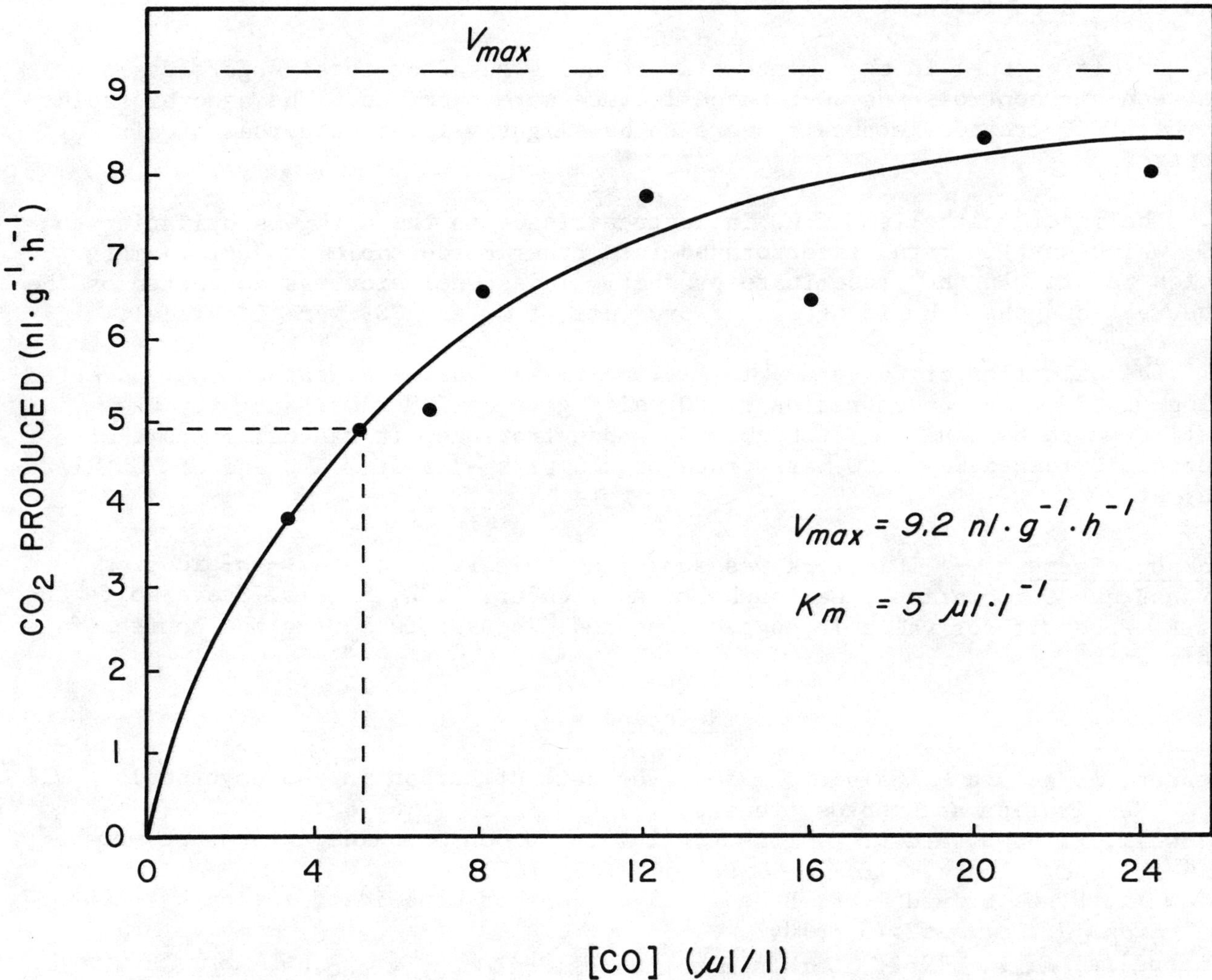

Figure 6. Dependence of the rate of CO oxidation to CO_2 on the CO concentration.

Conclusions

1. No differences in the respiration rates were observed between air controls and CO-treated lettuce.

2. No differences in the amino acid, reducing sugar or total sugar levels between air controls and CO-treated lettuce were observed. The ascorbic acid level of CO-treated lettuce appears to be slightly lower than that of air control.

3. The major metabolism of CO in lettuce tissue in the dark was oxidation to CO_2 which was, in turn, incorporated into other acid-stable products, among which malate was the predominate product. This conclusion was supported by the observations that the incorporated products of CO and CO_2 were identical.

4. The oxidation of CO to CO_2 in lettuce tissue was dependent on the concentration of CO. The concentration of CO which gave half maximal activity was estimated to be 5 μl/l. At higher CO concentrations, it was estimated that lettuce oxidized CO to CO_2 at a rate of 216 μl/kg-day at 2.5 C and of 700 μl/kg-day at 20 C.

Acknowledgement -- This work was supported in part by a grant-in-aid from TransFresh Corporation. We thank Dr. Adel Kader, L. L. Morris, Steve Wolfe and R. E. Woodruff for valuable suggestions and discussions during the course of this work.

References

Benson, A. A. and M. Calvin. 1950. The path of carbon in photosynthesis. VII Respiration and photosynthesis. *J. Exp. Bot.* 1:63-68.

Bidwell, R. G. S. and D. E. Fraser. 1972. Carbon monoxide uptake and metabolism by plants. *Can. J. Bot.* 50: 1435-1439.

Bidwell, R. G. S. and G. P. Bebee. 1974. Carbon monoxide fixation by plants. *Can. J. Bot.* 52:1841-1847.

Chappelle, E. W. 1962. Carbon monoxide oxidation by algae. *Biochim. Biophys. Acta.* 62:45-62.

Chappelle, E. W. and A. R. Krall. 1961. Carbon monoxide fixation by cell-free extracts of green plants. *Biochim. Biophys. Acta.* 49:578-580.

Dubois, M., K. A. Gilles, J. K. Hamilton, P. A. Rebers and F. Smith. 1956. Colorimetric method for determination of sugars and related substances. *Anal. Chem.* 28:350.

Fenn, W. O. 1970. CO burning in tissues. *Ann. N. Y. Acad. Sci.* 174:64-71.

Hegeman, G. 1980. Oxidation of carbon monoxide by bacteria. *Trends Biochem. Sci.* October:256-259.

Kader, A. A., G. A. Chastagner, L. L. Morris, and J. M. Ogawa. 1978. Effects of carbon monoxide on decay, physiological responses, ripening, and composition of tomato fruits. *J. Am. Soc. Hort. Sci.* 103:665-670.

Kortschak, H. P. and L. G. Nickell. 1973. Photosynthetic carbon monoxide metabolism by sugarcane leaves. *Plant Sci. Letters* 1:213-216.

National Academy of Sciences. 1977. In: Carbon monoxide. Committee on medical and biologic effects of environmental pollutants, Division of Medical Sciences, Assembly of Life Sciences, National Research Council, Washington, D. C.

Somogyi, M. 1952. Notes on sugar determination. *J. Biol. Chem.* 195:19-23.

Stokes, D. M., J. W. Anderson and K. S. Rowan. 1968. The isolation of mitochondria from potato-tuber tissue using sodium metabisulphite for preventing damage by phenolic compounds during extraction. *Phytochemistry* 7: 1509-1512.

Terada, M., Y. Watanabe, M. Kunitomo and E. Hayashi. 1978. Differential rapid analysis of ascorbic acid and ascorbic acid 2-sulfate by dinitrophenylhydrazine method. *Anal. Biochem.* 84:604-608.

Wolfe, S. K. 1980. Use of CO- and CO_2-enriched atmospheres for meats, fish and produce. *Food Tech.* March 1980:55-58.

Woodruff, R. E. 1977. Use of carbon monoxide in modified atmospheres for fruits and vegetables in transit. p.52-54. In: D. H. Dewey (ed.) Controlled atmospheres for the storage and transport of perishable agricultural commodities. Hort. Rept. 28, Mich. State Univ., East Lansing.

Wu, Sue-Ne. 1978. Ascorbic acid levels in crisphead lettuce (*Lactuca sativa* L) during storage. M.S. Thesis, University of California, Davis.

Yemm, E. W. and E. C. Cocking. 1955. The determination of amino-acids with ninhydrin. *Analyst* 80:209-213.

PHYSIOLOGICAL AND BIOCHEMICAL EFFECTS OF CARBON MONOXIDE ON FRUITS AND VEGETABLES

Adel A. Kader
Department of Pomology
University of California
Davis, CA 95616

During the previous CA conference, the status of knowledge as to the physiological effects of carbon monoxide (CO) on fresh fruits and vegetables was reviewed by Kader et al. (11) and Woodruff (24). In this report, emphasis will be placed on reviewing studies completed since 1977 especially those related to the physiological and biochemical effects of CO on fresh fruits and vegetables.

CO Production and Metabolism by Plants

CO is an air pollutant and its major source is the incomplete combustion from engines and automobile exhaust. Plants can be a minor source of CO. Bauer et al. (1) measured CO production by leaves of several woody plants and found a light-dependent CO production rate of up to 16×10^{-7} g/h. dm^2 leaf area. On the basis of assuming a mean CO production rate of 1.6×10^{-7} g/h. dm^2, they estimated a world-wide CO production by plants of 0.7×10^{14} g/year. Gladon and Staby (5) reported that CO is not emanated to any large extent (0 to 0.74 µl/Kg-hr) from tomato fruits, but is retained within the internal atmosphere. CO (28-41 µl/l) was found during all stages of fruit development, but no set pattern of CO concentration was evident. Whether CO, at such low levels, plays any role in plant physiology is not known. Also, these naturally-occurring CO levels are well below the levels used in postharvest applications (1-10% or 10,000 to 100,000 µl/l). The fate of CO in plant tissues will be discussed by Yang at this conference.

CO as a Fungistatic Gas

CO has been shown to be a fungistatic gas which suppresses but not completely inhibits fungal growth; its effectiveness is pathogen dependent and is greatly enhanced when added to reduced-O_2 atmospheres. During the past four years many reports have been published (2, 3, 8, 9, 10, 14, 15, 16, 17, 18, 19, 20, 21, 22, 23) on postharvest decay control in several commodities by CO. A level of 5% CO appears to be the minimum required for achieving decay control. El-Goorani and Sommer (4) recently reviewed available information on this subject. Further discussion of the pathological effects of CO will be presented by Sommer at this conference.

CO Effects on Fruit Ripening

CO is known to mimic ethylene effects and it can enhance fruit ripening when added to air. Kader et al. (8) reported that 5 or 10% CO added to air increased CO_2 and C_2H_4 production rates and hastened ripening of mature-green tomatoes. However, when 5 or 10% CO was added to 4% O_2 atmosphere, which is the effective decay control treatment, its effect on ripening was very small. After transfer to air at 20°C, fruits previously exposed to CO exhibited slightly higher CO_2 and C_2H_4 production rates and faster ripening than control fruits.

Morris *et al*. (14) found a CA combination of 4% O_2 + 2% CO_2 + 5% CO to be optimal for delaying maturation and ripening, maintaining good quality and retarding decay during storage of mature-green tomatoes for 4, 7, or 10 weeks at 12.7°C, followed by ripening at 20°C. However, fruits held for 10 weeks did not attain full red color when subsequently ripened at 20°C. Fruits exposed to CO with 4% O_2 with or without 2% CO_2 ripened faster than fruits held without CO after 10 weeks storage at 12.7°C (Table 1), but no effect on ripening was noted after 4 weeks, and only a slight effect was observed after 7 weeks.

Table 1. Effects of CA $\pm$ CO on ripening of mature-green 'Jackpot' tomatoes held at 12.7°C for 4, 7, or 10 weeks[1].

Composition of atmosphere %			Mean ripeness score[2] attained during CA storage		
O_2	CO_2	CO	4 weeks	7 weeks	10 weeks
21	0	0	3.7a	5.9a	6.0a
4	0	0	1.2b	2.8b	4.5c
4	0	5	1.0b	2.8b	5.4b
4	2	0	1.1b	1.3d	3.6d
4	2	5	1.1b	2.0c	4.3c

[1]From Morris *et al*. (14).

[2]USDA ripeness classes: 1=green to 6=red.

Although CO + air stimulated CO_2 and C_2H_4 production by peaches during holding at 5°C for 14 days followed by 3 days at 20°C, the effect on ripening was very small. No significant differences were observed in flesh firmness or color (9). Also, CO added to 4% O_2 $\pm$ 5% CO_2 had no effect on ripening rate of peaches.

Kiwifruits are extremely sensitive to C_2H_4 which induces their softening even during storage at 0°C. This may limit the use of CO as a supplement to CA for long-term storage of kiwifruits. Sommer *et al*. (21) found that kiwifruits held in air + 10% CO were softer (0.27 Kg) than those held in air (0.59 Kg) for 158 days at 0°C. This was also true for fruits held under 2.5% O_2 + 10% CO (0.41 Kg) relative to those held under 2.5% O_2 (0.86 Kg). However, the benefits of added CO in fungal suppression are greater than its detrimental effect on firmness of kiwifruits. CO should be used on fruits only in combination with reduced O_2 which is effective in retarding ripening (6) and in counteracting the C_2H_4-mimicing effect of CO.

CO Effects on Respiration

We have shown that CO (1, 3, or 5%) added to air or 2% O_2 reduced respiration rate of lettuce during holding at 2.5°C (7). However, lettuce previously subjected to CO had higher respiration rates during the subsequent holding in air at 10°C. Kramer et al. (12) stated that the beneficial effects of CO on storage-life of slices of apples, peaches and potatoes were attributable largely to reduced respiration rates and prevention of browning. On the other hand, CO added to air resulted in higher respiration rates of mature-green tomatoes (8) and peaches (9) as shown in Table 2. The effects of CO added to 4% O_2 on respiration rate were minimal.

Table 2. Effect of CO on respiration rate of 'Fay Elberta' peaches during storage at 5°C for 14 days followed by 3 days at 20°C.

Treatment	Days in storage									
	1	3	5	7	9	11	14	15	16	17
	ml CO_2/Kg-hr									
Air control	4.1	3.5	4.4	4.0	4.9	4.6	6.4	32.6	35.6	33.0
Air + 11% CO	5.8	5.4	6.1	7.9	9.7	9.1	8.5	33.3	36.2	32.2

CO Effects on Composition and Quality

McGill (13) stated that storing fresh leafy vegetables under an atmosphere of 1-5% CO and 1-10% O_2 retarded enzymatic deterioration changes. Woodruff (25) described a method for packaging cut vegetables (lettuce, cabbage, celery, broccoli, cauliflower, parsley, green onions) in polyethylene bags and modifying the atmosphere within these packages to include >25% O_2 and >3% CO. Under such conditions, cut vegetables retained their desirable color, flavor, and appearance much better than those held in air at temperatures below 7°C.

CO at very high concentrations (95%) is an effective enzyme inhibitor since it combines with iron- and copper-containing prosthetic groups or coenzymes which are required for the activity of certain enzymes. The effect of CO on inhibition of polyphenol oxidase activity and consequently tissue browning has been shown on mushrooms, lettuce, and fruit slices (12). Further work is needed to study the effects of CO on the activity of key enzymes in the metabolism of harvested fresh fruits and vegetables.

Kader et al. (8) reported that tomato fruits subjected to 5 or 10% CO + 4% O_2 maintained their sugar and acid content better than control fruits. No differences were observed between the two treatments in total ascorbic acid content or retention during storage and subsequent ripening. No significant differences in composition (sugars, acids, total carotenoids, and total ascorbic acid) were observed among peaches held in air or 4% O_2 + 5% CO_2 with or without 11% CO (9). Sommer et al. (21) found no effect for CO added to air or 2.5% O_2 on soluble solids content, but it reduced titratable acidity of kiwifruits. El-Goorani and Sommer (3) found no off-flavors attributable to 9% CO in apples or oranges, but strawberries developed moderate off-flavor under CA or CA +

9% CO which was attributed to reduced O_2 + elevated CO_2 and not added CO. Additional research is needed to evaluate the possible effects of CO on compositional changes and flavor quality of fresh fruits and vegetables which may benefit from its use as a fungistatic gas.

Literature Cited

1. Bauer, K., W. Seiler, and H. Giehl. 1979. CO production by higher plants. Z. Pflanzenphysiol. 94(3):219-230 (in German with English summary).

2. Buchanan, J. R., N. F. Sommer, and R. J. Fortlage. 1980. Production of aflatoxins (B_1 and G_1) in low O_2 and low O_2 + CO atmospheres. HortScience 15(3):424 (abstract).

3. El-Goorani, M. A. and N. F. Sommer. 1979. Suppression of postharvest plant pathogenic fungi by carbon monoxide. Phytopathology 69:834-838.

4. El-Goorani, M. A. and N. F. Sommer. 1981. Effects of modified atmospheres on postharvest pathogens of fruits and vegetables. Hort. Rev. 3:412-461.

5. Gladon, R. J. and G. L. Staby. 1979. Carbon monoxide formation in tomatoes. J. Amer. Soc. Hort. Sci. 104:74-76.

6. Kader, A. A. 1980. Prevention of ripening in fruits by use of controlled atmospheres. Food Technol. 34(3):51-54.

7. Kader, A. A., P. E. Brecht, R. Woodruff, and L. L. Morris. 1973. Influence of carbon monoxide, carbon dioxide, and oxygen levels on brown stain, respiration rate, and visual quality of lettuce. J. Amer. Soc. Hort. Sci. 98:485-488.

8. Kader, A. A., G. A. Chastagner, L. L. Morris, and J. M. Ogawa. 1978. Effects of carbon monoxide on decay, physiological responses, ripening, and composition of tomato fruits. J. Amer. Soc. Hort. Sci. 103:665-670.

9. Kader, A. A., M. A. El-Goorani, and N. F. Sommer. 1979. Effect of CO $\pm$ elevated CO_2 and/or reduced O_2 levels on postharvest behavior and quality of peaches. HortScience 14(3):471 (abstract).

10. Kader, A. A., J. M. Labavitch, F. G. Mitchell, and N. F. Sommer. 1980. Quality and safety of pistachio nuts as influenced by postharvest handling procedures. Calif. Pistachio Assoc. Annu. Rept. (1980):44-52.

11. Kader, A. A., L. L. Morris, and J. A. Klaustermeyer. 1977. Physiological responses of some vegetables to carbon monoxide. Mich. State Univ. Hort. Rept. 28:197-202.

12. Kramer, A., T. Solomos, F. Wheaton, A. Puri, S. Sirivichaya, Y. Lotem, M. Fowke, and L. Ehrman. 1980. A gas-exchange process for extending the shelf life of raw foods. Food Technol. 34(7):65-74.

13. McGill, J. N. 1969. Storage of fresh leafy vegetables. U.S. Patent No. 3,453,119 (July 1, 1969).

14. Morris, L. L., S. F. Yang, and D. H. Mansfield. 1981. Postharvest physiology studies. Calif. Fresh Market Tomato Adv. Bd. Annu. Rept. (1980-81):85-105.

15. Morris, L. L., S. F. Yang, and L. L. Strand. 1979. Postharvest physiology studies. Calif. Fresh Market Tomato Adv. Bd. Annu. Rept. (1978-79):59-66.

16. Morris, L. L., S. F. Yang, and L. L. Strand. 1980. Postharvest physiology studies. Calif. Fresh Market Tomato Adv. Bd. Annu. Rept. (1979-80):97-115.

17. Ogawa, J. M., G. A. Chastagner, and A. A. Kader. 1978. Effect of storage environment on development of postharvest diseases of temperate fruit. Third Int'l. Cong. Plant Path., Muchen, W. Germany, August 16-23, 1978, p. 27 (abstract).

18. Ogawa, J. M., B. T. Manji, and D. C. Janecke. 1978. Effects of modified oxygen and carbon monoxide environments on Monilinia decay of sweet cherries. Proc. Amer. Phytopath. Soc. 70th Annu. Meet., Tucson, Arizona, p. 173 (abstract).

19. Sommer, N. F., R. J. Fortlage, and J. R. Buchanan. 1980. Relation of oxygen to postharvest pathogen suppression by carbon monoxide. HortScience 15(3):424 (abstract).

20. Sommer, N. F., R. J. Fortlage, J. R. Buchanan, and A. A. Kader. 1981. Effect of oxygen on carbon monoxide suppression of postharvest pathogens of fruits. Plant Dis. 65(4):347-349.

21. Sommer, N. F., A. A. Kader, J. R. Buchanan, F. G. Mitchell, and R. J. Fortlage. 1980. Modified atmospheres with carbon monoxide for suppression of rot of perishable fruits and vegetables in storage and transit. Proc. Trop. Reg. Amer. Soc. Hort. Sci. (in press).

22. Sommer, N. F., F. G. Mitchell, and J. R. Buchanan. 1979. Controlled atmospheres and added carbon monoxide to slow softening and suppress storage rot and surface mold of kiwifruits. HortScience 14(3):461 (abstract).

23. Wolfe, S. K. 1980. Use of CO- and CO_2-enriched atmospheres for meats, fish, and produce. Food Technol. 34(3):55-58, 63.

24. Woodruff, R. E. 1977. Use of carbon monoxide in modified atmospheres for fruits and vegetables in transit. Mich. State Univ. Hort. Rept. 28:52-54.

25. Woodruff, R. E. 1980. Process and package for extending the life of cut vegetables. U.S. Patent No. 4,224,347 (Sept. 23, 1980).

THE ROLE OF CA, INCLUDING CO, IN PROLONGING THE STORAGE LIFE OF TOMATOES

L. L. Morris, D. Mansfield and L. Strand
Department of Vegetable Crops
University of California
Davis, CA 95616

Background

Research on the effects of reduced O_2 and increased CO_2 on tomatoes dates to the work of Kidd and West who studied tomatoes, as well as apples, in some of their early work. It is well established that ripening can be slowed by low O_2 levels and that increased CO_2 can be harmful. In the early days of consumer packaging it was observed that film packages of tomatoes required vent holes, and it is known that applications of wax coatings can retard or even prevent normal tomato ripening. Until the late 1940's, there was a widely held belief among tomato shippers that rail cars loaded with tomatoes required open vents during transit. This assessment, no doubt, had merit but was probably due more to avoidance of chilling temperatures rather than avoidance of harmful levels of O_2 or CO_2.

During the past few years we have studied the effects of modified atmospheres on tomatoes including the addition of CO. The interest stems largely from two assumptions: (1) it could be desirable to retard coloring, softening and decay of fruits harvested in a ripening condition in order to achieve better distribution, even under desirable temperature, and; (2) it could be desirable to hold tomatoes from a period of good supply to one of scarcity as can occur going from warm weather into cold weather. A third objective may become increasingly viable, and that is to reduce the costs of refrigeration by use of modified atmospheres to reduce the need for temperature control.

Our emphasis during the past 2 years has been toward prolonging storage up to several weeks in order to extend the marketing period. These results will be published elsewhere so we will limit our co ments to the more general aspects. In rief, our results indicate that mature green tomatoes can be successfully stored at 12.8 C for up to 7 weeks under 4% O_2, 2% CO_2 and 5% CO and still retain an adequate marketing life at acceptable quality for 1 to 2 weeks at 20 C under air.

Procedures

Emphasis was placed on duration of storage at 12.8 and on composition of the atmosphere. Some attention was given to stage of maturity on ripeness. The role of pathogen level and the effect of atmospheres on the pathogen growth on agar media were also considered. Two varieties of tomatoes were involved but direct comparison between varieties has not been made.

The CA treatments included O_2 levels of either 3 or 4%; CO_2 levels of 0, 2, 5%; and CO at 0, 5, or 8%. The fruits were held in large steel drums and the desired atmospheres were maintained by different procedures. In the first test the atmospheres were static and adjustments made by replacement of O_2 or removal of CO_2. In the second test the containers were flushed with the desired mix when changes in levels made this necessary. In the third test, the drums were supplied with a constant flow of the desired mixture.

We observed the ripening rates, quality of color developed, decay present, taste quality when table ripe, firmness, and total shelf life.

Results and Discussion

Fruit Maturity. Although definitive comparisons of fruit maturities have not been made, it seems reasonable to assume that storage life can be maximized by starting with green fruits with a definite green life. Fruits that have started to ripen can be stored successfully under CA but their potential marketing period after removal is likely to be short. At present, we believe that fruits subjected to CA storage should be of a maturity of score 2 to 3.

Effect on Postharvest Maturation and Ripening. At 12.8 C, low O_2 results in a significant reduction in postharvest maturation and ripening rates as compared with air. This is true for storage of 4, 7, or 10 weeks. Addition of 2% CO_2 to the 4% O_2 mixture slowed ripening somewhat more. In contrast, the addition of 5% CO to the CA mixture resulted in more ripening in most comparisons.

Effect on Visual Quality. When the fruits reached the table ripe stage they were scored for visual quality. Four weeks of storage resulted in visual quality nearly as good as non-stored fruits and differences among treatments were slight. However, after 7 weeks of storage the ripened fruits from all CA treatments ripened without storage. The fruits from the CA treatments showed abnormal color at transfer after 7 weeks of storage but were of normal appearance when rated as table ripe. After 10 weeks of storage all CA treatments appeared abnormal in color upon transfer and they failed to achieve the deep red color of a non-stored fruit. However, the visual quality of all CA fruits was better than that of the air-stored fruits. The presence of CO seemed to enhance the visual quality somewhat.

Effect on Decay. Two tests were carried out in 1980. The first was with early summer fruits and the second was with late fall fruits that had received chilling in the field before harvest. With the summer-grown fruits, there were no significant differences among treatments at the time of transfer following either 4 or 7 weeks storage. However, after 10 weeks of storage, all CA treatments were significantly lower in decay than the air controls. Those CA treatments with 5% CO showed less decay than their appropriate controls but the differences were not statistically significant. With the fall-grown fruits, decay was very extensive (over 70%) with the air-stored samples and was low with the fruits held under CA. Added CO gave a further reduction that was significant. When the fruits were innoculated, added CO resulted in a significant reduction in decay -- not only at transfer but after an additional 6 days at 20 C.

Effect on Taste Quality. Results in 1979 indicated that tomatoes could be held for 6 weeks under CA and remain acceptable from a taste standpoint when table ripe. Tests run in 1980 with fruits held for 10 weeks showed some advantages to CA storage as compared with air storage in regard to overall taste quality. However, the level of overall quality was in the undesirable range for this prolonged storage period for both CA and air stored fruits. Although the differences are not significant at the 5% level there seemed to be somewhat improved retention of sweetness under CA and less development of off flavors. We believe that CA storage should be limited to 6 or 7 weeks at 12.8 C.

Conclusions

A potential exists for the storage of tomato fruits in 4% O_2 at 12.8 C for a period up to 7 weeks. The addition of 2% CO_2 to the storage atmosphere retarded postharvest maturation and ripening without influencing decay incidence or quality parameters during a 7 week storage period. Decay incidence was a function of storage time, fruit maturity, atmosphere composition and inoculum level of *Botrytis cinerea*. The addition of 5% CO to the atmosphere reduced the incidence of decay during storage without otherwise influencing shelf life, visual quality, taste quality, firmness or color achieved at the table ripe stage. Storage for up to 7 weeks resulted in fruits of acceptable visual and taste quality. But fruits stored for 10 weeks neither developed normal red color nor ripened to acceptable taste quality. Commercial applications of prolonged storage should await pilot-scale trials.

Acknowledgements: We wish to thank Billy Manji in the Department of Plant Pathology at the University of California, Davis, for providing *Botrytis cinerea* culture. The 1979 test was carried out by Mr. F. Adamicki (Poland) while on study leave at Davis.

CARBON-MONOXIDE SUPPRESSION OF POSTHARVEST DISEASES OF FRUITS AND VEGETABLES

N. F. Sommer
University of California, Davis, California

M. K. El-Kazzaz
University of Tanta, Egypt

J. R. Buchanan and R. J. Fortlage
University of California, Davis, California

and M. A. El-Goorani
University of Alexandria, Egypt

Controlled atmospheres (CA) have only a modest ability to suppress postharvest diseases. Any reduction or delay in disease development has often been attributed to delay in the onset of senescence of the host commodity rather than to a direct effect of the CA upon the pathogen. The growth rate of the fungus is a better indication of the direct effects of the atmosphere on the fungus. Measurements of the growth rates abundantly indicate that important benefits might accrue from adding a fungistatic gas (2, 3).

The success in using carbon-monoxide (CO) to inhibit certain browning reactions led to interest in its possible fungistatic properties. Added interest followed the finding by Kader and Ogawa (4) that a much greater fungistatic effect was achieved when O_2 was low. CO might stimulate ripening of some commodities, which is a serious disadvantage, but has not been fully evaluated. Consequently tests were conducted in culture and with kiwifruits and sweet cherries in storage or in simulated transit to further explore the possibility that CO might be the fungistatic gas needed.

In Vitro Fungal Growth

The direct effect of atmospheric gases upon the physiology of fungi is commonly determined by measuring growth rate. Different postharvest pathogens vary considerably in their responses to O_2, CO_2, and CO as single gases or in combination.

Oxygen: Fungi often grow best when the O_2 concentration is that in normal air, i.e., about 21%. Such fungi commonly show little reduction in growth rates as O_2 is lowered until a concentration of about 2% is reached. At concentrations below 2%, growth rates drop very rapidly. Under anaerobic conditions, fungal pathogens cease to grow, but many can live and undergo anaerobic respiration for a considerable period.

A few pathogens, such as the sour rot organism *Geotrichum candidum* Link ex Pers., grow better at 2-4% O_2 than in air (2, 9). Other postharvest fungi that grow better at low O_2 include Phytophthora cactorum (Leb. & Cohn) Schroet. (1, 2), P. citrophthora (R. E. Sm. & Br.) Leonian (5), and P. nicotiana var. parasitica (Dast.) Waterh (2,5). At least some isolates of Monilinia fructicola (Wint) Honey appear to grow best at O_2 between 4 and 10%. Rhizopus stolonifer (Ehrenb. ex Fr.) Lind has been shown able to grow at extremely low O_2, and growth was greater at 2.3% O_2 than in air, but no growth occurred when O_2 was rigorously excluded (2, 3, 7, 8).

Carbon dioxide: Although CO_2 is fixed by many fungi and growth may be suppressed in its absence, fungi cannot use CO_2 as the sole source of carbon. As concentrations of CO_2 increase above the normal 0.03% of air, the effect of CO_2 is commonly stimulatory, but at some point it becomes toxic. At 5% CO_2, common in controlled-atmosphere storages, the effect on fungal growth is often not very different from that of normal air or is only slightly inhibitory; however, the very modest suppression of many pathogens at 5% CO_2 is sometimes greater if O_2 is also low (3).

Carbon monoxide: Data in Tables 1 and 2 show how interactions of CO with O_2 and CO_2 suppress growth of fungi in culture. Low O_2 dramatically enhanced the suppression by CO of most fungi tested (2, 6). Only Whetzelinia sclerotiorum (Lib.) Korf & Dumont was suppressed nearly equally whether O_2 was low or not. Penicillium digitatum Sacc. and P. italicum Wehmr. were also considerably suppressed by 9% CO in air, and benefits were greater when CO_2 was also present. The small response of Rhizopus stolonifer to CO and to CO with low O_2 or 5% CO_2 may be associated in some way with the ability of this fungus to grow at very low O_2 and to tolerate 5% CO_2. We do not have evidence that the CO at concentrations up to 9% has any fungicidal effect; it is evidently fungistatic only.

Table 1. Percent reduction in fungal growth in atmospheres at 12.5 C. Growth in air = 100%.

	Air + 9% CO	Air + 5% CO_2 + 9% CO	2.3% O_2	2.3% O_2 + 9% CO	2.3% O_2 + 5% CO_2	2.3% O_2 + 5% CO_2 + 9% CO
Alternaria alternata	<10	<10	<10	40	<10	60
Botrytis cinerea	<10	<10	<10	35	<10	45
Geotrichum candidum	<10	<10	<10	50	<10	50
Rhizopus stolonifer	<10	15	10	15	20	20
Whetzelinia sclerotiorum	80	80	45	90	40	85

Table 2. Percent reduction in growth of Penicillia in various atmospheres at 12.5 C. Growth in air = 100%.

	Air + 9% CO	Air + 5% CO_2 + 9% CO	2.3% O_2	2.3% O_2 + 9% CO	2.3% O_2 + 5% CO_2	2.3% O_2 + 5% CO_2 + 9% CO
P. expansum	20	30	<10	75	20	65
P. digitatum	30	60	<10	95	<10	90
P. italicum	40	75	35	95	30	90

Disease Suppression

Components of a synthetic atmosphere suppress postharvest diseases by affecting the physiology of the fungus or by affecting the physiology of the host commodity. In addition to lowering the fungal growth rate, such factors as the suppression of synthesis or the actions of certain fungal toxins or enzymes would likely slow the development of the disease lesion.

Ripening is the physiological change which is susceptible to influence by modified atmospheres, and may have a striking effect on disease resistance. The response of commodities may change from near immunity to high susceptibility within several days. The effects of modified atmospheres on defense mechanisms are largely unknown. Those defense mechanisms that depend on the synthesis of new compounds seem likely to be suppressed by these atmospheres; a situation that would depress respiration rates of the commodities. For example, it is well documented that low O_2 atmospheres slow wound-healing in potatoes resulting in increased danger of rot.

Botrytis Rot of Kiwifruit

California-grown kiwifruits stored for 3-6 months at 0 C may suffer severe losses from *Botrytis cinerea* Pers. ex Fr. Inoculation may occur as fruits are injured during harvest and handling. Lesions initially localized at the stylar- or calyx-ends are common and suggest prior colonization of adherent styles or sepals. Experimental evidence does not yet permit evaluation of the importance of these structures in the development of the disease in storage.

Botrytis rot in storage commonly appears after about three months at 0 C, at which time the fruits may have softened considerably. Loss of individual infected fruits is only a part of the problem: often more important is that the fungus has the ability to produce aerial mycelium that can penetrate nearby sound fruits. A fruit so penetrated will rot, and the mycelium from that fruit will penetrate adjacent fruits. This multiple fruit "nest" will continue to enlarge, and, given sufficient time, all fruits in a container will be involved.

Tests of the use of CO in atmospheres for disease suppression were designed primarily to determine whether lesion development is suppressed and whether the development of aerial mycelium is prevented. The addition of 9% CO to air resulted in only a very modest reduction in the rate of lesion development (Table 3), a result anticipated from previous studies (2, 6). If O_2 was low, on the other hand, CO greatly reduced the rate of lesion development. The presence of CO_2 further slowed lesion development. At the end of five months, the disease lesions, although viable, were relatively very small in the CO low-O_2 atmosphere. However, the most striking benefit of using CO with low-O_2 was the prevention of the formation of aerial mycelium during the period of the experiment, thus completely eliminating fruit-to-fruit infection.

Table 3. Lesion diameters (cm) in kiwifruits needle-inoculated with conidia of *Botrytis cinerea*. Measurements were taken after three months at 0 C in indicated atmospheres.

Atmosphere	Lesion diameter (cm)
Air	5.6 $\pm$ 1.2
Air + 9% CO	4.0 $\pm$ 0.3
CA (2% O_2, 5% CO_2)	1.4 $\pm$ 0.3
CA + 9% CO	0.5 $\pm$ 0.2

Although the suppression of Botrytis rot was highly effective, the effects of CO on fruits held in 0 C storage for five months was to soften the fruit excessively. Data in Table 4 show that fruit held in CA (2.5% O_2 plus 5% CO_2) plus 10% CO were essentially as soft as fruit held in air. When the CO_2 was increased to 7.5%, the CO-induced softening was counteracted. Unfortunately, this increased CO_2 concentration caused fruit damage.

Table 4. Firmness of kiwifruits after five months storage at 0 C in various atmospheres. Pressure measured using Magness-Taylor Tester with a 3/8-inch tip.

Composition (%)			
O_2	CO_2	CO	Pressure (lbs)
Air			2.2 $\pm$ 0.4
Air	5		8.1 $\pm$ 0.3
Air	7.5		9.0 $\pm$ 2.7
2.5	5		10.8 $\pm$ 2.7
2.5	7.5		10.6 $\pm$ 2.1
2.5	5	10	3.1 $\pm$ 0.9
2.5	7.5	10	8.4 $\pm$ 2.3

Botrytis and Brown Rots of Sweet Cherries

The two principal postharvest diseases usually found in sweet cherries are Botrytis rot and brown rot (*Monilinia fructicola*). Botrytis rot may develop in refrigerated storage or transit and is commonly found after sweet cherries have been stored at 0 C or higher for several weeks before marketing. It may be mistaken for brown rot when the fruits are marketed. *Monilinia fructicola*, on the other hand, grows very little at 5 C or lower. Consequently, fruits may appear sound upon removal from refrigeration, only to be found heavily diseased a few days later.

Data in Table 5 show that lesions of Botrytis rot were dramatically suppressed by 10% CO, particularly when included in CA. Data in Table 6 show that no brown rot lesions developed in the fruits during 21 days at 5 C because of suppression by cold. When fruits were placed at 20 C, however, visible lesions were detected the second day in fruits held in air during the 21-day storage. By contrast, fruits held in CA plus 10% CO atmospheres during storage did not exhibit lesions until the fourth day. Other atmospheres resulted in intermediate delays in lesion development.

Table 5. Mean diameter of *Botrytis cinerea* lesions in cherries as influenced by atmosphere. Each fruit was needle-inoculated with conidia before refrigerated storage.

Atmosphere	Early Burlat[z] (15 d at 5 C) (mm)	Bing[z] (21 d at 5 C) (mm)
Air	18.6[a]	41.7[a]
Air + 10% CO	11.3[b]	26.2[b]
CA (2.3% O_2 + 5% CO_2)	10.4[c]	26.2[b]
CA + 10% CO	0.6[d]	3.3[c]

[z]Numbers followed by the same letter are not significantly different (P = 0.05) according to Duncan's Multiple Range Test.

Table 6. Development of brown rot in inoculated 'Bing' cherries held 21 days at 5 C followed by seven days at 20 C.

Atmosphere	% diseased fruits after indicated days at 20 C						
	1	2	3	4	5	6	7
Air	0	0.8	1.7	15	40	57	73
Air + 10% CO	0	0	0.8	10	25	38	51
CA (2.3% O_2 + 5% CO_2)	0	0	1.7	14	35	57	71
CA + 10% CO	0	0	0	2	8	27	46

The effects of atmospheres containing CO on respiration and ethylene (C_2H_4) evolution were studied with 'Bing' sweet cherries held at 0 C (Table 7). The addition of CO to air did not significantly increase respiration, as measured by CO_2 evolution, above that of fruits in air. Similarly, the addition of CO to 2.3% O_2 did not significantly increase CO_2 evolution above that of fruits in 2.3% O_2. On the other hand, C_2H_4 evolution was approximately tripled by adding CO to air or to 2.3% O_2.

Table 7. Effect of atmospheres containing CO on CO_2 and ethylene (C_2H_4) evolution from 'Bing' cherry fruits held in air or 2.3% O_2 at 0 C for 21 days.

Treatment	Total CO_2^z (mg/kg-hr)	Total $C_2H_4^z$ (μg/kg-hr)
Air	2.81a	0.0048b
Air + 10% CO	2.83a	0.0158a
2.3% O_2	1.67b	0.0042b
2.3% O_2 + 10% CO	1.74b	0.0150a

[z]Numbers followed by the same letter are not significantly different (P = 0.05) according to Ducan's Multiple Range Test.

The color of cherries in CO-containing atmospheres was slightly darker than in the comparable atmosphere without CO (Tables 8, 9) but was significantly so only between air plus CO and air. Fruits were slightly softer after storage in CO atmospheres. Neither soluble solids nor acidity were altered significantly by the atmosphere.

Table 8. Effect of CO treatments on color, firmness, soluble solids and acidity of 'Early Burlat' cherries held in air and controlled atmosphere (CA) in the presence or absence of 10% CO for 15 days at 5 C, followed by two days in air at 20 C.

Treatment	Color[xz] 5 C	Color[xz] 20 C	UC firmness[z] tester reading (g) 5 C	UC firmness[z] tester reading (g) 20 C	Soluble[y] solids content (%) 5 C	Soluble[y] solids content (%) 20 C	Titratable[y] acidity (% citric acid) 5 C	Titratable[y] acidity (% citric acid) 20 C
Air	20.38a	18.00a	248b	159a	12.5	13.1	0.66	0.83
Air + 10% CO	15.48b	15.17a	164c	147a	13.7	13.8	0.64	0.89
CA (2.3% O_2 + 5% CO_2)	22.46a	20.00a	281a	178a	13.7	13.7	0.70	0.87
CA + 10% CO	20.73a	20.61a	249b	150a	13.7	13.5	0.66	0.92

[z]Numbers followed by the same letter are not significantly different (P = 0.05) according to Duncan's Multiple Range Test.

[y]Juice of 24 fruits taken at random.

[x]Measurements made with Gardner XL-23 colorimeter. Lower numerical values indicate darker red color.

Table 9. Effect of CO treatment on color, firmness, soluble solids and acidity of 'Bing' cherries held in air and 2.3% O_2 in the presence or absence of 10% CO for 21 days at 0 C, followed by two days in air at 20 C.

Treatment	Color[xz] 5 C	Color[xz] 20 C	UC firmness[z] tester reading (g) 5 C	UC firmness[z] tester reading (g) 20 C	Soluble[y] solids content (%) 5 C	Soluble[y] solids content (%) 20 C	Titratable[y] acidity (% citric acid) 5 C	Titratable[y] acidity (% citric acid) 20 C
Air	19.60a	16.90a	409a	372a	17.9	17.6	0.97	0.94
Air + 10% CO	19.83a	14.30b	370a	345a	17.5	16.9	0.97	0.90
2.3% O_2	19.03a	16.20ab	384a	345a	18.5	17.0	0.95	0.91
2.3% O_2 + 10% CO	18.56a	14.23b	359a	366a	17.4	16.9	0.99	0.91

[z]Numbers followed by the same letter are not significantly different (P = 0.05) according to Duncan's Multiple Range Test.

[y]Juice of 24 fruits taken at random.

[x]Measurements made with Gardner XL-23 colorimeter. Lower numerical values indicate darker red color.

Conclusions

The addition of CO to CA may provide a means of suppressing important post-harvest diseases. However, the effectiveness of the gas varies widely among pathogens. Further, the tendency for CO to stimulate C_2H_4 evolution may prevent use with kiwifruits or other commodities highly sensitive to C_2H_4. The poisonous properties of CO pose added limitations to widespread use.

Literature Cited

1. Covey, R. P., Jr. 1970. Effect of oxygen tension on the growth of Phytophthora cactorum. Phytopathology 60:358-359.

2. El-Goorani, M. A. and N. F. Sommer. 1979. Suppression of postharvest plant pathogenic fungi by carbon monoxide. Phytopathology 69:834-838.

3. El-Goorani, M. A. and N. F. Sommer. 1981. Effects of modified atmospheres on postharvest pathogens of fruits and vegetables. Hort. Rev. 3:412-461.

4. Kader, A. A., G. A. Chastagner, L. L. Morris, and J. M. Ogawa. 1978. Effects of carbon monoxide on decay, physiological responses, ripening, and composition of tomato fruits. J. Amer. Soc. Hort. Sci. 103:665-670.

5. Klotz, L. J., L. H. Stolzy, and T. A. DeWolfe. 1963. Oxygen requirements of three root-rotting fungi in a liquid medium. Phytopathology 53:302-305.

6. Sommer, N. F., R. J. Fortlage, J. R. Buchanan, and A. A. Kader. 1981. Effect of oxygen on carbon monoxide suppression of postharvest pathogens of fruits. Plant Dis. 65(4):347-349.

7. Wells, J. M. 1967. Growth and production of pectic and cellulolytic enzymes by *Rhizopus stolonifer* in low-oxygen atmospheres. Phytopathology 57:1010. (Abstr.)

8. Wells, J. M. 1968. Growth of *Rhizopus stolonifer* in low-oxygen atmospheres and production of pectic and cellulolytic enzymes. Phytopathology 58:1598-1602.

9. Wells, J. M. and D. H. Spalding. 1975. Stimulation of *Geotrichum candidum* by low oxygen and high carbon dioxide atmospheres. Phytopathology 65:1299-1302.

ETHYLENE RELATIONSHIPS IN STORAGE OF SOME VEGETABLES

P. Toivonen, J. Walsh, E.C. Lougheed and D.P. Murr
Horticultural Science Department, University of Guelph
Guelph, Ontario NIG 2WI

The presence of potentially harmful volatiles, including ethylene, in controlled atmosphere (CA) storages for apples was recognized early by researchers (Smock, 1979) and this recognition was evident in a practical sense in the installation of scrubbing devices, usually containing brominated, activated charcoal, in some storage rooms (Smock, 1979). Because these devices were not particularly effective in controlling either volatile levels or storage scald, their use diminished. This disinterest probably paralleled the birth of the notion that ethylene had little or no effect at low storage temperatures and CA conditions (Fidler, 1964), and was nurtured by the discovery by Smock (1955) of chemicals which would control storage scald, the most important and supposedly volatile-related disorder of apples. This notion has been shown to be at least suspect (Smock, 1979) and likely incorrect (Liu, 1977), and work is now going on to develop an effective means of controlling ethylene levels in storage (Herregods, 1977; Liu, 1977; McKee and Dilley, 1980; Saltveit, 1980).

There has not been the same interest in volatile accumulation in vegetable storage for several reasons. The major reason may be that the major volatile (ethylene) is not produced in large quantities by vegetables (excluding those fruits such as tomatoes classed as vegetables), and a secondary reason is that not many vegetables are stored in CA in which volatiles might accumulate. However, the literature does contain references to the effect of volatiles upon vegetables in storage. Although low pressure storage offers an elegant method for control of volatiles in storage (McKeown et al., 1978), its use does not appear economically practical for vegetables.

Burton (1952) showed that the previously purported effect of carbon dioxide on controlling sprouting of potatoes was not due to carbon dioxide but to volatile products induced within the potatoes and accumulating in the atmosphere, which resulted in sprout inhibition. From this knowledge Burton developed the use of nonyl alcohol as a sprout inhibitor. Herregods (1977) has studied the effect of volatiles upon harvested vegetables and also reviewed some of the pertinent literature. Furry et al. (1973) reported that "volatile" products of atmospheric generators harmed cabbage in CA storage. It may be that one reason that CA storage of vegetables is not particularly effective is the accumulation of, as yet, unidentified volatiles.

In vegetable storage, as with fruits, ethylene is probably the volatile present in largest amounts and is most easily identified (Isenberg, 1979). Ethylene has long been associated with isocoumarin production in carrots (Carlton et al., 1961), senescence and abscission of leaves of vegetables (Furry et al., 1979; Pendergrass et al., 1976), and disorders in lettuce (Rood, 1956; Morris and Kader, 1977). Its presence is related also to suppression of sprouting (Burton, 1952) and break of dormancy (Rosa, 1925; Rylski et al., 1974) in potatoes, and toughening of asparagus through fibre development (Haard et al., 1974). Of practical interest in the report that CA storage of cabbage in New York was most successful when a scrubber for ethylene was installed (Furry et al., 1979a).

The work reported here concerns ethylene effects on some vegetables with the full realization that we may indeed have effects attributed to ethylene which may be due to other "volatiles", as yet unidentified.

An attempt was made to determine a loss estimate due solely to ethylene for cabbage and Brussels sprouts that were stored for 84 days in air at 1°C. One treatment involved the exposure of the produce to apple volatiles (presumed to be mainly ethylene) from a storage room containing apples and the other involved exposure to apple volatiles which had been scrubbed to remove ethylene (Table 1). At the end of the storage period there was minimal disease development on both cabbage and Brussels sprouts, so the basic effects seen could be attributed to ethylene. A significant increase in trim loss in cabbage and Brussels sprouts was found with an average ethylene level of approximately 4 ppm versus little damage at the scrubbed ethylene level of approximately 0.75 ppm ethylene. Brussels sprouts were harmed much more by ethylene than cabbage.

CA storage of cabbage (3% O_2, 5% CO_2) allowed the accumulation of ethylene to much higher levels than found in simulated air-storage in a sealed chamber (20% O_2, 0% CO_2). Despite levels of ethylene approaching 100 ppm or more, there were no apparent ethylene effects on the cabbage. Cabbage stored in CA was in much better condition and exhibited less trim loss than cabbage stored in "air" at 1°C after 5 months (Table 2). Removal of ethylene from air storage with "Purafil" reduced deterioration and trim losses in cabbage. In contrast, ethylene scrubbing had no visible affect on cabbage stored in CA. However, a taste test indicated that the high levels of ethylene were associated with development of a bitter flavor in cabbage held in CA and ethylene scrubbing reduced the development of this flavor. The off-flavor diminished considerably over time in air at 20°C.

Results with celery stored in CA (3% O_2, 2% CO_2) and in air (20% O_2, 0% CO_2) were quite similar to the tests with cabbage (Table 3).

However, no development of off-flavors with high levels of ethylene occurred in controlled atmosphere storage of celery. During the storage of broccoli in CA (3% O_2, 10% CO_2) ethylene accumulated to about five times that in air storage, but never exceeded 5 ppm in CA. For the short, 39-day storage test, there was no difference in quality between broccoli stored in CA and in air (Table 4). The produce was still of excellent marketable quality.

Two storage trials with cauliflower produced similar results. In both cases CA (3% O_2, 10% CO_2) chambers had extremely elevated ethylene levels which were three to five times those found in the broccoli tests. No apparent deleterious effects of ethylene were found in the CA-stored cauliflower (Table 5). Cauliflower stored in CA retained a high degree of green color in the wrapper leaves and the curd was maintained in a clean white condition. In contrast, air-stored cauliflower at 1°C displayed senescence and abscission of the wrapper leaves and dark specking occurred over most curds.

In summary, it appears that ethylene, and possibly other unidentified volatiles, can cause measurable and significant losses in cabbage and Brussels sprouts; Brussels sprouts, broccoli and cauliflower may show loss of quality in air storage at 1°C with the ethylene levels that occur in a sealed storage. In CA an accumulation of high levels of ethylene can occur; however, no visible effects of these potentially noxious levels were noted. This may be due to competitive inhibition of ethylene action by the high CO_2 levels. Nevertheless the bitter flavor which was noted in CA-stored cabbage may have been due to ethylene levels that approached 100 ppm or more. No significant off-flavor was found in broccoli or cauliflower stored in CA despite previous reports that off-flavors develop at lower levels of O_2 (Lipton and Harris, 1974; Lipton et al., 1967). These workers reported that off-flavors disappeared if the produce was aerated over a period of time and this was probably what occurred in the present work.

It appears that ethylene can be of importance in both air storage and in CA storage of vegetables. Although the effect of ethylene may be reduced in CA conditions, it or other volatiles may have significant deleterious effects.

REFERENCES CITED

Abeles, F.B. 1973. Ethylene in Plant Biology. Academic Press. New York.

Burton, W.G. 1952. Studies on the dormancy and sprouting of potatoes. III. The effect upon sprouting of volatile metabolic products other than carbon dioxide. New Phytologist. 51:154-163.

Carlton, B.C., L.E. Peterson and N.E. Tolbert. 1961. Effects of ethylene and oxygen in production of a bitter compound by carrot roots. Plant Physiol. 36:550-552.

Fidler, J.C. 1964. Controlled atmosphere storage of apples. Ditton Lab. Memoir 78, East Malling, England (In Smock, 1979).

Furry, R.B., F.M.R. Isenberg and M.C. Jorgenson. 1979a. Post-harvest controlled atmosphere requirements of cabbage intended for long duration storage. Paper #79-6011. Joint summer meeting ASAE and CSAE, June, 1979.

Furry, R.B., J.R. Hicks, M.C. Jorgenson, and P.M. Ludford. 1979b. Effects of ethylene on controlled atmosphere storage of cabbage. Paper #79-4537, Winter meeting ASAE, Dec., 1979.

Furry, R.B., F.M.R. Isenberg, M.C. Jorgensen and J.R. Carroll. 1973. Pilot studies on the use of catalytically generated atmospheres for storage of cabbage, *Brassica oleracea* L. ASAE Paper #73-3506, Dec. 1973 (As referred to in Furry et al., 1979b).

Haard, N.F., S.C. Sharma, R. Wolfe and C. Frenkel. 1974. Ethylene induced isoperoxidase changes during fiber formation in post harvest asparagus. J. Food Science. 39:452-456.

Herregods, M. 1977. Disorders caused by some volatiles other than ethylene, found in storage rooms. Acta. Hort. 62:247-256.

Isenberg, F.M. 1979. Controlled atmosphere storage of vegetables. Hort. Rev. 1:337-394.

Lipton, W.J. and C.M. Harris. 1974. Controlled atmosphere effect, on the market quality of stored broccoli (Brassica oleracea L., Italica Group). J. Amer. Soc. Hort. Sci. 99:200-205.

Lipton, W.J., C.M. Harris and H.M. Couey. 1967. Culinary quality of cauliflower stored in CO_2-enriched atmospheres. Proc. Amer. Soc. Hort. Sci. 91:852-859.

Liu, F.W. 1977. The ethylene problem in apple storage. Proc. 2nd Nat. CA Res. Conf., Hort. Rpt. #28 C.D.A. D.H. Dewey, Ed., Mich. State U.: 86-96.

McKee, M.W. and D.R. Dilley. 1980. Reducing the level of ethylene in controlled atmosphere apple storages by catalytic oxidation. HortScience 13 (3), (Sec. 2):423.

McKeown, A.E., E.C. Lougheed and D.P. Murr. 1978. Comparability of cabbage, carrots and apples in low pressure storage. J. Amer. Soc. Hort. Sci. 103:749-752.

Morris, L.L. and A.A. Kader. 1977. Physiological disorders of certain vegetables in relation to modified atmospheres. Proc. 2nd Nat. CA Res. Conf. Hort. Rpt. 28 (D.H. Dewey, Ed.) Mich. State U: 142-147.

Pendergrass, A., F.M.R. Isenberg, L.L. Howell and J.E. Carroll. 1976. Ethylene-induced changes in appearance and content of Florida-grown cabbage. Can. J. Plant Sci. 56:319-324.

Pratt, H.K. and J.D. Goeschl. 1969. Physiological roles of ethylene in plants. Annu. Rev. Plant Physiol. 44:541-

Rood, P. 1956. Relation of ethylene and post harvest temperature to brown spot of lettuce. Proc. Amer. Soc. Hort. Sci. 68:296-303.

Rosa, J.T. 1925. Shortening the rest period of potatoes with ethylene gas. Potato News Bull., 2:363-365.

Rylski, I., L. Rappaport and H.K. Pratt. 1974. Dual effects of ethylene on potato dormancy and sprout growth. Plant Physiol. 53:648-662.

Saltveit, M.E. 1980. An inexpensive chemical scrubber for oxidizing volatile organic contaminants in gases and storage room atmospheres. HortScience. 15:759-760.

Smock, R.M. 1955. A new method of scald control. Amer. Fruit Gr. 75 (1):20.

Smock, R.M. 1979. Controlled atmosphere storage of fruits. Hort. Rev. 1:301-306.

ACKNOWLEDGEMENTS

The authors acknowledge with gratitude the financial support of Agriculture Canada, The Ontario Ministry of Agriculture and Food, and the National Science and Engineering Research Council (Canada), and the help of Mr. A.W. Loughton, Director, Ontario Ministry of Agriculture and Food, Horticulture Experiment Station, Simcoe, Ont., Mr. M. Valk, Ontario Ministry of Agriculture and Food, Muck Research Station, Bradford, Ontario, and growers who contributed produce. As well the authors acknowledge the many papers and personal contacts which cannot be otherwise recognized herein.

Table 1. Percentage increase in losses in cabbage and Brussels sprouts due to exposure to unscrubbed apple volatiles.

	Cabbage			Brussels Sprouts
	April Green	Evergreen Ballhead	Ultra Green	Silver-star
Fresh weight	0.3	0.4	0.3	0.5
Trim	7.0	4.0	21.0	40.0
Total chlorophyll	25.0	30.0	66.0	71.0

Table 2. The effect upon cabbage of scrubbing ethylene from CA (3% O_2, 5% CO_2) and "air" at 1°C during the 1980-81 season.

	Treatment			
	CA	CA + Scrub	"Air"	"Air + Scrub"
Weight change	+ 0.9%	+ 1.0%	0.0%	+ 0.3%
Trim loss	14.6%	14.5%	20.2%	16.7%
Total chlorophyll (mg/g fr. wt.)	0.015	0.013	0.004	0.010
Taste	Bitter	Not Bitter	Slightly Bitter	Not Bitter

Table 3. The response of celery to CA (3% O_2, 2% CO_2) and "air" storage at 1°C for the 1980-81 season.

	Treatment	
	CA	"Air"
Weight loss	0.5%	0.9%
Trim loss	22.4%	39.1%
Chlorophyll content (mg/g fr. wt.)	0.015	0.011
Taste	Not Bitter	Not Bitter

Table 4. The response of broccoli to CA (3% O_2, 10% CO_2) and air storage at 1°C in 1980.

	Treatment	
	CA	"Air"
Taste[z]	15	14
Visual score[y]	1.0	1.0

[z]Accumulated score in a taste ranking of the two treatments.

[y]0-4 scale, 0 being excellent and 4 poor.

Table 5. The effect upon cauliflower of CA (3% O_2, 10% CO_2) and "air" storage at 1°C for the 1980 season.

	Treatment	
	CA	"Air"
Wrapper leaf chlorophyll content[z]	0.056	0.021
Wrapper leaf abcission	none	in all samples
Curd condition[y]	2.3	5.3
Weight loss	2.6%	5.0%

[z]mg/g fresh wt.

[y]0-11 scale, 0 being excellent and 11 very poor.

EFFECTS OF ATMOSPHERE AND ETHYLENE ON CABBAGE METABOLISM DURING STORAGE

J.R. Hicks, P.M. Ludford and J.F. Masters
Vegetable Crops Department
Cornell University
Ithaca, New York

The effects of ethylene on vegetative tissues are well documented. Pendergrass et al. (6), were the first to investigate the effects of this hormone on cabbage (Brassica oleracea). Furry et al (3), have reported on the ethylene concentration that accumulated in commercial CA cabbage storages and the effectiveness of ethylene scrubbers. Recently we reported on the effects of continuous low ethylene levels on cabbage stored under CA conditions (5). This present study was to investigate effects of a slightly higher level of ethylene on a variety that has better storage characteristics.

Cabbage (cultivar Bartola) was commercially harvested in early November 1979 and transported to Ithaca. Treatments were initiated November 15 and consisted of normal atmosphere (air) or CA (5% CO_2 + 2 1/2% O_2), each with 0, 1, or 5 ppm ethylene. N_2, CO_2, O_2 and C_2H_4 were combined to attain the appropriate atmosphere which was continuously flushed through the approximately 20 liter chambers containing 2 heads each. Treatments were replicated 3 times. Each container was sealed at the start of the experiment and remained unopened until removal for analysis.

After removal from storage, each cabbage was cut and an "inner" sample (5 cm^2) was taken from the seven leaves immediately above the growing tip. An "outer" sample (5 cm^2) was taken from the seven outermost, sound leaves from the top of the head. Samples were immediately frozen in liquid N_2 and held frozen until they were freeze-dried, ground in a Wiley Mill and analyzed. Organic acids were extracted by the method of Fernandez-Flores (2) et al (0.25g freeze-dried material). Sugars were extracted with hot 80% ethanol, passed through an anion-cation exchange column and derivatized (TMS) by the method of Sweeley et al, (7). Analysis of both acids and sugars was by gas chromatography with a flame ionization detector, separation was with a OV-101 column and identification was by retention time and mass spectroscopy.

In agreement with previous reports (1,4,6) CA storage was effective in maintaining the green color in cabbage. The effect of ethylene on the color of cabbage stored in air was very pronounced, as shown by the increasing "L" and decreasing "a" values with increasing C_2H_4 levels (Table 1). Under CA conditions the levels of C_2H_4 used did not affect either of these color parameters markedly. There was a time X treatment interaction for the "a" values in air (Fig. 1). These changed very little during storage in air alone, but with 1 ppm C_2H_4 , there was a slow decline until April and then a sharper drop during May and June. The addition of 5 ppm C_2H_4 resulted in lower "a" values after 1 month storage than other treatments, followed by a steady increase through March after which there was another sharp decrease.

		Hunter Values		
Atmos	C_2H_4	"L"	"a"	"b"
Air	0	76.5^{bc}	9.7^{a}	3.7^{a}
	1	77.4^{ab}	8.1^{b}	3.0^{bc}
	5	78.2^{a}	7.2^{c}	3.6^{ab}
CA	0	74.1^{d}	10.3^{a}	3.4^{ab}
	1	75.2^{cd}	9.5^{a}	2.4^{c}
	5	74.1^{d}	10.2^{a}	2.9^{bc}

Table 1. Hunter Color Difference Meter readings for cabbage stored for 1 to 7 months in air or CA with or without ethylene. Values are an average of 18 samples. Mean separation is by Duncan's Multiple Range test at the 5% level. There was a significant interaction with time for the "a" values.

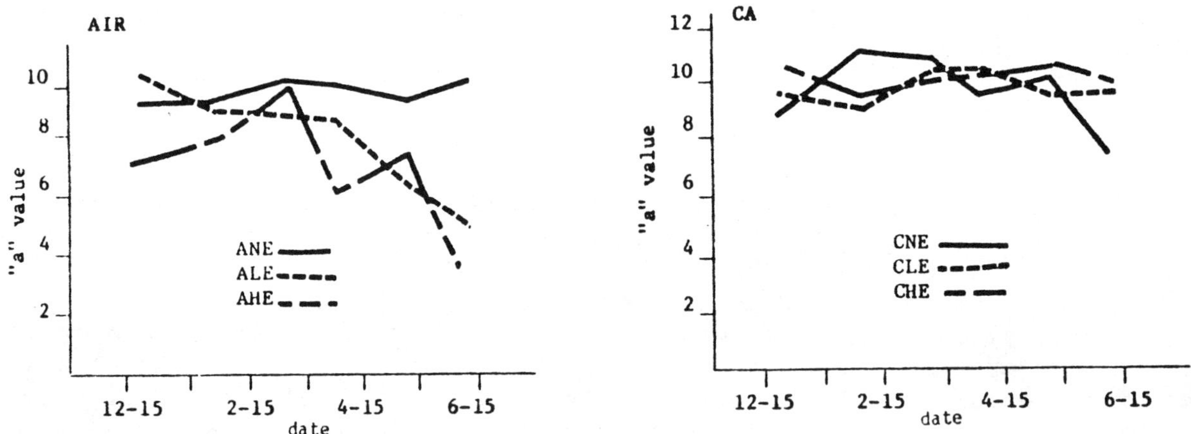

Fig. 1. Effect of storage on green color retention in cabbage.
Air alone (ANE), + 1 ppm ethylene (ALE), + 5 ppm ethylene (AHE)
CA alone (CNE), + 1 ppm ethylene (CLE), + 5 ppm ethylene (CHE)
"a" values represent Hunter Color Difference Meter readings taken from the outermost sound leaf of the intact head. Each number is the average of 6 samples.

There was no reduction in fresh weight loss with Bartola due to CA storage (Table 2). This is in agreement with the results reported for Hidena (5) but contrary to those of Bohling and Hansen (1). In addition, there was no effect of C_2H_4 on weight loss of cabbage stored in air. Under CA storage there was less weight loss when ethylene was present, particularly in the 1 ppm C_2H_4 treatment.

Table 2: Percent loss in fresh weight for cabbage stored from 1-7 months in CA or air with or without ethylene. Each number is an average of 36 samples. Mean separation is by Duncan's Multiple Range Test at the 5% level.

	% Weight Loss		
	Ethylene	PPM	
Atmos	0	1	5
Air	2.83^{ab}	2.36^{abc}	2.57^{ab}
CA	3.06^{a}	1.81^{c}	2.22^{bc}

There was a decrease in % dry weight (approximately 2.5%) during the storage period. The effects of atmosphere and C_2H_4 are shown in Table 3. Although the differences are not nearly the magnitude of those with time a slight promotion in dry weight loss was found with C_2H_4 under both air and CA storage.

Table 3. Percent dry weight for cabbage

	% Dry Weight		
	Ethylene	PPM	
Atmos	0	1	5
Air	9.11^{ab}	8.69^{b}	8.80^{b}
CA	9.39^{a}	9.15^{ab}	9.17^{ab}

It is apparent from figure 2 that the loss in dry weight was not due to utilization of sugars. There was no effect of time, atmosphere, or C_2H_4 on levels of glucose, fructose or total sugars. However, although sucrose in the initial samples accounted for less than 10% of the total sugar, concentration doubled during 6 months of storage in CA with C_2H_4 (Figure 3). Increases in sucrose levels in air storage were not as large as those in CA. Storage in air alone resulted in considerable variation in sucrose from month to month with peak levels being attained in February and April. On the other hand the air plus C_2H_4 treatments were much more consistent from month to month and although these treatments reached maximum sucrose levels in February, the levels were not as high as the air control.

Malic acid content increased with time in storage regardless of the storage atmosphere (Figure 4). In air storage the amount of increase seemed to be related to exogenous ethylene levels. Under CA conditions malic acid levels remained relatively steady until May, though the CA + 5 ppm C_2H_4 stored cabbage contained more malic acid than other CA treatments through March after which there was a pronounced drop. After March malic acid in CA plus C_2H_4 treated cabbage began to increase.

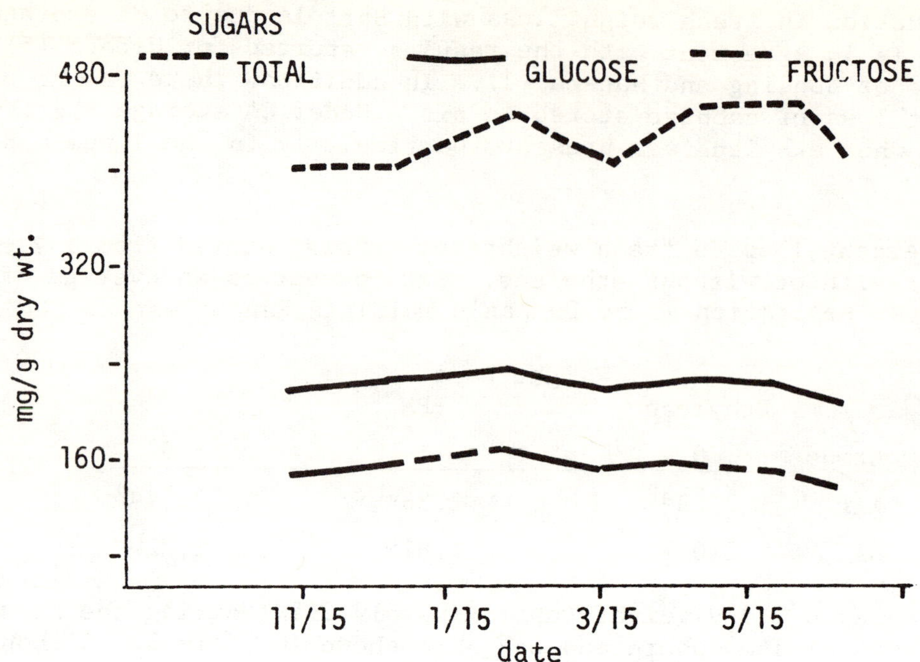

Fig. 2. Effect of storage time on glucose, fructose and total sugar levels in cabbage. Each point is the average of 36 samples across ethylene and atmosphere treatments.

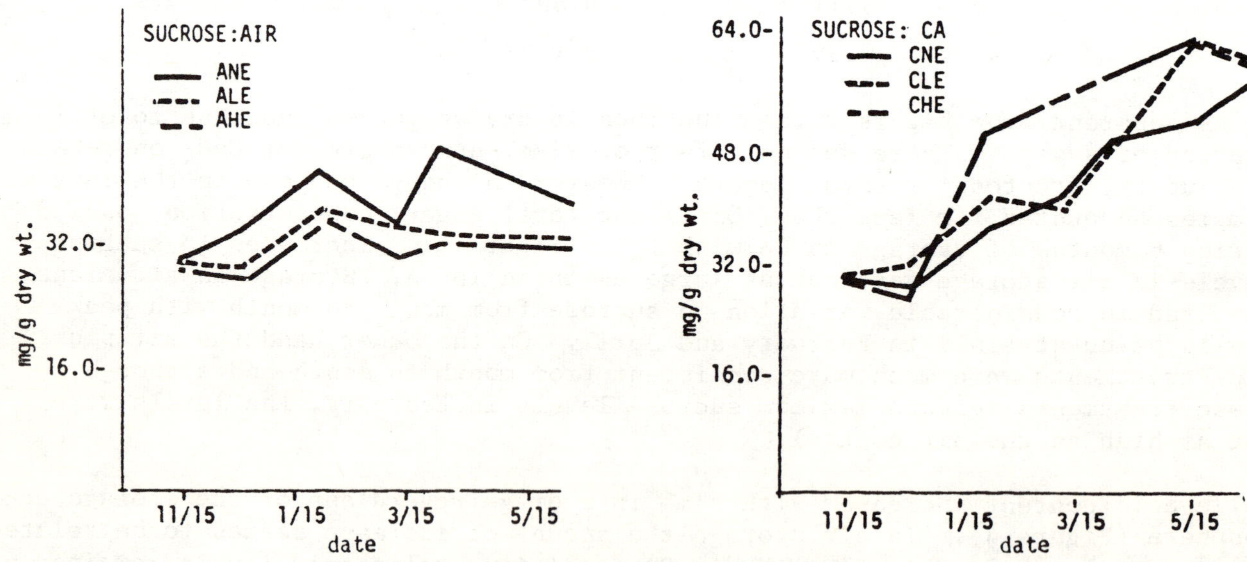

Fig. 3. Effect of storage time, atmosphere, and ethylene levels on sucrose content of cabbage (see legend Fig. 1). Each number is the average of 6 samples.

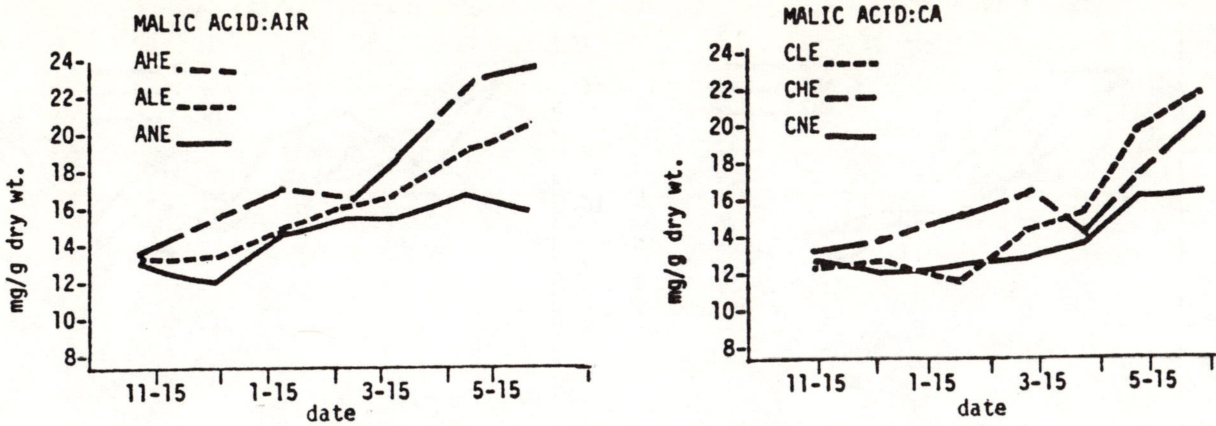

Fig. 4. The effect of storage time, atmosphere and ethylene on malic acid levels in cabbage (see legend Fig. 1.)

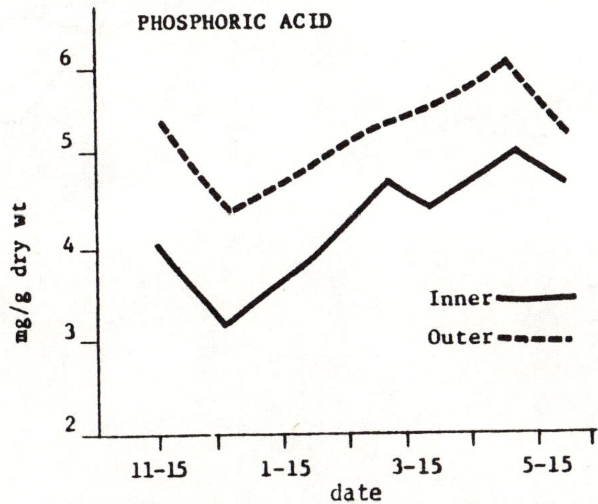

Fig. 6. Effect of storage time and leaf position on phosphoric acid content of cabbage. Each value is an average of 18 samples.

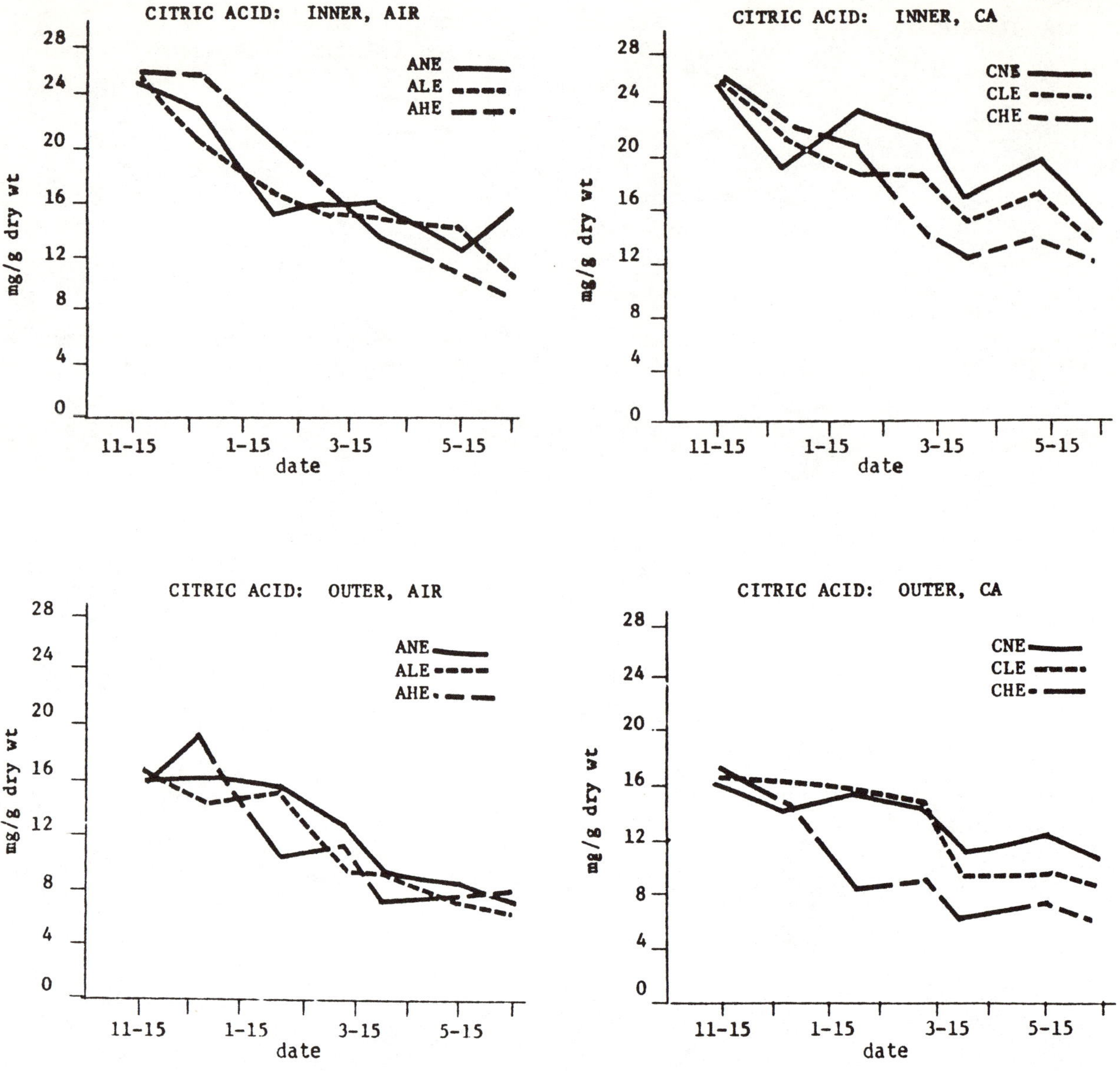

Fig. 5. The effect of storage time, atmosphere, ethylene and position on citric acid levels in cabbage. Inner samples were from leaves just above the apex. Outer samples were from outermost leaves on the head (see legend Fig. 1). Numbers are the average of 3 reps.

Citric acid was the only parameter examined to show an interaction involving leaf position in the head as well as atmosphere and time. Citrate was found at higher levels in inner tissue than outer (Figure 5). Without C_2H_4, the rate of citrate loss was more rapid in air than in CA. Ethylene appeared to have a more pronounced effect on citric acid in cabbage stored in CA than those stored in air. High C_2H_4 appeared to enhance citrate concentration in the inner portion of the head during early storage, and while this effect lasted only a month in CA it persisted until March in the air sample.

Phosphoric acid was the only metabolite measured that was present in greater amounts in the outer tissue than the inner (Figure 6). After an initial drop during the first month of storage there was a general increase in the level of phosphoric acid which was not induced by either atmosphere or ethylene.

The cabbage stored in air showed visual effects to exogenous ethylene at both concentrations. The 5 ppm treatment was considerably worse in terms of loss of green color and the presence of decay. Cabbage stored in CA showed very little, if any, visual response to the added ethylene. In addition, as in previous reports (4), CA delayed both the loss of green color and the development of decay.

The lack of response to CA in retention of sugars is somewhat confusing. It has been noted on numerous occasions that cabbage from CA storage has a sweeter taste than that from conventional storage, particularly later in the storage season. The increase in sucrose during CA storage may have had a more marked influence on taste than total sugar content. It is also surprising that after approximately 7 months storage there was no noticeable loss in sugar, particularly in the air + ethylene treatments. It is possible that starch was being broken down in order to maintain a constant sugar level. In any case, both the sugar and acid contents in this report are based on dry weight which did change during storage and seemed to be somewhat responsive to both atmosphere and ethylene.

In contrast to results reported with the variety Hidena, in the air treatment there was very little effect of ethylene on the citrate content although under CA storage ethylene had a very pronounced effect. In fact the high rate of ethylene in CA resulted in citrate levels similar to those found in air with or without ethylene. Maleate content was influenced by ethylene regardless of storage atmosphere.

The pattern on phosphoric acid content and the fact that it was higher in the outer than the inner leaves seem to indicate that it might be a breakdown product and related to senesence of the tissue. However, since neither CA or ethylene were effective in altering the levels this is doubtful.

The Bartola variety used in this study is a very good storage variety which may account for some of the differences between this report and an earlier one (5). In addition the storage atmospheres were imposed soon after harvest which may also be important. The fact that there was no visual effect of ethylene under CA conditions is contrary to observations made on the use of ethylene scrubbers in commercial storage rooms. However, it has also been noted that a flow-through experimental CA system where there is constant atmosphere change results in better quality (appearance) than a static system. It is quite possible that both the purafil scrubbers being used commercially and the experimental flow-through systems are removing volatiles other than ethylene that are detrimental to cabbage. On the other hand, both the sugar and acid determinations show that ethylene is influencing the metabolism of cabbage stored under CA conditions.

References

1. Bohling, H. and H. Hansen. 1977. Storage of white cabbage (Brassica Oleracea var. Capitata) in controlled atmospheres. Acta Horticulturae 62: 49-54.

2. Fernandez-Florez, E., D. A. Kline and A. R. Johnson. 1970. GLC determination of organic acids in fruits as their trimethylsilyl derivatives. J. AOAC 53: 17-20.

3. Furry, R. B., J. R. Hicks, M. C. Jorgensen and P. M. Ludford. 1979. Effects of ethylene on controlled atmosphere storage of cabbage. ASAE Paper no. 79-4537, Dec. 1979.

4. Geeson, J. D. and K. M. Browne. 1980. Controlled atmosphere storage of winter white cabbage. Ann. Appl. Biol. 95: 267-272.

5. Hicks, J. R. and P. M. Ludford. 1981. Effects of low ethylene levels on storage of cabbage. Acta Horticulturae Vol. No. 116 in press.

6. Pendergrass, A., F. M. R. Isenberg, L. L. Howell, and J. E. Carroll. 1975. Ethylene-induced changes in appearance and hormone content of Florida grown cabbage. Can. J. Plant Sci. 56: 319-324.

7. Sweeley, C. C., R. Bentley, M. Makita, and W. W. Wells. 1963. Gas-liquid chromatography of trimethylsilyl derivatives of sugars and related substances. J. Am. Chem. Soc. 85: 2497-2507.

LOW OXYGEN, HYPOBARIC STORAGE AND ETHYLENE SCRUBBING

David R. Dilley[1], Peter L. Irwin[1] and M. Wayne McKee[2]

[1]Department of Horticulture, Michigan State University

East Lansing, Michigan and [2]Hyde Park, New York

Some earlier studies (3,8,11) indicated that ethylene did not stimulate ripening of apples during low temperature storage. Hypobaric storage which maintains a low O_2 and ultra-low ethylene environment prevents apples from ripening and they remain free of superficial scald (1,5). Now after more than a quarter of century of controversy among postharvest researchers, it has been resolved that removing ethylene from the storage atmosphere retards fruit ripening and diminishes the incidence of superficial scald (14,18). The conflicting results among researchers can largely be attributed to differences in maturity of the fruit at harvest, ethylene levels in the fruit and storage atmosphere, concentration of oxygen and carbon dioxide in the storage atmosphere and, to some extent, on differences among cultivars.

The concept of employing a low oxygen and low ethylene atmosphere to preserve apples and pears during storage is not new. It has been known since the work of Kidd and West in England in the 1930's that limiting the oxygen supply to the fruits retarded ripening and that ethylene is responsible for fruit ripening. Hansen (12) demonstrated the requirement of oxygen for ethylene production by pear fruits. However, it was not until the discovery of hypobaric storage by Burg and Burg (4), which allowed fruit to be stored under hyponormal ethylene levels, that it was proven experimentally that ethylene was obligatory for ripening. Subsequently, many studies have borne this out (1,5,19).

Commercial development of hypobaric storage for fruits has been slow due to high costs and technical problems. Alternative means to remove ethylene from the storage atmosphere have given mixed results (9,16). It is important to recognize that while it may be possible to remove most of the ethylene from around the fruit it is difficult to remove ethylene from inside the fruit except by hypobaric ventilation. And, it is the ethylene inside rather than outside the fruit that is involved in ripening metabolism. Thus, it is necessary to restrict ethylene production by the fruit and inhibit the action of ethylene.

The role of ethylene in fruit ripening and the practical considerations of maturity, ripening and storage are depicted in Fig. 1 (7). Professor Biale (2) originated this relationship and over the ensuing years many postharvest researchers have contributed to extending and validating this important relationship for specific fruits and situations.

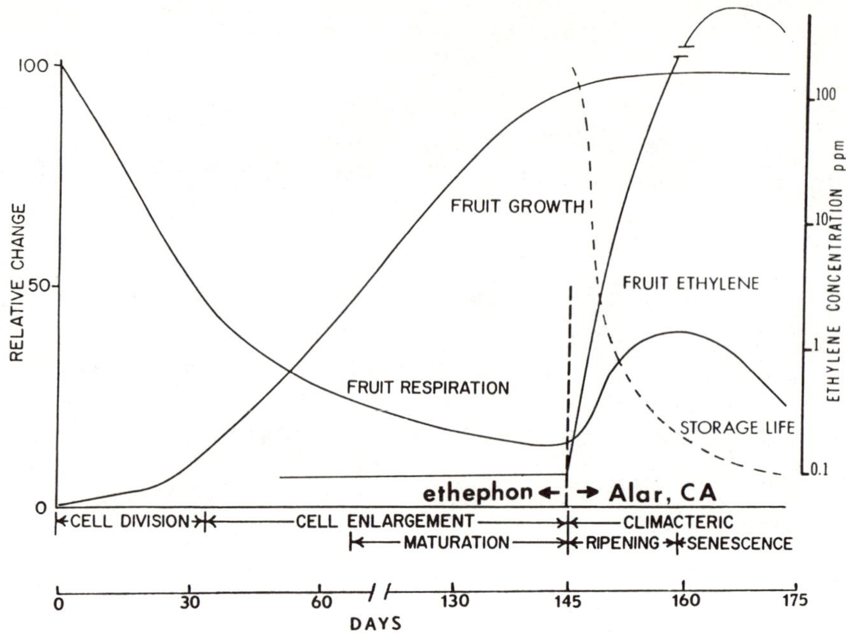

Fig. 1. Growth and development of apple and pear fruits in relation to the effects of ethylene, ethephon, alar and controlled atmosphere storage on ripening and storage life.

Our studies at MSU (10) established that ethylene initiates a fundamental redirection of metabolism, resulting in the synthesis of new enzymes. These catalyze the manifold biochemical processes of ripening, such as: cell wall softening, loss of chlorophyll, decrease in acidity, and development of flavor and aroma. These ripening processes are brought about through qualitative and quantitative changes in metabolism and will not occur without ethylene action. New biochemical pathways are initiated, and the metabolic rate, exemplified by the respiratory climacteric, increases dramatically as shown in Fig. 1.

This describes important features of apple and pear fruit growth and development that are essential to understand the processes of maturation and ripening and the role of growth regulators and storage technology. The time scale roughly approximates the situation for Red Delicious and is used for comparative purpose only.

The nature of the fruit growth is important in relation to development and ripening. The 30 to 80 million cells that comprise an apple or pear fruit are laid down during cell division during the first month after bloom. These cells subsequently enlarge and continue to expand as long as the fruits remain on the tree. The growth rate during the last few weeks is about 1% per day, and this accompanied by a significant increase in the intercellular air space between the cells. This, in large measure, is responsible for the gradual decrease observed in flesh firmness before harvest. It is important to recognize that this decrease in flesh firmness is a result of growth rather than ripening, and imparts to the fruits the desired textural characteristics of crispness, as opposed to the tough and wood-like texture of immature fruits.

There are several other features of Fig. 1 that have important bearing on harvesting and storage. Note particularly the precipitous and logarithmic increase in ethylene which marks the beginning of the ripening process. During the last few weeks of the maturation period the ethylene level is very constant and mostly below 0.1 ppm in the fruit. Then it begins to accumulate in the intercellular air space and the seed cavity. When the concentration rises much above 0.1 ppm it stimulates further ethylene production in an autocatalytic manner responsible for the log-linear increase. When this occurs, the course of ethylene production can not be reversed. And to slow down ethylene production and the events it initiates in ripening requires prompt establishment of controlled atmosphere conditions. Storage life diminishes rapidly as fruits gain capacity to produce ethylene. The ability of CA storage to retard ripening diminishes as the ripening process is initiated.

Carbon dioxide in the CA atmosphere is a natural inhibitor of ethylene action but it becomes ineffective when ethylene accumulates to about 1 ppm in the fruit. Oxygen is required for fruits to make ethylene, and low temperature slows down the rate of the chemical reactions of metabolism. These considerations serve as the basis for CA storage. They explain why CA storage is most successful, in terms of quality retention, when this technique is applied to fruits before they begin to produce much ethylene. CA storage can thereby delay the "time-line" of ripening, as indicated by the arrow in Fig. 1.

Hypobaric storage provides a good experimental system to determine the effect of O_2 concentration on storage behavior of fruits. This is because the O_2 partial pressure varys directly with the absolute pressure of air. Results of Michigan studies indicate that preclimacteric apples of most cultivars can be stored at 0^oC at 25 mm-Hg (equivalent to 0.55% O_2) very satisfactorily (Table 1). An exception was the 'McIntosh' cultivar which suffered from an apparent low O_2/low temperature disorder similar in symptoms to that observed when fruits are stored in CA at 0^oC. This disorder can be attenuated by warming McIntosh to 20^oC in air for 2 days after 2 months of hypobaric storage at 25 or 50 mm-Hg at 0^oC without affecting ripening.

In more than 12 years of hypobaric storage testing of apples in Michigan at O_2 concentrations as low as 0.5% the only low O_2 damage that has been encountered was with overmature 'Red Rome' apples. In this experiment 'IdaRed' and 'Red Rome' fruits were harvested late at median ethylene levels of 53 and 41 ppm, respectively. After cooling to 0^oC the fruits were placed in hypobaric chambers and ventilated with air at 760, 160, 80, 40 and 20 mm-Hg. After 6 months, the fruits were returned to air at 0^oC at atmospheric pressure for 1 month and then evaluated.

Table 1. Influence of very low pressure storage at 0^oC on behavior of selected apple cultivars.

Cultivar	Conditions at harvest: harvest date (1977)	flesh firmness lbs.	internal ethylene ppm	Flesh firmness (lbs - 7/16") after 8 months storage at 0^oC: 760 mm-Hg 0	7	14	25 mm-Hg* 0	7	14 days at 20^oC
McIntosh	9/13	15.2	0.05	7.8	7.6	4.2	14.9	14.8	12.8
Empire	9/20	16.6	0.08	8.4	5.3		15.8	15.8	
Jonathan	9/20	15.2	2.6	8.3	10.0	8.8	14.6	13.7	11.2
IdaRed	9/23	17.8	0.45	10.7	9.5	11.5	13.4	13.9	13.6
Red Delicious	9/23	16.7	1.3	11.0	11.5	10.3	15.4	16.1	14.5
Golden Delicious	9/23	17.7	0.04	8.8	7.2	6.6	14.6	13.6	13.8
Red Rome	9/23	21.5	0.05	11.8	11.8	7.2	18.0	17.0	15.6

*Storage at 25 mm-Hg in air is equivalent to about 0.55% O_2.

As seen in Table 2 only fruits stored at 40 mm-Hg did not soften appreciably. Only slight retention of firmness was obtained at 80 mm-Hg and the O_2 level at this pressure is equal to about 2%. Fruits at 20 mm-Hg had a fermented flavor and the 'Red Romes' had external and internal low O_2 damage. This data indicates that 40 mm-Hg is the lower limit for safe hypobaric pressure of ripe apples. This is equal to about 1% O_2. The interesting aspect is that hypobaric ventilation retarded softening of these very late harvested over-mature fruits. This indicates that even though the fruits had gained essentially full capacity to produce ethylene that low O_2 storage coupled with removing ethylene was benefical. Moreover, it may mean that ethylene may play a role in fruit ripening beyond that of initiating the synthesis of the ripening enzymes.

Table 2. Effect of hypobaric storage pressure on ripening of post-climacteric[1] 'IdaRed' and 'Red Rome' apples during 6 months at 0^oC.

Storage pressure (mm-Hg)	IdaRed firmness (lbs)	IdaRed ethylene (ppm)	Red Rome firmness (lbs)	Red Rome ethylene (ppm)
760	10.2	114	12.1	163
160	10.7	84	12.6	143
80	11.2	99	13.8	134
40	13.6	72	16.5	169
20[2]	11.8	99	14.6	141

[1]Flesh firmness at harvest on Oct. 30, 1979 was 15.3 and 18.6 lbs for 'IdaRed' 'Red Rome', respectively; internal ethylene was 53 and 41 ppm, respectively.

[2]Both cultivars had fermented flavor and Red Rome had internal and external low O_2 injury.

Hypobaric storage investigations at MSU with apples have led to some interesting observations. Fruits harvested and stored before entering the ethylene climacteric can be kept in the preclimacteric stage without any ripening changes (5,6). Moreover, storage under hypobaric conditions at .05 to 0.1 atmosphere was found to uncouple ethylene synthesis from ethylene action. 'Empire' and 'IdaRed' apples were harvested at various maturity and ripening stages and were stored at 0^oC at equivalent O_2 tensions of about 8 mm-Hg at total pressures of 760 and 40 mm-Hg and in air at 760 mm-Hg. After 3 months of hypobaric storage at 40 mm-Hg, regardless of maturity at harvest, no significant softening was observed in fruits held for 8 days at 20^oC with or without exogenous ethylene treatment at 10 ppm for 24 hours. Fruits stored for 3 months at equal O_2 tension but at atmospheric pressure softened during the 8 day post-storage period at 20^oC. After 6 months of hypobaric or low O_2 storage only the fruits which had begun to produce much ethylene when harvested softened appreciably. Whereas preclimacteric fruits from hypobaric storage gain normal ability to produce ethylene upon removal from storage (6) they are unresponsive to endogenous or exogenous ethylene in terms of developing ripening characteristics. Holding such fruits in air at 0^oC for several weeks after removing them for hypobaric storage 'conditions' the fruits for normal ripening upon transferring them to 20^oC.

The ethylene synthesis/action uncoupling phenomenon of hypobaric storage was investigated to determine if the effect was due to hypobaric conditions or to low oxygen. Experiments were conducted with 'Empire' apples harvested before the ethylene climacteric. Ethylene production by fruits from hypobaric and low O_2 CA for various durations is shown in Table 3. Hypobarically stored fruits gain ethylene production capacity more slowly than fruits from low O_2 CA. And, fruits from low O_2 CA were similar when N_2 or He was used in the CA atmosphere indicating that low O_2 rather than diffusivity *per se* was the primary determinant. When the fruits from hypobaric and low O_2 CA were 'conditioned' in air at 0^oC for several weeks, normal ripening occurred.

Studies in 1980 revealed that the lack of ethylene responsiveness of hypobaric and 1% O_2 CA stored apples was associated with enzyme synthesis. Malic enzyme was used as a marker of protein synthesis. The increase in water soluble polyuronides was used as a biochemical parameter of ripening because it is positively correlated with cell wall softening. After 5 months of hypobaric storage, the malic enzyme activity was only about 1/5 that of the air stored fruit which increased during the storage period (Table 4).

The malic enzyme activity of the hypobaric stored fruit was also reduced to less than half the initial level (Table 4). The fruits stored in 1% O_2 CA in He or N_2 also had reduced malic enzyme levels but not so low as the hypobarically stored fruits. Upon transferring the fruits from hypobaric or 1% O_2 CA to air at 20^oC the malic enzyme activity increased but less for hypobaric than for the 1% O_2 CA fruits. The change in water soluble polyuronides shows a similar response in hypobaric and 1% O_2 CA and low levels are associated with retention of flesh firmness.

of inclusions an effect of ethylene alone, ethylene in conjunction with low O_2 or a combination of all three components? In regard to the first question raised, it has been found that concentrations as low as 50 ppb ethylene will stimulate the appearance of these white core inclusions as well as causing some flesh softening. It was observed that the severity and incidence of the white inclusions was time-ethylene concentration dependent. Figures 2 and 3 illustrate this point. These injury-type symptoms appeared after four weeks of storage in all ethylene concentrations tested (50-1000 ppb). The severity and overall occurrence was greater in the treatments with the higher ethylene concentrations. The ethylene concentration effect was also seen when evaluating flesh firmness. After 8-10 weeks in storage there was a rapid decrease in flesh firmness in the 1000 and 500 ppb ethylene treatments. In the 100 and 50 ppb ethylene treatments there was a loss in firmness as compared to the CA control but it was not nearly as dramatic as in the other two treatments previously mentioned.

In an effort to obtain an answer to the second question (what interaction results in core inclusions), eight storage treatments were evaluated: air $\pm$ 1 ppm C_2H_4, 2% O_2 $\pm$ 1 ppm C_2H_4, 5% CO_2 $\pm$ 1 ppm C_2H_4 and CA (2% O_2 + 5% CO_2) $\pm$ 1 ppm C_2H_4. In all cases, when ethylene was present, fruit softening was enhanced; however, injury symptoms (white core inclusions) were only observed when high CO_2 and C_2H_4 were combined (the 5% CO_2 plus 1 ppm ethylene and CA plus ethylene treatments). It appears from these results that the starch inclusions seen during storage are a result of biochemical interactions between carbon dioxide and ethylene. The severity of these injury symptoms was accentuated by the presence of low oxygen even though in the presence of low oxygen and ethylene no injury symptoms were observed. The basis of this interaction between ethylene and carbon dioxide remains unknown at this time.

In conclusion, it has been found that the presence of ethylene under CA conditions in concentrations as low as 50 ppb can negate the positive benefits of CA storage by causing enhanced flesh softening and altering the internal appearance of the fruit. The basis of this alteration seems to involve an interaction between carbon dioxide and ethylene which either affects the degradation and/or synthesis of starch. Further studies are continuing to elucidate the basis of this interaction.

References Cited

1. Arpaia, M. L. 1980. The effect of ethylene and modified atmospheres on the storage performance of kiwifruits (*Actinidia chinensis*, Planchon). M.S. Thesis, University of California, Davis, 50 pp.

2. Arpaia, M. L., F. G. Mitchell and G. Mayer. 1980. The effect of ethylene and high carbon dioxide on the storage of kiwifruit. HortScience 15:423 (abstract).

3. Forsyth, F. R. and C. A. Eaves. 1975. Ripening of apples in CA storage, low or high ethylene levels and medium or high humidity levels. *In*: Colloques Internationaux CNRS #238. Facteurs et Régulations de la Maturation des Fruits. p. 67.

4. Liu, F. W. 1977. Varietal and maturity differences of apples in response to ethylene in CA storage. J. Amer. Soc. Hort. Sci. 102: 93-95.

Conditioning the fruit in air at $0^{o}C$ following hypobaric or 1% O_2 CA restores the fruits ability to synthesize malic enzyme, increase in water soluble polyuronides and to ripen (Table 5).

Alcohol dehydrogenase activity was found to increase more during hypobaric and 1% O_2 CA storage than during storage in air. This may indicate substrate (ethanol) induced en_yme synthesis due to storage at the low O_2 levels.

Table 5. Effect of post-storage conditioning of 'Empire' apples in air at 0° on ripening at $20^{o}C$ following hypobaric and 1% O_2 CA storage.

Storage treatment[1]	Days at $20^{o}C$	Flesh firmness (lbs.) minus[1]	Flesh firmness (lbs.) plus[1]	Water soluble polyuronides ug/g minus	Water soluble polyuronides ug/g plus	Malic enzyme eu/g minus	Malic enzyme eu/g plus
Hypobaric	0	16.1	15.9	43	49	50	65
	12	15.2	13.6	76	95	59	94
1% O_2 CA	0	16.1	15.3	57	52	48	69
	12	13.7	12.5	84	90	90	93

[1] Fruits were stored in hypobaric or 1% O_2 CA storage at 0° for 8 months (minus) or for 7 months and then conditioned for 1 month in air at $0^{o}C$ (plus).

Several recent studies (13,17) and events caused re-evaluation of scrubbing ethylene from the CA storage atmosphere because some benefits were found. It has been recognized for many years that only a few ppm of ethylene in the atmosphere is sufficient to cause fruit ripening. Moreover, ethylene can accumulate to several hundred ppm in the CA atmosphere and there was no good means in 1979 to reduce the concentration much below 50 ppm in large storage rooms.

At the beginning of the 1979 fruit storage season a survey of the C_2H_4 levels in commercial CA storages in the Hudson Valley was conducted. It was observed that in rooms where an atmosphere generator such as Arcat or COB-1 had been used, the C_2H_4 levels were lower than levels prevailing in other CA storage rooms. In addition, one storage operator used a COB-1 CA generator without fuel combustion but operated by the "preheat mode" with electricity, whereby the room atmosphere was circulated over the hot catalyic oxidizer at about $600^{o}F$. This reduced the C_2H_4 level in the room atmosphere from 180 ppm to about 9 ppm as it passed over the catalyst at 15 cfm. This observation led to testing other atmosphere generators operated in the preheat mode for the removal of C_2H_4 from commercial CA storages.

Two Hudson Valley storage rooms were selected for experimentation; room 1, at Clintondale, N.Y. was of 16,640 cu. ft. capacity and contained 6,460 bu. of 'McIntosh'; room 2 capacity was 34,476 cu. ft. and contained 12,157 bu. of

'McIntosh' and was at Clermont, New York. A 20,000 bu. CA room containing 'Red Delicious' and 'Jonathan' was also tested at Belding, Michigan. The initial C_2H_4 levels were 360, and 440, and 150 ppm in the 3 rooms, respectively. The fruits contained in these storage rooms had passed through the ethylene climacteric as evidenced by these high ethylene levels. Thus, any reduction achieved in ethylene level at this stage of storage would probably not result in any improvement in retention of fruit firmness or quality. The experiment was conducted to test the scrubbing efficiency of commercial CA generators.

The C_2H_4 concentration in the room atmosphere was measured in each room by using a portable detector[1]. Oxygen and CO_2 levels were measured by Orsat analyzers. The atmosphere generator was brought to ca. 600°F at which time the CA storage room atmosphere was circulated over the hot catalyst. Oxygen and CO_2 levels were checked frequently to ascertain that this operation did not adversely effect these parameters. Ethylene was measured at the inlet and outlet of the generator. The temperature of the catalyst and the discharge temperature of the atmosphere were measured with a thermocouple thermometer. Measurements were taken up to 3 times each day and the generator operated continuously for the duration of experiment.

In one test an Arcat[2] CA generator lowered the C_2H_4 in the storage atmosphere from 360 ppm to 45 ppm in a 48 hr period. The concentration of C_2H_4 at the generator discharge at 48 hrs was lower than that of the storage room indicating that the machine was still effectively removing C_2H_4 from the storage atmosphere. This was also the case at the end of the trial when the generator was detached from the room, suggesting that it would be possible to lower the C_2H_4 level even further. The Arcat catalytic bed temperature varied between 590°F and 640°F and generally the higher the bed temperature the greater the rate of C_2H_4 removal.

In a second test a COB-1[3] unit reduced the C_2H_4 concentration from 450 ppm to 7.1 ppm in ca. 70 hrs. The ethylene level was still being lowered when test was terminated after 70 hrs. Catalytic bed temperature varied between 560°F and 590°F and again the higher bed temperatures tended to remove C_2H_4 better. The O_2 level increased from 4.2% to 4.9% between hour 30 and when the test was terminated after 70 hrs.

A CA room at Belding, Michigan was also scrubbed with a Arcat CA generator in a similar manner. This unit was operated for two scrubbing cycles. The ethylene level at the start of cycle 1 was 150 ppm and this was reduced to 40 in ca. 72 hrs. The scrubber was then turned off and the ethylene level rose to 145 ppm in ca. 7 days after which the scrubber was reactivated and this brought

---

[1]Model 12-110 'Snoopy' Ethylene Detector, Bio-Gas Detector Corp., Okemos, Michigan.

[2]Atlantic Research Corporation, Alexandria, Virginia.

[3]Food Storage Systems, Yakima, Washington.

ethylene level to ca. 35 ppm in about 72 hrs. The inlet and outlet connectors of the Arcat to the room, were not strategically located to ensure good flushing of the room atmosphere.

These and other tests in the laboratory at Michigan State University demonstrated the capability of catalytic oxidation of ethylene in low O_2 atmospheres. Practical considerations of storage volume, ethylene production rate, storage ethylene levels, atmosphere exchange rate, preheater and catalyst design, effluent gas cooling and energy utilization were evaluated in designing a high capacity ethylene scrubbing systems. Table 6 provides data essential for design considerations for ethylene scrubbing. For example to keep the CA storage atmosphere ethylene level below 0.4 ppm if the fruit were producing ethylene at a rate of 0.2 ul/kg-hr would require a scrubbing rate of 5.6 CFM/1000 bu of apples.

The data in Table 6 indicate that a scrubber of high volumetric flow rate and ethylene removal capacity is required. Moreover, the scrubbing system must be hermetically sealed to prevent oxygen gain in the storage room. In 1980 a prototype high-capacity ethylene scrubber was constructed at MSU and tested on a commercial CA storage. The system consisted of a large blower capable of several hundred CFM, an electrical heater, a monolithic catalyst similar to that employed for engine exhaust emission control, a heat exchanger to cool the ethylene-free atmosphere on return to the storage atmosphere and the necessary electrical controls. The scrubber was tested on a 20,000 bu CA storage of 'Jonathan' and 'Red Delicious' operated at 2% O_2 and ca 1% CO_2 at 0^oC.

Table 6. Ventilation rate as a function of ethylene production rate to maintain low ambient ethylene levels in fruit storages.

Ethylene production rate	Ventilation rate (ft^3 min^{-1} 1000 bu^{-1})* storage atmosphere ethylene (ul 1^{-1})					
(ul kg^{-1} hr^{-1}	.05	0.1	.2	.4	.8	1.6
0.05	11.2	5.6	2.8	1.4	.7	.4
.10	22.4	11.2	5.6	2.8	1.4	.7
.20	44.8	22.4	11.2	5.6	2.8	1.4
.40	89.6	44.8	22.4	11.2	5.6	2.8
.80	179.	89.6	44.8	22.4	11.2	5.6
160	358.	179.	89.6	44.8	22.4	11.2

*ft^3 min^{-1} = ___ul ethylene kg^{-1} hr^{-1} x 10^3bu x $\frac{1}{\text{storage ethylene ul } l^{-1}}$ x .01122175

ml ul^{-1} lbs bu^{-1} kg lb^{-1} ft^3 ml^{-1} hr min^{-1}.

Figure 2 shows that the scrubber rapidly removed ethylene from the atmosphere down to the low levels required to retard ripening and prevent scald. Intermittant operation of the scrubber through February kept the ethylene level below 1 ppm after which scrubbing was discontinued due to a heater failure. Ripening of fruits in the ethylene-scrubbed room was much less advanced than similar fruits in CA storage without ethylene removal.

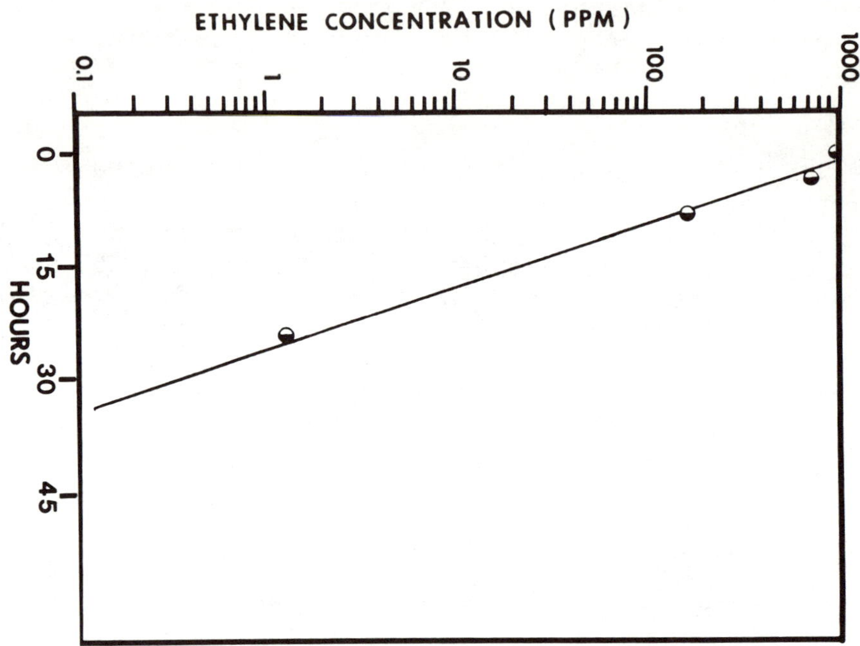

Fig. 2. The removal of ethylene from a CA storage room with a catalytic oxidizer operating at about 400°F. Flow rate was 130 CFM. Subsequent to this ethylene pull down the ethylene level was kept below 1 ppm. The room contained 20,000 bu. of Red Delicious and Jonathan. The high initial ethylene level was from incompletely oxidized propane from the pull down with a CA generator.

The favorable test results have encouraged the design and manufacture of a unit that will have capability to scrub both oxygen and ethylene from the CA storage atmosphere.

CA storage of 'McIntosh' in 1% O_2 CA at 3°C began on a commercial basis in Canada in 1980 based on studies of Lidster et al (15). Successful storage requires that the fruits be harvested just as they enter the ethylene climacteric. Fruits harvested after ripening has begun are subject to physiological breakdown. Perhaps scrubbing ethylene from the 1% O_2 CA may allow fruits of more advanced maturity to be stored successfully and achieve fruits of better eating quality.

In summary ethylene does promote ripening of apples during CA storage even at oxygen levels of 2%. Ripening at low oxygen CA can be markedly attenuated by removing ethylene from the atmosphere. The effectiveness of ethylene removal in retarding ripening diminishes as the fruits gain capacity to produce ethylene autocatalytically. Inhibition of ripening following long-term hypobaric or 1% O_2 CA storage of preclimacteric apples is apparently due to an uncoupling of ethylene synthesis from ethylene action related to a diminution of activity or amount of enzymes involved in ripening. Enzyme activity and ripening is restored by conditioning the fruits in cold air for several weeks following hypobaric or 1% O_2 CA storage. A high capacity ethylene scrubber based on catalytic oxidation is described for use on large volume CA storage rooms.

Literature Cited

1. Bangerth, F. 1975. The effect of ethylene on the physiology of ripening of apple fruits at hypobaric conditons. p. 183-188, In Facteurs et regulation de la maturation des fruits. Coll. Intern. du CNRS 238. Paris

2. Biale, J.B. 1960. The postharvest biochemistry of tropical and subtropical fruits. Advances in Food Research 10:293-354.

3. Blanpied, G.D. O. Cadun and T. Tamura. 1972. Ethylene in apple and pear experimental CA chambers. J. Amer. Soc. Hort. Sci. 97:207-209.

4. Burg, S.P. and E.A. Burg. 1966. Fruit storage at subatmospheric pressures. Science 153:314-315.

5. Dilley, D.R. 1972. Hypobaric storage - a new concept for preservation of perishables. 102nd Ann. Rpt. Mich. State Hort. Soc. 102:82-89.

6. Dilley, D.R. 1977. The hypobaric concept for controlled atmosphere storage. p. 29-37 In D.H. Dewey (ed.) Hort. Rpt. 28 Proc. 2nd Nat. CA Res. Conf. Mich. State Univ.

7. Dilley, D.R. 1981. Assessing fruit maturity and ripening and techniques to delay ripening in storage. 110th Ann Rpt Michigan State Hort. Soc. p. 132-146.

8. Fidler, J.C. 1956. The effect of ethylene on the rate of respiration of of apples as a function of temperature. Proc. 8th International Congress of Botany, Paris, 1954. Sect. 11-12, 392.

9. Forsyth, F.R., C.A. Eaves and H.J. Lightfoot. 1969. Storage quality of McIntosh apples as affected by removal of ethylene from the storage atmosphere. Can. J. Plant Sci. 49:567-572.

10. Frenkel, C., I. Klein and D.R. Dilley. 1968. Protein synthesis in relation to ripening of pome fruits. Pl. Physiol. 43:1146-1153.

11. Gerhardt, F. and H.W. Siegelman. 1955. Storage of pears and apples in the presence of ripened fruit. Agr. and Food Chem. 3:428-433.

12. Hansen, E. 1942. Quantitative study of ethylene production in relation to respiration of pears. Botan. Gaz. 103:543-558.

13. Knee, M. 1977. Ethylene removal during storage. E. Malling Res. Sta. Ann. Rpt 1977. p. 150.

14. Knee, M. and S.G.S. Hatfield. 1981. Benefits of ethylene removal during apple storage. Annuals Applied Biol. 98:157-165.

15. Lidster, P.D., K.B. McRae and Katherine A. Sanford. 1981. Response of 'McIntosh' apples to low oxygen storage. J. Amer. Soc. Hort. Sci. 106:159-162.

16. Lougheed, E.C., E.W. Franklin, S.R. Miller and J.T.A. Proctor 1973. Firmness of McIntosh apples as affected by Alar and ethylene removal from the storage atmosphere. Can. J. Plant Sci. 53:317-322.

17. Lui, F.W. 1977. Varietal and maturity differences of apples in response to ethylene in controlled atmosphere storage. J. Amer. Soc. Hort. Sci. 102:93-95.

18. Lui, F.W. 1979. Interaction of daminozide, harvesting date and ethylene in CA storage on 'McIntosh' apple quality. J. Amer. Sco. Hort. Sci. 104:599-601.

19. Streif, J. and F.B. Bangerth. 1976. The effect of different partial pressures of oxygen and ethylene on the ripening of tomato fruits. Scientia Hort. 5:227-237.

THE ETHYLENE PROBLEM IN MODIFIED ATMOSPHERE STORAGE OF KIWIFRUIT

Mary Lu Arpaia, F. Gordon Mitchell, Adel A. Kader and Gene Mayer
Department of Pomology
University of California
Davis, CA 95616

The kiwifruit, *Actinidia chinensis*, has recently become a horticultural commodity of international repute. Part of this interest stems from the potential six-month storage life of the fruit. Because of worldwide interest in kiwifruit marketing, it has been necessary to characterize the requirements for its optimum long-term storage. The major problem that hinders long-term storage is the flesh softening which occurs during 0 C storage. Within 8-10 weeks after harvest the fruit may soften to a point which severely limits its subsequent handling and marketing. Beyond proper temperature management (0 C), perhaps the most important factor which will affect the successful storage of kiwifruit is the presence of ethylene. It has been repeatedly shown that ethylene (as low as 30 ppb) can influence the rate of softening of the fruit at 0 C (6, 8). Subsequently, it was found that the use of controlled atmospheres could significantly reduce the fruit softening problem, so that optimum quality could be maintained throughout the storage period. Two percent oxygen plus 5% carbon dioxide appears to be the best storage atmosphere to use (7).

In relation to ethylene contamination of CA storage facilities, it has been shown by many researchers (3, 4, 5, 9) that the presence of ethylene seems to influence the quality of apples after storage. It seemed reasonable, therefore, based on these observations and the kiwifruits' sensitivity to ethylene in air storage, to investigate the possible ramifications of ethylene's presence under CA conditions. Consequently, it was found that the presence of 1 ppm ethylene during CA (2% O_2 + 5% CO_2) storage of kiwifruits also affected fruit quality (1, 2).

The effect of ethylene was two fold. First, ethylene caused fruit softening after 10-12 weeks storage which paralleled the softening seen during air storage. Secondly, ethylene's presence influenced the visual quality of the fruit (Figure 1). This second effect of ethylene involved the appearance of large white inclusions in the core of the fruit and smaller inclusions near the seeds after six weeks of storage. Microscopic examination of these inclusions showed that they contained large amounts of starch granules. At the time of harvest the fruit may contain approximately 5% of its dry weight as starch. Whether these core inclusions are newly synthesized starch or not remains unknown.

In conjunction with this study the effect of high CO_2 pretreatments were evaluated. Small white inclusions (also containing large starch granules) were observed near the seeds in the CO_2 pretreated fruit, but not in the core of the fruit. The incidence of these inclusions near the seeds was greater in fruits which were transferred to CA storage (2% O_2 + 5% CO_2) following the 20% CO_2 pretreatment.

Two questions which arose from these observations were: a) what is the lowest ethylene level permissible during CA storage and b) is the development

Table 3. Internal ethylene of Empire apples following storage for various durations in hypobaric and low oxygen storage.*

Storage treatment	Ethylene (ppm)				
2 Weeks	0	2	5	9	Days at 20°C
LPS 40	.39	.16	5.0	167	
CA-He	.12	.12	31.	196	
CA-N_2	.11	.15	12.8	190	
Control	25.	78.	212.	250	
2 Months	0	4	8		Days at 20°C
LPS 40	.30	114	251		
CA-He	.51	117	261		
CA-N_2	.71	90	272		
Control	17.	212	370		
5 Months	0	3	7	11	Days at 20°C
LPS 40	.22	.16	51	235	
CA-He	1.30	11.	177	265	
CA-N_2	1.26	13.	191	251	
Control	60.	230	356	519	

*Oxygen partial pressure equal to 8 mm Hg.

Table 4. Effect of hypobaric and low 0_2 storage duration on malic enzyme activity of 'Empire' apple fruits (eu/g fr. wt.).

Storage conditions			Storage duration						
P_{Total}	P_{O_2}	Inert gas	2 weeks		2 months		5 months		
(mm-Hg)			2	12	0	12	0	12	Days at 20° C
40	8	N_2	44	174	49	155	17	59	
760	8	He	47	162	66	174	32	94	
760	8	N_2	58	150	42	190	39	79	
760	152	N_2	102	145	59	110	130	112	

5. Liu, F. W. 1978. Effects of harvest date and ethylene concentration in controlled atmosphere storage on the quality of 'McIntosh' apples. J. Amer. Soc. Hort. Sci. 103:388-392.

6. McDonald, Barry and Derek Snowball. 1980. Cool storage of kiwifruit. Cool Storage Sub-committee of Kiwifruit Exporters Assn. Seminar, New Zealand, pp. 9-18.

7. Mitchell, F. G., M. L. Arpaia and G. Mayer. 1981. Modified atmosphere storage of kiwifruits (*Actinidia chinensis*). Proceedings of the National CA Research Conference, Oregon State University, Corvallis, pp. 235-238.

8. Reid, M. S. and S. Harris. 1977. Factors affecting the storage life of kiwifruit. The Orchardist of New Zealand 50(3):76-77, 79.

9. Stoll, K., F. Hauser and D. Datwyler. 1975. Maturation des Pommes en Chambre froide à Atmosphere Controlée sous L'infuence D'une Réduction du Taux de L'éthylène dans L'atmosphère. *In*: Colloques Internationaux CNRS #128. Facteurs et Régulations de la Maturation des Fruits, p. 81.

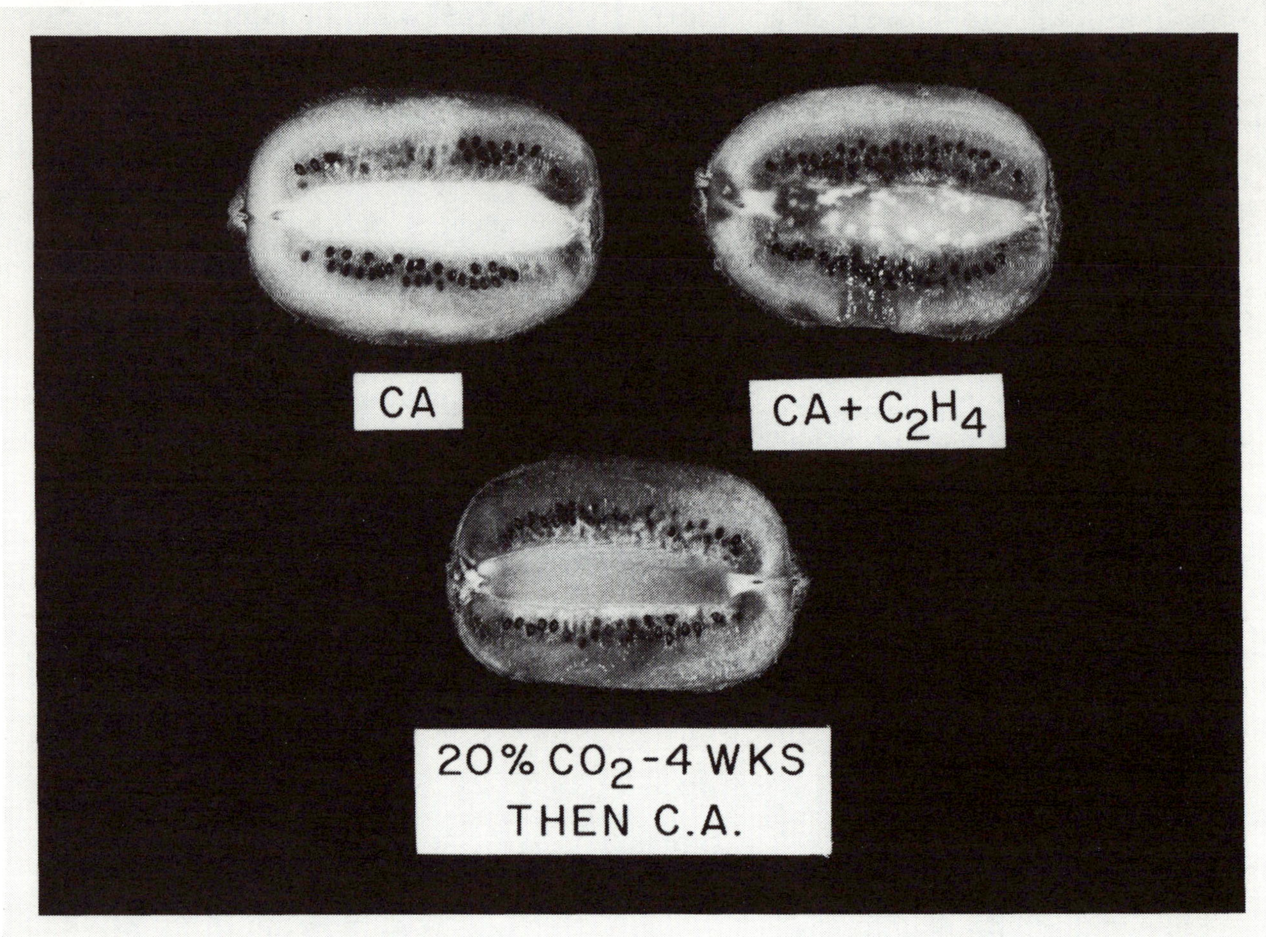

Fig. 1. The effect of CA (2% O_2, 5% CO_2) plus C_2H_4 and CO_2 (20%) pretreatment on the internal appearance of kiwifruit. Note the large white inclusions in the core of the ethylene treated fruit and the grainy appearance around the seeds in the ethylene and CO_2 pretreated fruit.

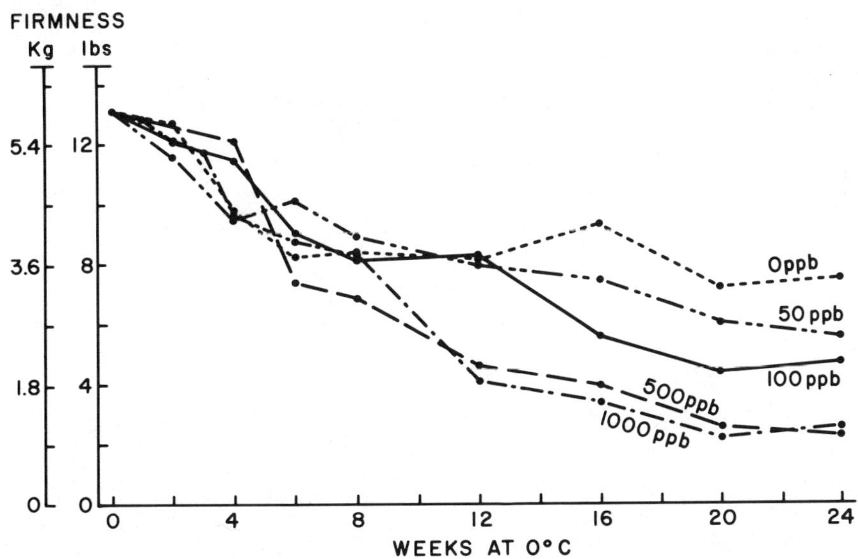

Fig. 2. The effect of ethylene on fruit softening during CA (2% O_2, 5% CO_2) storage of kiwifruit at varying concentrations: 0 ppb (----------), 50 ppb (—— -- ——), 100 ppb (————————), 500 ppb (—— —— ——) and 1000 ppb (—— - ——).

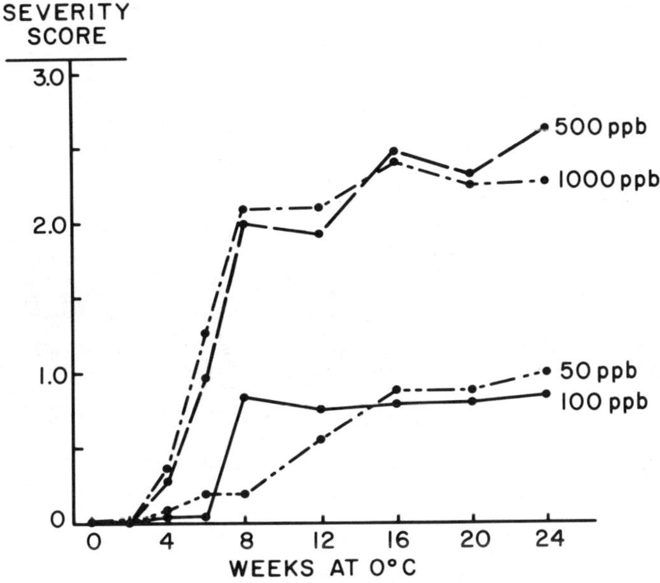

Fig. 3. Average score of severity of white core inclusion in kiwifruit treated with varying levels of ethylene: 50 ppb (—— -- ——), 100 ppb (————————), 500 ppb (—— ——) and 1000 ppb (—— - ——). A score of 0 = no inclusions, 1 = slight incidence of inclusions, 2 = moderate incidence of inclusions and 3 = severe incidence of inclusions.

Low Ethylene CA Storage for Apples

G. D. Blanpied, J. R. Turk, and J. B. Douglas
Pomology Department, Cornell University, Ithaca, NY

The internal ethylene concentration of apples held in a low ethylene environment is directly related to the rate of ethylene production by the apples (Fig. 1). On several occasions we have observed rather high concentrations of internal ethylene when McIntosh apples were held in low ethylene CA. Data for individual apples given in Table 1 are typical of these cases. To retard the rate of McIntosh flesh softening in CA, an internal ethylene concentration of 1 ppm or less must be maintained (Fig. 2). In low ethylene CA, the May internal ethylene concentrations for early and mid-season Empire apples were below 1 ppm and the apples were very firm; but internal ethylene of late picked Empires was 1.36 ppm and these apples were soft. When ethylene production in late picked Empires was suppressed with AVG, the apples were very firm (Table 2).

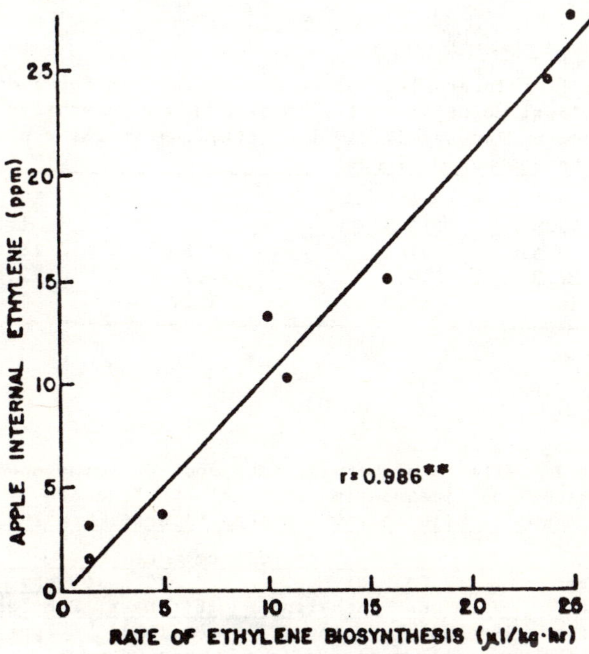

Figure 1. Relationship between internal ethylene and the rate of ethylene biosynthesis by 'McIntosh' apples held at 19°C.

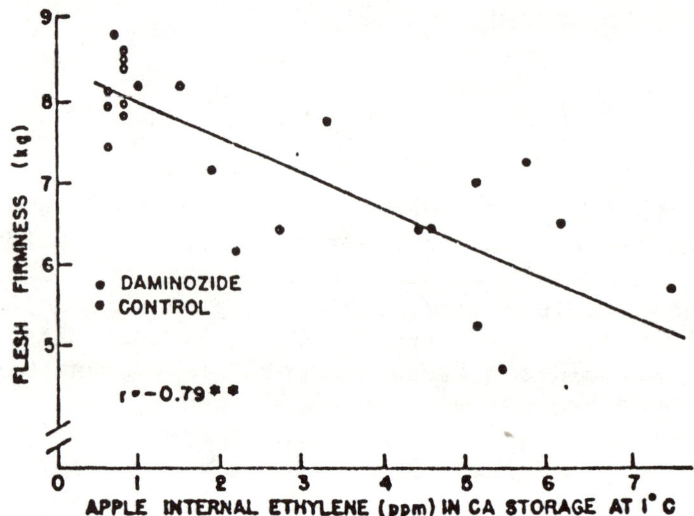

Figure 2. Relationship between flesh firmness and internal ethylene of 'McIntosh' apples held in CA storage until late February.

Table 1. Internal ethylene concentrations (ppm) of individual McIntosh apples removed from a low ethylene CA room on January 21, 1981. (Ethylene in the storage environment was 0.91 ppm).

4.1	4.1	26.1	13.1
12.8	16.3	25.6	1.5
3.5	7.6	27.6	16.6
27.3	4.1	24.7	21.8
26.2	19.2	1.2	

Table 2. Effects of harvest date and AVG (ethylene inhibitor) on firmness and internal ethylene of Empire apples held in low ethylene CA until May 1981.

Harvest date	Orchard treatment: Control Firmness (lbs)	Control Ethylene (ppm)	AVG - 150 ppm Firmness (lbs)	AVG - 150 ppm Ethylene (ppm)
Oct 2	15.3	0.05	15.7	0.13
Oct 8	15.0	0.06	15.3	0.02
Oct 15	10.9	1.36	14.5	0.02

The rate of ethylene production and concomitant internal ethylene may increase during the storage period in low ethylene CA (Fig. 4). If this occurs, the rate of ethylene production may exceed the capacity of the ethylene scrubber, as illustrated in Figure 5 for our ethylene scrubbed 1400 bushel capacity CA room. When this occurred, we could have added more ethylene scrubbing capacity to the CA room, but we rejected this option because internal ethylene was well above 1 ppm (see Table 1), the threshold to induce flesh softening, prior to the accelerated accumulation of ethylene in the CA room.

After steps have been taken to minimize the rate of ethylene production of the apples, it will be necessary to prevent the accumulation of ethylene above 1 ppm in the CA storage room. Without ethylene scrubbing, ethylene will accumulate in the CA room, diffuse into the apple flesh, and accelerate flesh softening. This occurred with AVG Empire apples stored in normal CA (Fig. 3). We have tested several types of ethylene scrubbers during the past two years, looking for a suitable system to prevent the accumulation of ethylene in the CA room.

Dr. F. W. Liu measured McIntosh ethylene production in low ethylene CA during several seasons. His data indicated the maximum rate of ethylene production occurred at the end of the storage period, May, and was approximately 2 µl per kilogram hour, which translated to 33 ml of ethylene per 1000 bushels per hour. An ethylene scrubber for our 1400 bushel CA room would therefore need the capacity to remove 0.77 ml or approximately 1 ml of ethylene each minute from an atmosphere containing less than 1 ppm ehtylene.

Figure 3 summarized the previous comments. Retardation of flesh softening was dependent upon low ethylene in CA and low rates of ethylene production by the apples. Apples were soft unless both of these conditions were met. We will now discuss methods of reducing the rate of ethylene production by apples. Scrubbing ethylene from the storage atmosphere will be discussed at the end of the paper.

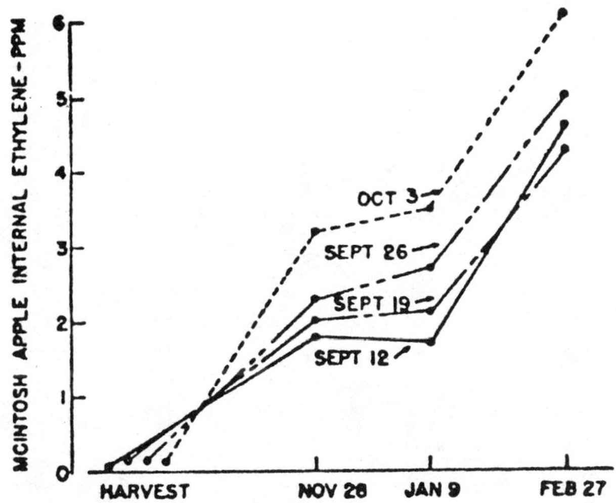

Figure 4. Internal ethylene of non-Alar McIntosh apples harvested on four dates and held in experimental low ethylene CA chambers.

		Ethylene in CA storage	
		Normal	Low
Apple Ethylene Production	Normal	9.8a	10.9b
	Low	9.8a	14.5c

Figure 3. Firmness (lbs) of Empire apples held in CA storage until May 1981.

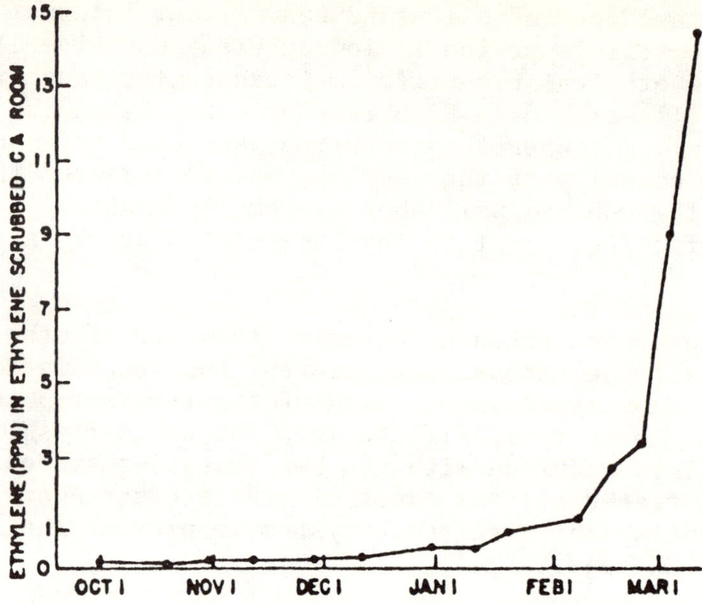

Figure 5. Concentration of ethylene in an ethylene scrubbed CA room filled with 1400 bushels of Alar McIntosh. 1980-81.

Table 3. Internal ethylene concentrations of individual 'McIntosh' apples held in two storage chambers at 1°C with 2.5% O_2 and 2.5% CO_2.

Orchard		Storage	C_2H_4 in chamber		Harvest date 1980			
Tree	Alar	chamber	atmosphere (ppm)		Sept 12	Sept 19	Sept 26	Oct 3
					(apple internal C_2H_4 - ppm)			
1	Yes	A	0.2	range [1/]	0.7-1.1	0.7-2.4	0.6-4.1	0.8-7.0
				average	0.8	1.0	1.5	2.7
2	No	B	0.9	range	1.0-7.3	1.2-9.4	1.2-8.8	2.8-9.3
				average	3.3	4.6	4.5	5.5

1/range for 10 apples from each harvest date.

Table 4. Effect of harvest date and ethylene inhibitor (AVG) on internal ethylene of Empire apples at harvest and on ethylene production at 70° after removal from low ethylene CA in February and internal ethylene during CA in May.

Orchard AVG (ppm)	Harvest date 1980 Oct 2	Oct 8	Oct 15
	(internal ethylene at harvest - ppm)		
0	0.08	0.21	0.31
300	0.09	0.20	0.22
	(Feb. ethylene production - µl/kg.hr)		
0	0.18a	0.35a	7.05b
300	0.09a	0.13a	0.13a
	(May internal ethylene during CA - ppm)		
0	0.05a	0.06a	1.36b
300	0.10a	0.02a	0.01a

The later the harvest date, the higher was the internal ethylene of McIntosh held in low ethylene CA. Alar, an ethylene biosynthesis inhibitor, reduced the internal ethylene (Table 3). Although the ethylene inhibitor AVG and harvest date did not significantly influence Empire internal ethylene at harvest, both influenced ethylene production in February and internal ethylene in May (Table 4). Late picked apples that had not been orchard sprayed with AVG had significantly more ethylene after harvest. Although Alar significantly reduced ethylene in air and in CA storage, the effect of the CA atmosphere was much more dramatic (Table 5). The concentration of internal ethylene of apples stored in low ethylene air was ten or more fold higher than internal ethylene of comparable apples stored in low ethylene CA. This observation may explain why ethylene scurbbing has not been effective for air stored apples. In summary, our studies showed ethylene production/internal ethylene concentrations were decreased by early harvest, orchard application of an ethylene inhibitor, and storage in a low oxygen environment. The response to the last of these variables was much greater than the response to the other two variables.

The relationship between scrubber efficiency and scrubber CFM required to maintain a concentration at 0.5 ppm ethylene in a CA room of apples producing 1 ml of ethylene per minute is shown in Table 6. The ethylene scrubbers we have tested ranged from ten to almost 100 percent efficiency. We used 10 ppm ethylene in some of our early tests, but changed to much low concentrations when we realized it was more difficult to remove 1 ml of ethylene from an atmosphere containing 1 ppm or less (Table 7).

Ozone has been used for decades to destory ethylene gas. Our tests showed ozone destruction of ethylene in air storage might be practical, but not under the low oxygen conditions of CA (Table 8). Several years ago we discovered that our catalytic oxygen burner (Arcat) would destroy ethylene if hydrogen gas was substituted for propand as the fuel. When we repeated the tests this spring, we found that metering the hydrogen to maintain a catalyst bed temperature of 600-650°F resulted in excellent ethylene scrubbing efficiency (Table 9), but the CFM of our unit was too small (20 CFM) to maintain the ethylene concentration below 1 ppm in our CA room.

Table 5. Internal ethylene and flesh firmness of McIntosh apples held in flowing atmospheres at 33°F until late February 1981.

Storage atmosphere	Atmosphere changes per hour	Orchard Alar	Average internal ethylene (ppm)[1/]	Flesh firmness (lbs)
Air	4	no	33.6b	9.4
Air	4	yes	30.6a	10.3
2½% O_2 + 2½% CO_2	1	no	3.3b	14.1
2½% O_2 + 2½% CO_2	1	yes	1.0a	17.7

[1/] Averages for 3 trees, 4 harvest dates, and 3 periods of storage.

Table 6. Effect of ethylene scrubber efficiency on calculated CFM required to sustain removal of 1000 µl ethylene/minute from a CA room atmosphere.

Atmospheric ethylene in scrubber		Scrubber efficiency	Scrubber CFM required to sustain removal of 1000 µl ethylene each minute
Intake (ppm)	Outlet (ppm)	(%)	(cfm)
0.500	0.000	100	71
0.500	0.125	75	94
0.500	0.250	50	141
0.500	0.375	25	282
0.500	0.450	10	706

Table 7. Ethylene removal from an empty 5460 cu. ft. CA room. Fifty pounds of Purafil held in 168 CFM scrubber. 1981.

Atmospheric ethylene (ppm)		Time	Average ul ethylene removed each minute
Start	Finish	(minutes)	per pound of Purafil
13.30	6.90	90	220
6.90	2.70	105	124
2.70	0.47	180	38

Table 8. Results of preliminary tests with ethylene scrubbing with ozone and Purafil.

Transformer setting for ozone generator	Atmosphere flow rate (liters/min)	Ethylene in scrubber discharge (ppm)	Ethylene removal (µl/min)
(10 liter ozone generator - air + 10.1 ppm ethylene)			
0	4.7	10.1	0
40	4.7	9.8	1.4
70	4.7	3.7	30.1
95	4.7	1.5	40.4
(10 liter ozone generator - 3.1% O_2 + 2.1 ppm ethylene)			
95	2	0.1	4.0
95	4	0.5	6.2
95	6	1.1	6.1
95	8	1.3	5.6
95	10	1.5	5.5
(Purafil tube, 31 inches X 1½ inches dia. - 3.1% O_2 + 2.1 ppm ethylene)			
—	4	0.0	8.2
--	7	0.0	14.1
—	10	0.0	20.5

Table 9. Destruction of atmospheric ethylene in a heated[1/] catalyst bed of Arcat oxygen burner. 1981.

Arcat catalyst bed temperature (°F)	Atmospheric ethylene into Arcat (ppm)	from Arcat (ppm)	removed by Arcat (ul/min)
500	0.470	0.190	160
600	0.360	0.018	195
800	0.390	0.018	213
900	0.390	0.014	215
650	0.680	0.012	382

1/Electric heaters used to raise catalyst bed to 500°. Heating of catalyst bed above 500° accomplished by metering hydrogen gas into the machine.

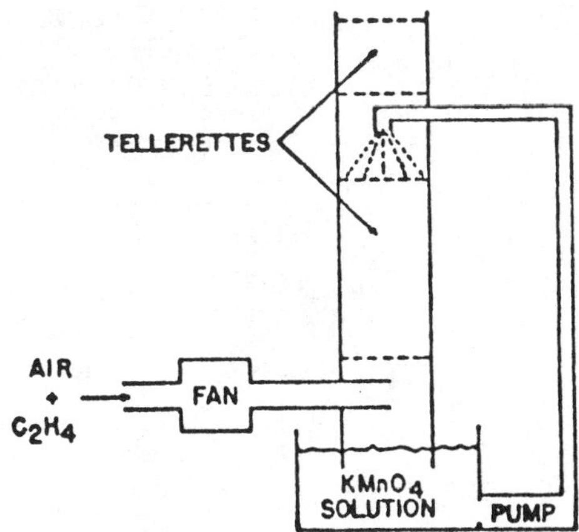

Figure 6. Counter-flow ethylene scrubber tower.

Several tests were run with a counter-flow scrubber tower (Fig. 6) in which a saturated solution of potassium permanganate ($KMnO_4$) was pumped (4 gallons/minute) through a 60° full-cone nozzle. The 10 inch diameter tower contained 6 feet of 1-inch tellerretes to maximize the droplet fomation needed for contact between the solution and the air. Air was blown up the tower at various flow rates. One foot of tellerettes above the nozzle minimized droplet carry-over. Results with this system were extremely erratic. Frequently ethylene removal from the air stream was greatest after the pump was stopped and the $KMnO_4$ dried on the tellerettes. The erratic results, the more efficient scrubbing by the dried $KMnO_4$, and problems associated with the $KMnO_4$ blown out the top of the tower precluded further testing of this system.

We also conducted several tests with potassium permanganate ($KMnO_4$) absorbants of ethylene gas. Although some of the ethylene absorbants we made in the lab were more effective than commercially available absorbants, the data for the lab-made absorbants are not included in this report because these absorbants are not commercially available. The results of some test with two commercially available absorbants, Oxi-Scrubb and Purafil, are presented in Tables 10 and 11. These tests were run subsequent to the observation that Purafil was much more effective than Oxi-Scrubb in removing ethylene from a CA room filled with apples.

To quantitatively measure the ethylene absorption rates by Oxi-Scrubb and Purafil, we used two types of scrubber cabinets in an empty CA room: a commercially made cabinet with 787 CFM, and a home-made plywood cabinet with 167 CFM. Both absorbants were tested in both cabinets, all tests were run twice. The flow rate into the empty CA room of a one percent ethylene-in-air mixture was adjusted until the ethylene in the CA room held steady for 12 hours at a concentration between 0.5 and 0.6 ppm. Under these steady conditions, ethylene was absorbed at the same rate that it was metered into the room. Ethylene absorption rates were similar in both cabinets (Table 10). Purafil absorbed ethylene three to four times faster than Oxi-Scrubb. The faster and greater total ethylene absorption by Purafil may be attributed, at least in part, to its greater amount of $KMnO_4$ per pound and the small pellet size (Table 11), which gave it more surface per pound.

Table 10. Rates of ethylene removal from an empty CA room (5460 cu. ft.) maintained at 0.5-0.6 ppm ethylene for 12 or more hours. Averages for duplicate runs. 1981.

Ethylene absorbant			Ethylene removal/minute	
Trade name	lbs.	Scrubbers[1/]	total µl	µl/lb. of absorbant
Oxi-Scrubb	75	A	250	3.33
Oxi-Scrubb	125	B	510	4.08
Purafil	50	A	645	12.90
Purafil	100	B	1265	12.65

1/A - homemade, plywood scrubber with 169 cfm
B - commercially made metal scrubber with 787 cfm

Table 11. Some characteristics of two ethylene absorbants.

Absorbant	Density (gms/100 cc)	$\%KMnO_4$	Total ethylene absorption (ml/lb.)	Size distribution (mm) of pellets >4.0	3.4-4.0	2.4-3.4	2.0-2.4	<2.0
				(percent by weight)				
Oxi-Scrubb	87.3	1.31	207	14.0	23.1	62.1	0.8	0.0
Purafil	73.7	3.55	646	0.0	0.2	60.7	24.1	15.0

Our ethylene scrubbing tests indicated a $KMnO_4$ ethylene absorbant may be the most practical system to scrub ethylene from the atmosphere in a CA room. Our calculations indicated a requirement of 0.2 pounds of Purafil per bushel per season. This quantity of material represents a substantial addition to the cost of CA storage. Although this additional cost is not prohibitive, we hope a less expensive ethylene scrubber will be developed. The benefits derived from a system combining procedures to reduce ethylene production by the apples and ethylene scrubbing show great promise for attaining the goal of marketing high quality apples very late in the storage season.

Effect of AVG on Storage Behavior of Apples

Kenneth L. Olsen
USDA-ARS, Tree Fruit Research Laboratory, Wenatchee, Washington

Summary

'Red Delicious' fruit were not affected by 500 ppm AVG applied 14 days prior to harvest and held in regular commercial storage. However, 'Golden Delicious' treated with AVG retained more firmness even though storage was with predominantly non-treated fruit. There was some indication of a slower starch hydrolysis in AVG treated 'Golden Delicious' apples.

Introduction

Increased apple production has stimulated research aimed at prolonging storage life by reducing the postharvest ripening rate of the fruit. Maturation and ripening involves many biochemical changes which appear to be under an integrated hormonal control. The preclimacteric minimum in the respiratory curve of apples is generally accepted as the point of physiological maturity. This is followed by an increase in the rate of respiration and an increase in ethylene production. Ripening is stimulated in apples harvested near the preclimacteric minimum and exposed to an exogenous supply of ethylene. Ethylene is a ripening hormone and is generally considered as the natural stimulator of the ripening process in climacteric fruit (8). The autocatalytic production of ethylene assures that once the production of this hormone is initiated, the fruit becomes dedicated to ripening.

Reports of rhizobitoxine and its analogs as inhibitors of ethylene production (11,7) have resulted in their experimental use on intact fruit. Bangerth (1) reported that a series of AVG (aminoethoxyvinylglycine) sprays of 500 ppm applied starting 39 days from harvest resulted in firmer 'Golden Delicious' at harvest and after 65 days at $2^{o}C$. 'King of the Pippin' apples were as firm at harvest as fruit sprayed 27 days before picking, but the treated fruit retained more firmness after 6 to 11 days at $20^{o}C$. Preharvest sprays of AVG on 'McIntosh', 'Spencer', and 'Spartan' delayed or prevented ripening at $25^{o}C$ after harvest. However, little difference was observed in treated and control fruit stored together at $0^{o}C$ (5). 'Red Delicious' were sprayed with 450 ppm AVG 12 days before harvest (13). All fruit had essentially the same firmness at harvest but the treated fruit was about 0.5 kg firmer than the controls after 3 and 6 months in storage. Treatment of 'Anjou' and 'Bartlett' pears with L-2-amino-4-(2-aminoethoxy)-trans-3-butenoic acid (AAR) delayed the softening of 'Anjou' but not 'Bartlett' during ripening (12).

The following tests were conducted to determine the residual effect of AVG on fruit held in storage in presence of normal physiological concentrations of ethylene. 'Golder Delicious' apples were used from three and 'Red Delicious' from two commercial orchards.

Method

AVG was applied on 'Golden Delicious' at 250 and 500 ppm in one orchard with three replications and at 500 ppm in two others with 6 replications. 'Red Delicious' were treated also in the second two orchards. The first harvest was made 14 days after application of the AVG followed by a second harvest 1 week later.

The fruit was brought to the laboratory and three matched samples were prepared from each replication. One sample was analysed immediately and the others were stored in trays. The 'Golden Delicious' were placed in polyethylene liners and the 'Red Delicious' were stored without liners for examination in January and May. The fruit examined in January was held in regular storage and the fruit examined in May was held in a commercial CA storage.

Examinations included circumference and weight for size. Relative chlorophyll content was determined by light transmittance (O.D. 740-690) with a Difference Meter (4). The external color was compared using an Agtron E-5 which spins the apple and gives a reading in a green/red or red/green ratio. A visual color evaluation was also made to the nearest 5% of surface area covered with good red color for 'Red Delicious' or on a scale (1-7) of dark green to golden color for 'Golden Delicious'. Destructive tests included firmness as measured by a Magness-Taylor pressure tester on the individual fruit. A composite sample of juice was then prepared from 2 longitudinal pie shaped sections from each apple using a home vegetable juicer. Soluble solids of the juice was determined on a hand refractometer. An aliquot of juice was neutralized with standard 0.1 N base to determine the total titratable acidity. At harvest time each apple was cut in half and treated with an iodine solution for the starch test. Visual ratings ranged from 1 to 6, i.e. starch present throughout the cross section of the fruit to complete hydrolysis of starch. A rating of 2 indicated starch free just in the core area. After storage, scald ratings were made on the 'Red Delicious' at both examinations.

Results

Maturation of 'Red Delicious' was not affected by the AVG application made 14 days before harvest in 1979. No significant differences were found in either orchard in any of the measurements described.

In 'Golden Delicious' there was a significant retention of firmness during storage by fruit in two of the orchards with the AVG treatment (Table 1). In the third orchard, there was a similar trend but with an average firmness of 0.5 kg higher in the treated fruit, the differences were significant only at the 10% level. The initial firmness in this orchard was considerably higher and held better in storage than normal. This would tend to mask the effect of the treatment as observed in the other orchards.

The only other difference recorded in 'Golden Delicious' from the AVG application was a significant retention of starch in fruit from orchard No. 2 (Table 2). It had the earliest bloom date and seemed the most advanced fruit at harvest.

Discussion

Development of the fruit and changes in storage followed a fairly standard pattern during the test period. Most measures were significantly different on fruit from the two picking dates and again at the different examination periods out of storage. On 'Red Delicious' there was no apparent affect from the treatment alone or in interaction with either harvest or examination time. The normal softening pattern of 'Red Delicious' follows closer to a linear pattern than that of 'Golden Delicious' over the storage life of the fruit (2). Assuming that the AVG also entered the 'Red Delicious' fruit, it appears questionable that AVG had any other effect than limiting ethylene production. Softening of this variety exhibited no difference between fruit from the control and treated trees.

In 'Golden Delicious' softening procedes very rapidly over the first 3 to 4 months in delayed CA or regular cold storage (3). Very little softening occurs late in the storage period and the fruit may even become a little firmer at a later examination. This is due to moisture loss and the tissue actually becomes slightly tough. Under this softening pattern, a significant retention of firmness was found in treated fruit. This occurred in spite of the presence of over 50 ppm ethylene after the storages were closed in the fall.

The presence of exogenous ethylene was the same for both varieties. Since some difference was noted in the response of 'Golden Delicious' to the AVG treatment, a question is raised as to a possible secondary effect of AVG in this variety. An effect of AVG on protein synsthesis has been reported by Mattoo et al (9). The retention of firmness in the 'Golden Delicious' was not comparable to what can be achieved by high CO_2 treatment or rapid CA (10). However, storage of either variety after treatment with AVG in rooms devoid of ethylene might give entirely different results.

It has been suggested that an increase in respiration may precede the increase in ethylene production in climacteric fruit (6). Although this is generally not accepted, it is another indication of our incomplete understanding of the conversion in a fruit from the vegetative to a ripening phase. It appears that AVG and similar analogs of rhizobitoxine may serve as good experimental tools in helping to sort out the complexity of the chemical reactions in fruit undergoing maturation and ripening. Effects of timing of the application of AVG to fruit on and off the tree in relation to the preclimacteric minimum and the results from storage tests under different atmospheres should prove interesting with different cultivars of apples.

References

1. Bangerth, F. 1978. The effect of substituted amino-acid on ethylene biosynthesis, respiration, ripening, and preharvest drop of apple fruits. J. Amer. Soc. Hort. Sci. 103:401-404.

2. Bartram, Richard, Harold A. Schomer, Charles F. Pierson, and Kenneth L. Olsen. 1969. High quality 'Red Delicious' apples for late-season marketing. Wash. State Univ. Ext. Ser. EM 3033.

3. Bartram, Richard, Charles F. Pierson, Kenneth L. Olsen, and Ronald B. Tukey. 1971. 'Golden Delicious' apples for high quality and maximum storage life. Wash. State Univ. Ext. Ser. EM 3496.

4. Birth, G. S., and K. H. Norris. 1965. The Difference Meter for Measuring Interior Quality of Foods and Pigments in Biological Tissues. U.S.D.A. Agri. Res. Ser. Tech. Bull. No. 1341.

5. Bramlage, W. J., D. W. Greene, W. R. Autio, and J. M. McLaughlin. 1980. Effects of aminoethoxyvinylglycine on internal ethylene concentrations and storage of apples. J. Amer. Soc. Hort. Sci. 105:847-851.

6. Koseyachinda, Suraphong, and Roy E. Young. 1975. Ethylene production in relation to the initiation of respiratory climacteric in fruit. Plant & Cell Physiol. 16:595-602.

7. Lieberman, M., A. T. Kunishi, and L. D. Ownes. 1975. Specific inhibitors of ethylene production as retardants of the ripening process in fruits. In Facteurs et Regulation de la Maturation des Fruits, Colloques Int. C.N.R.S. No. 238, Paris 1974, pp. 161-170.

8. Lieberman, M. 1979. Biosynthesis and action of ethylene. Annu. Rev. Plant Physiol. 30:533-591.

9. Mattoo, Autar K., James D. Anderson, Edo Chalutz, and Morris Lieberman. 1979. Influence of anol ether amino acids, inhibitors of ethylene biosynthesis, on aminoacyl transfer RNA synthetases and protein synthesis. Plant Physiol. 64:289-292.

10. Olsen, Kenneth L. 1980. Rapid CA and low oxygen storage of apples - A new concept for long storage of Golden Delicious and more effective storage of Red Delicious apples. Washington State Horticultural Association Proceedings. 76:121-125.

11. Owens, L. D., M. Lieberman, and A. T. Kunishi. 1971. Inhibition of ethylene production by rhizobitoxine. Plant Physiol. 48:1-4.

12. Wang, Chen Yi, and Walt M. Mellenthin. 1977. Effect of aminoethoxy analog of rhizobitoxine on ripening of pears. Plant Physiol. 59:546-549.

13. Williams, Max W. 1980. Retention of fruit firmness and increase in vegetative growth and fruit set of apples with aminoethoxyvinylglycine. HortSci. 15:76-77.

USE OF AVG TO CONTROL RIPENING OF APPLES

William J. Bramlage and Wesley R. Autio
Department of Plant and Soil Sciences
University of Massachusetts, Amherst, MA 01003
(Presented by W.J. Bramlage)

Initial studies showed that aminoethoxyvinylglycine (AVG), an inhibitor of ethylene biosynthesis, delayed the ripening of apples and pears (1, 2, 5, 7). We thus initiated tests to determine how AVG use might alter current postharvest handling of apples.

In 1979-80 we found that preharvest sprays with 500 ppm AVG delayed or prevented postharvest ripening of several apple cultivars at room temperature, but that after 3 or 4 months of air storage at 0°C there was no discernible effect of AVG on fruit ripening (3). These results were extended in 1980-81, with 3 experiments designed to answer specific questions about use of AVG.

The results of 1979-80 suggested that the effects from a given AVG concentration increased when it was applied to late maturing cultivars. To test this observation, limb treatments of 0, 125, 250, 500, or 1000 ppm AVG were applied 1 week before harvest to Early McIntosh, McIntosh, Cortland, and Delicious trees, cultivars that ripen from early-August to mid-October.

Ripening was monitored by alternate day measurement of internal ethylene concentrations of harvested fruit kept in a well-ventilated laboratory at 70-75°F. The same fruit were used continuously unless rot developed. Fruit were judged to be ripe when internal ethylene concentration reached 1 ppm.

Results are summarized in Figure 1, in which the mean number of days for fruit to reach 1 ppm internal ethylene concentration after harvest is plotted against AVG concentration. Overall, there was a highly significant quadratic relationship between AVG concentrations and days required for fruit to ripen. However, there was also a highly significant interaction between the effects of AVG concentrations and the days from bloom to ripening (i.e., time to maturity) for a given cultivar. Substantial delay of Early McIntosh ripening required 500 to 1000 ppm AVG, while ripening of Cortland and especially Delicious were substantially delayed by 125 ppm AVG. Whereas 1000 ppm delayed ripening of Early McIntosh by about 10 days, it delayed ripening of McIntosh by about a month, while only 125 ppm delayed ripening of Cortland by about 10 days and ripening of Delicious by about 3 weeks. In all cultivars peak internal ethylene levels were reduced as AVG concentration increased, and the percent reduction by a given concentration was approximately the same for all cultivars, although early maturing cultivars produced much more ethylene than later maturing ones.

This test showed that optimum AVG concentration for delaying apple ripening varied with cultivar, and that in general greater concentrations were required and less response was obtained for an earlier maturing than for a later maturing cultivar.

In the 1979-80 test, results were about the same whether AVG was applied 3 or 18 days before harvest of McIntosh and Spartan (3). To test the importance of time of application and to see if early application might influence maturation processes as well as the onset of ripening, limb treatments with 500 ppm AVG were applied to trees of Puritan, an early maturing cultivar, on June 26, July 11, or July 28. Ten fruit were harvested from each limb on August 1, 4, 7 and 10 and were used to assess fruit firmness, soluble solids, peel chlorophyll, titratable acidicy, and internal ethylene. AVG treatments all delayed ripening and reduced ethylene peak height (Figure 2). However, the fruit from the earliest treatment ripened significantly faster than those from the 2 later treatments (Table 1). No application, even the one 6 weeks before harvest, had a significant effect on firmness, soluble solids, peel chlorophyll, or titratable acidity (Table 1). It therefore appears that although greater response was obtained from treatments within a couple of weeks of the onset of ripening, treatment could be made earlier without an evident effect on maturation processes of the fruit.

The third experiment was designed to test whether the loss of AVG effect during storage that was seen the previous year was due to storage ethylene levels, or to an effect of low temperature. McIntosh and Delicious tree limbs were sprayed with 500 ppm AVG 1 week before harvest. Only AVG-treated fruit were used, and these were stored in 3 different storage conditions: (a) 0°C, low ethylene; (b) 3.3°C, low ethylene; (c) 3.3°C high ethylene. (McIntosh is a cold-sensitive cultivar and Delicious is cold tolerant). "Low ethylene" was maintained well below 0.1 ppm by placing 150 g of Purafil in each box of fruit; these samples were the only fruit in a 200 bu. storage room. Relatively high ethylene was attained by placing additional ripe apples in a room with the samples, and also by occasionally injecting sufficient ethylene into the atmosphere to temporarily reach 10 ppm. Fruit were kept in storage for 23 weeks after harvest and then were assessed for internal ethylene, firmness, ground color (McIntosh only), and occurrences of disorders at room temperature. For McIntosh, there was no difference in internal ethylene between low ethylene and high ethylene storage, but internal ethylene was initially lower from the 0°C storage (Figure 3). In all cases, however, internal ethylene exceeded levels likely to be needed to trigger ripening, hence all samples had apparently initiated ripening during storage. Delicious fruit all had very low ethylene levels initially, but they gradually rose to about 0.5 ppm regardless of storage environment (data not shown).

After 1 and 3 weeks at room temperature, AVG-treated McIntosh stored in a low ethylene environment had less breakdown than ones stored in a higher-ethylene environment (Table 2). Furthermore, ones stored at 0°C were firmer, had a significantly greener ground color, and had less decay after 3 weeks at room temperature. Thus, exogenous ethylene had a somewhat deleterious effect on AVG-treated McIntosh, whereas lower temperature (0°C) helped conserve quality of these fruit. Fruit in all storages developed 60-70% browncore. This might account for the relatively small differences between fruit in low-and high-ethylene rooms: low-temperature stress may have produced sufficient endogenous ethylene to trigger ripening in the low-ethylene room.

For Delicious there was little effect of storage conditions on properties of the fruit after storage except that fruit from 3.3°C were softer than ones from 0°C storage. Apparently, Delicious did not initiate ripening in any of the storage conditions.

From these results, and those of others, the potential roles of AVG or another ethylene inhibitor in handling of apples can be envisioned. AVG certainly can delay ripening on the tree and have a stop-drop effect (6) and while the magnitude of delay has not been determined, for at least late-maturing cultivars it could be profound. After harvest and without refrigeration the delays for late-maturing cultivars were profound; the delay from high concentrations exceeded the limits of our tests in the absence of high concentrations of exogenous ethylene. Even for inherently fast-ripening early maturing cultivars, ripening of harvested fruit was slowed substantially. AVG therefore might be a very good marketing tool, producing slower ripening of freshly harvested apples as they pass through the marketing channels, and an important benefit might be in marketing of early maturing cultivars that normally ripen very fast. AVG might also reduce energy comsumption by reducing the need for refrigeration of apples destined for immediate marketing.

AVG would seem to have limited value for air-stored fruit, since an accumulation of ethylene in the storage atmosphere will override the AVG effect. However, if fruit are intended for only short storage in air, a significant effect might be obtained. Its usefulness in CA might be greater, assuming that pre-climacteric fruit were placed under CA conditions before ripening was initiated. In a CA atmosphere AVG-treated fruit will probably start to ripen more slowly, thereby slowing the buildup of ethylene in CA-storage.

The effects of AVG are similar to the effects of Alar (daminozide) in a number of ways (6). Its effects in conjunction with Alar would be interesting to test and it may be that AVG could contribute significantly to low-ethylene CA storage, especially in combination with Alar.

A potential concern is the apparent increase in brown core in AVG-treated fruit (3,4). Development of browncore is usually avoided by storage at 3.3°C, yet we obtained as much at this temperature as at 0°C (Table 2). If AVG increases cold-temperature sensitivity, serious problems could be encountered.

AVG is no longer available for use in significant quantities and its safety has not been established. Nevertheless, other ethylene-inhibiting compounds are becoming available for testing, and the results with AVG should provide models for testing these materials.

Literature Cited

1. Baker, J.E., M. Lieberman, and J.D. Anderson. 1978. Inhibition of ethylene production in fruit slices by a rhizobitoxine analog and free radicle scavengers. Plant Physiol. 61: 886-888.

2. Bangerth, F. 1978. The effects of substituted amino acids on ethylene biosynthesis, respiration, ripening, and preharvest drop of apple fruits. J. Amer. Soc. Hort. Sci. 103: 401-404.

3. Bramlage, W.J., D.W. Greene, W.R. Autio, and J.M. McLaughlin. 1980. Effects of aminoethyoxyvinylglycine on internal ethylene concentrations and storage of apples. J. Amer. Soc. Hort. Sci. 105: 847-851.

4. Chu, C.L., D.C. Elfving, and E.C. Lougheed. 1981. Aminoethoxyvinylglycine (AVG) effects on apple fruit storage. HortScience 16(3): 238.

5. Ness, P.J. and R.J. Romani. 1980. Effects of aminoethoxyvinlglycine and counter-effects of ethylene on ripening of Bartlett pear fruit. Plant Physiol. 65: 372-376.

6. Walsh, C.S. and T. Solomos. 1981. Effect of AVG and daminozide on ACC concentration, ethylene evolution, and ripening of apple fruits. HortScience 16(3): 239.

7. Wang, C.Y. and W.M. Mellenthin. 1977. Effect of aminoethoxy analog of rhizobitoxine on ripening of pears. Plant Physiol. 59: 548-549.

Table 1. Effects of 500 ppm AVG, applied on different dates, on the properties of Puritan apples after harvest. Data are means ot harvests on August 1, 4, 7, and 10.

Treatment date	Days to ripen	Peel chlorophyll (ug/g f.wt.)	Firmness (lbs)[z]	Soluble solids (%)	Titratable acidity (ml 0.1 N NaOH/5 ml)
June 26	3.9b[y]	93.7a	20.1a	12.4a	7.5a
July 11	6.4c	94.6a	19.9a	12.5a	7.4a
July 28	6.2c	90.0a	20.0a	11.6a	7.4a
Control	2.8a	92.2a	20.1a	11.8a	7.2a

[z]Firmness values were adjusted by covariance to account for different fruit sizes.

[y]Means in columns not followed by a common letter are significantly different at the 5% level.

Table 2. Effects of temperature and C_2H_4 concentration in storage on poststorage properties of apples that had been sprayed with 500 ppm AVG before harvest.

Cultivar	Storage conditions	Firmness (lbs. pressure)	% decay		% breakdown		% browncore	Ground color[z]
			1 wk.	3 wks.	1 wk.	3 wks.		
McIntosh	Prestorage	17.5a[y]	---	---	---	---	---	---
	Low C_2H_4[x], 0°	11.0b	2a	3a	1a	3a	63a	3.3a
	Low C_2H_4, 3.3°	10.3c	1a	9b	8b	15b	69a	2.6b
	High C_2H_4[w], 3.3°	9.7c	3a	9b	17c	25c	61a	2.3b
Delicious	Prestorage	20.2a	---	---	---	---	---	---
	Low C_2H_4[x], 0°	16.6b	1a	8a	1a	3a	---	---
	Low C_2H_4, 3.3°	14.8c	0a	9a	0a	4a	---	---
	High C_2H_4[w], 3.3°	13.8c	2a	9a	1a	5a	---	---

[z]Color scale: 1 = light yellow-green, to 5 = dark green, using Cornell color chart.

[y]Means in a column and cultivar not followed by a common letter are significantly different at the 5% level.

[x]Low C_2H_4 varied from 0.001 to 0.014 ppm atmospheric C_2H_4 during periodic measurements.

[w]C_2H_4 was injected into the atmosphere to attain 10 ppm concentration at intervals.

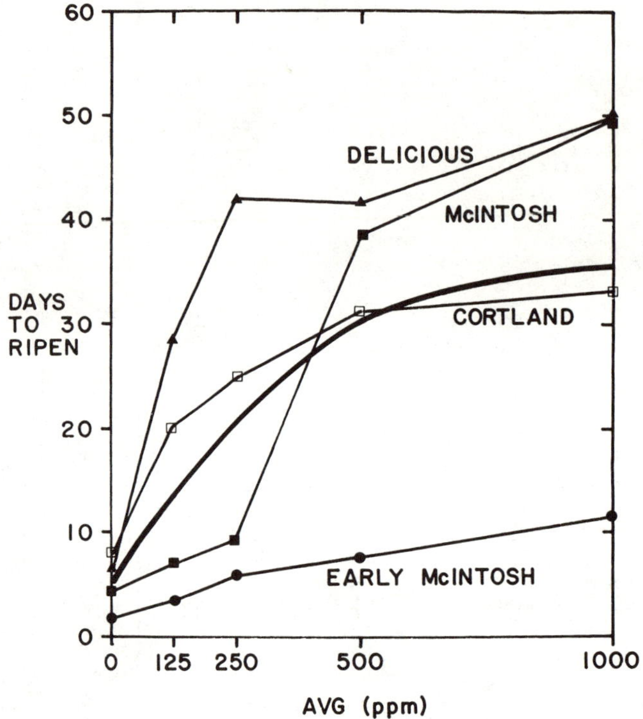

Figure 1. Days at room temperature required for fruit to reach 1 ppm internal C_2H_4 following AVG application 1 week before harvest. Each value is the mean within a treatment and cultivar for all the fruit that ultimately reached ripeness (1 ppm internal C_2H_4). The central line is the calculated quadratic relationship for all cultivars.

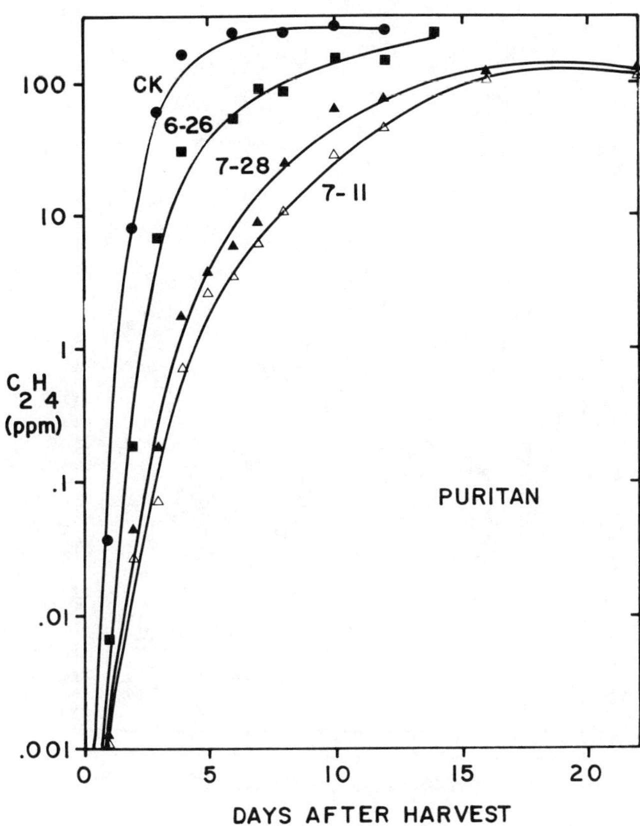

Figure 2. Internal C_2H_4 concentrations of Puritan apples, sprayed with 500 ppm AVG on indicated dates, at intervals after harvest. Values are means of harvests on August 1, 4, 7, and 10.

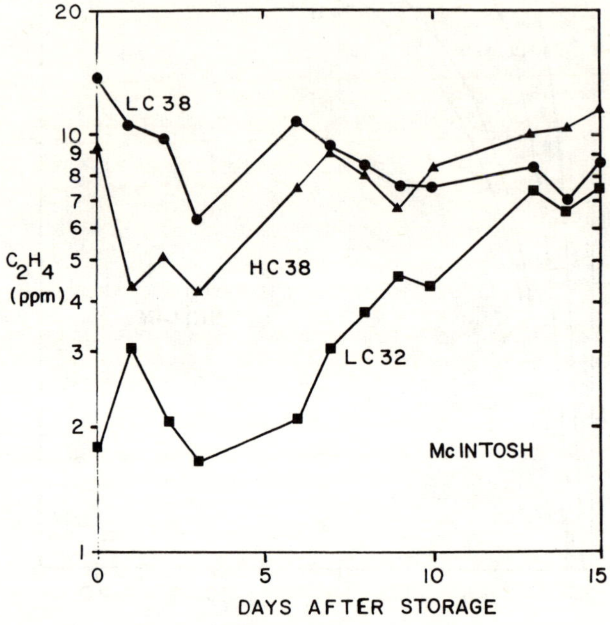

Figure 3. Internal C_2H_4 concentrations of McIntosh apples after storage for 23 weeks at relatively low (LC) or high (HC) C_2H_4 concentrations at 0°C (32) or 3.3°C (38).

CO_2 ATMOSPHERES FOR TRUCK SHIPMENTS OF STRAWBERRIES

John M. Harvey
U. S. Department of Agriculture
Agricultural Research Service
2021 South Peach Avenue
P. O. Box 8143
Fresno, California 93747

Introduction

Almost 358 million pounds of fresh California strawberries were shipped to market in 1980. Over 92% of this fruit was transported by truck and the remainder by air (1). Postharvest losses of California strawberries shipped to the New York metropolitan area were surveyed extensively in 1973 by our Eastern Market Pathology Laboratory (3). This study showed that about 5% of the losses had occurred when the berries reached retail level and an additional 18% occurred at the consumer level. About two-thirds of this loss was caused by decay and one-third by mechanical injuries. Although temperature is a predominant factor affecting postharvest losses (6, 10, 11), significant reductions in decay have been obtained by maintaining CO_2-enriched atmospheres in transit (2, 4, 5, 7, 8, 11).

Controlled Atmosphere Effects

In controlled tests, conducted under laboratory conditions, decay of strawberries was significantly reduced with CO_2-enriched atmospheres (Table 1). These tests simulated cross-country truck transit, subsequent retailing and consumer use.

Table 1. Effects of carbon dioxide on decay of strawberries during simulated transport and marketing.[1]

Holding condition	Percentage of decayed berries in lots held in indicated CO_2 atmospheres			
	0 (air)	10%	20%	30%
3 or 5 days at 41°F (5°C) in CO_2 atmosphere	11.4a[2]	4.5a	1.7a	1.3a
Plus 1 day at 60°F (15.5°C) in air	35.4b	8.5a	4.7a	4.0a
Plus 2 days at 60°F (15.5°C) in air	64.4c	26.2b	10.8a	8.3a

[1] Data from reference 5.
[2] Means without a letter in common differ at the 95% probability level.

After all holding periods fruit stored in CO_2-enriched atmospheres had less decay than fruit stored in air. Ten percent CO_2 was not as effective as 20 or 30% CO_2 in controlling decay. Decay differences due to CO_2 concentration were not statistically significant (95% probability level) immediately after "arrival at market" [3 or 5 days at 41°F (5°C)], but the differences became significant during subsequent holding at 60°F (15.5°C) in air. The CO_2 atmospheres also slowed ripening of the berries. The shelf life was extended, therefore, by the "in-transit" CO_2 treatment, and losses that would be sustained in the consumer's kitchen were reduced.

CO_2-enriched atmospheres also reduced the growth of Botrytis cinerea in culture. Botrytis is the principal cause of decay in California strawberries (Figure 1).

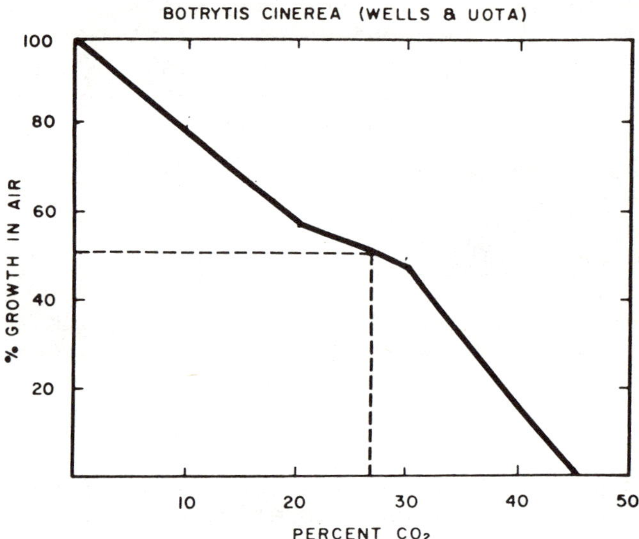

Figure 1. Growth of Botrytis cinerea in CO_2-enriched atmospheres (from reference 11).

At 26% CO_2, growth was reduced to one-half of that in air, and at 45% CO_2 growth almost ceased (11). CO_2-enriched atmospheres have an even more pronounced effect on spore germination of Botrytis cinerea — inhibition was over 90% in 16% CO_2.

Flavor evaluation of berries held in CO_2-enriched atmospheres [36 hours at 59°F (15°C)] showed that tasters consistently objected to berries held in 30% CO_2, if the evaluation was made immediately after treatment, but not if the evaluation was delayed for an additional 24 hours in air (4, 5). Concentrations of 20% CO_2 or less did not adversely affect flavor.

CO_2-Enriched Atmospheres in Commercial Shipments

So much for the effects of CO_2-enriched atmospheres in controlled laboratory tests — how effective are such atmospheres in commercial truck shipments. The atmospheres are maintained in transit by covering the pallet loads of berries with plastic bags, sealed to a gas-tight pallet floor (Figure 2). The bags are purged through a port with the desired amount of CO_2 and are then re-sealed. The initial charge of CO_2 plus that produced by respiration serves to maintain the enriched atmospheres during transit. Under commercial conditions, CO_2 atmospheres were found to average about 13% when applied at shipping point and about 10% when the pallet loads arrived at wholesale markets (9). However, there was considerable variation in the CO_2 levels within individual pallet loads, which indicated a need to carefully monitor the initial application of CO_2 and to assure a tight seal on the bag.

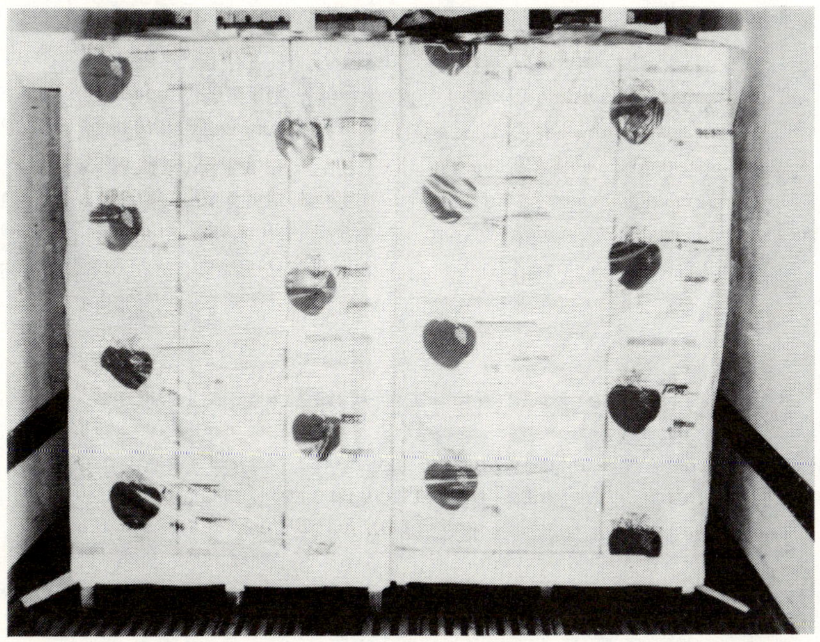

Figure 2. Film-covered pallets of strawberries with CO_2-enriched atmospheres loaded in truck.

Transit temperatures in film-covered pallet loads average about 2°F (1°C) warmer than those in open-pallet loads (Figure 3).

In full loads, of film-covered pallets, the thermostat can be lowered slightly to compensate for the restrictive effect of the cover on temperature maintenance. However, this technique could not be used in mixed loads of berries and other commodities, because of the danger of freezing produce that was not covered with film.

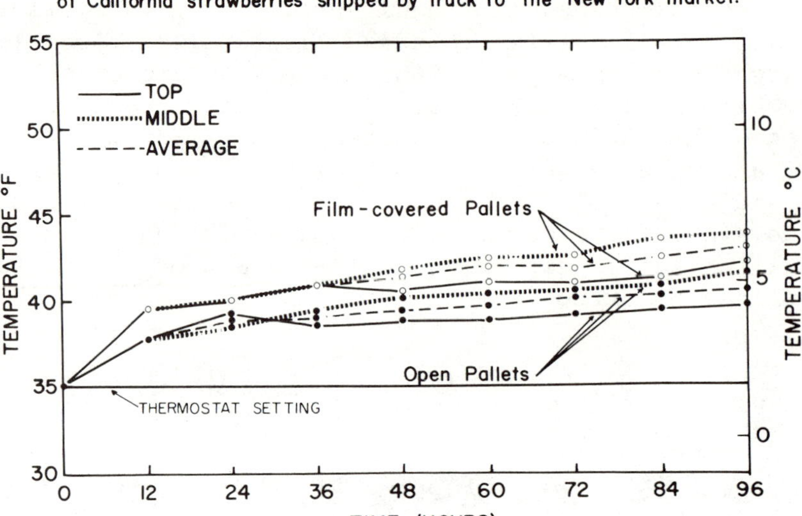

Figure 3. Average transit temperatures in film-covered and open pallet loads of California strawberries shipped by truck to eastern markets.

Decay of Berries in Film-Covered Pallets

Differences in decay due to atmosphere modification were negligible on arrival, but the differences became progressively greater after 1 and 2 days at 60°F (15.5°C) (Table 2).

When only those berries in film-covered pallets in which the CO_2 levels were above 10% were compared with those in open pallets, decay was significantly lower in the covered than in the open strawberry pallets, after both 1 and 2 days under simulated marketing conditions.

The interacting effects of transit temperatures and atmosphere modification on shelf life of strawberries are shown in Table 3.

Table 2. Average incidence of decay in California strawberries in film-covered and in open pallets shipped via truck to eastern markets.

Pallet type and atmosphere	Incidence of decayed berries		
	On arrival	After 1 day at 60°F (15.5°C)	After 2 days at 60°F (15.5°C)
	- - - - - - - - - - - - Percent	- - - - - - - - - - - -	
All test shipments:			
Film-covered	2	6	2*
Open	3	8	8*
Test shipments with CO_2 above 10%:[1]			
Film-covered	2	4*	14*
Open	2	7*	24*

[1]In film-covered pallets only.
*Difference significant at 90% level.

Table 3. Incidence of decay in California strawberries shipped by truck at various average transit temperatures with and without CO_2 atmospheres.

Average transit temperatures		Incidence of decay after 2 days at 60°F (15.5°C)		
		All CO_2 treatments	CO_2 over 10%	Open pallet
°F	°C	- - - - - - - - - - - - Percent	- - - - - - - - - -	
42 and above	5.5	27	14	33
40 to 41	4.5–5.0	25	17	33
38 to 39	3.5–4.0	21	10	29
37 and below	3.0	6	5	24

After holding 2 days at 60°F (15.5°C) at the market, decay percentages were lowest (5%) in berries from trays with transit temperatures that were 37°F (3°C) or lower and that were shipped in covered pallets with 10% or more CO_2 on arrival. Decay percentages were highest (33%) when transit temperatures were 40°F (4.5°C) or above and no atmosphere modification was provided. Even at these relatively high transit temperatures, which are not uncommon in commercial shipments, CO_2 atmospheres over 10% provided some reduction in decay (14–17%), compared to the open pallets.

The overall effect of CO_2 concentrations in transit on decay of berries held for 2 days at 60°F (15.5°C) is shown in Figure 4.

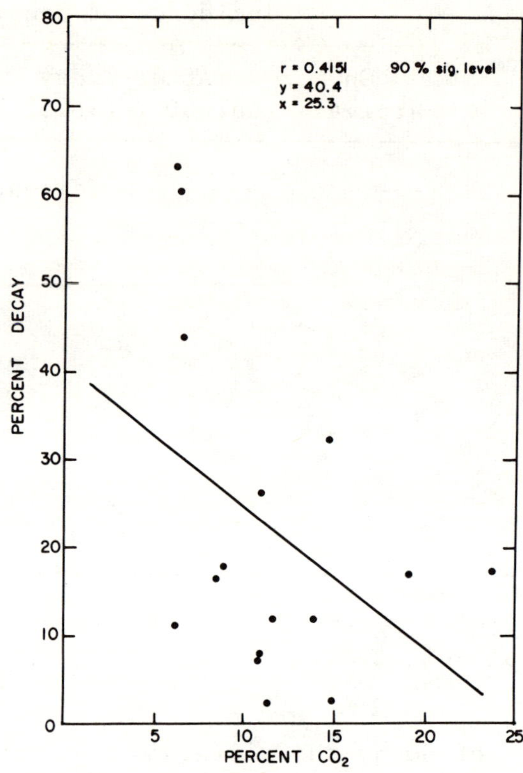

Figure 4. Effect of CO_2 concentration in transit on incidence of decay of California strawberries shipped by truck to eastern markets.

The incidence of decay decreased as the CO_2 concentration increased. Under the conditions of these experiments, decay was reduced by one-half at about 13% CO_2 in transit, compared to open pallets without CO_2.

Conclusions

Laboratory findings have demonstrated the advantages of CO_2-enriched atmospheres for decay reduction in fresh strawberries held under simulated transit conditions. CO_2 has a direct effect on suppression of fungus growth and spore germination. It also slows the ripening of berries.

Transport tests conducted under commercial conditions confirmed that CO_2 atmospheres were beneficial, if an adequate concentration of CO_2 was maintained in transit. The tests also showed the need for maintaining optimum transit temperatures. Losses were smallest when transit temperatures were at 37°F (3°C) or below and CO_2 atmospheres on arrival were 10% or more.

References

(1) Anonymous. 1981. California Strawberry Report. Federal-State Market News Service 67(36):1.

(2) Brooks, C., E. V. Miller, C. O. Bratley, and others. 1932. Effect of solid gaseous carbon dioxide upon transit diseases of certain fruits and vegetables. U.S. Department of Agriculture, Technical Bulletin No. 318, 60 p.

(3) Ceponis, M. J. and J. E. Butterfield. 1973. The nature and extent of retail and consumer losses in apples, oranges, lettuce, peaches, strawberries, and potatoes marketed in greater New York. U.S. Department of Agriculture, Marketing Research Report No. 996, 23 p.

(4) Couey, H. Melvin and John M. Wells. 1970. Low-oxygen or high-carbon dioxide atmospheres to control postharvest decay of strawberries. Phytopathology 60:47-49.

(5) Harris, C. M. and J. M. Harvey. 1973. Quality and decay of California strawberries stored in CO_2-enriched atmospheres. Plant Disease Reporter 57(1):44-46.

(6) Harvey, J. M., H. M. Couey, C. M. Harris, and F. M. Porter. 1965. Air transport of California strawberries: Factors affecting market quality in summer shipments - 1965. U.S. Department of Agriculture, Marketing Research Report No. 751, 12 p.

(7) Harvey, J. M., H. M. Couey, C. M. Harris, and F. M. Porter. 1966. Air transport of California strawberries: Factors affecting market quality in spring shipments - 1966. U.S. Department of Agriculture, Marketing Research Report No. 774, 9 p.

(8) Harvey, J. M., C. M. Harris, and F. M. Porter. 1971. Air transport of California strawberries: Pallet covers to maintain modified atmospheres and reduce market losses. U.S. Department of Agriculture, Marketing Research Report No. 920, 10 p.

(9) Harvey, J. M., C. M. Harris, W. J. Tietjen, and T. Serio. 1980. Quality maintenance in truck shipments of California strawberries. U.S. Department of Agriculture, Advances in Agricultural Technology W-12, 13 p.

(10) Maxie, E. C., F. G. Mitchell, and Arthur Greathead. 1959. Studies on strawberry quality. California Agriculture 13(2):11, 16.

(11) Wells, J. M. 1970. Modified atmosphere, chemical, and heat treatments to control postharvest decay of California strawberries. Plant Disease Reporter 54:431-434.

(12) Wells, J. M. and M. Uota. 1970. Germination and growth of five fungi in low-oxygen and high-carbon dioxide atmospheres. Phytopathology 60:50-53.

STORAGE OF PECAN KERNELS UNDER WHOLESALE AND RETAIL CONDITIONS

Stanley J. Kays
Department of Horticulture
University of Georgia
Athens, Georgia

While pecans are grown in Brazil, Mexico, Australia, Israel and South Africa, it is only in the United States that they represent a major nut tree crop. The average annual production in the United States is around 90 million kg (1) with a value of approximately 90 million dollars. Pecans have been experimentally planted throughout much of the United States, however, it is only in the South that the species has developed into a commercial concern. Leading states in production are Georgia, Texas, Alabama, Arkansas, Arizonia, Florida, Louisiana, Mississippi, New Mexico, North Carolina, Oklahoma and South Carolina.

Handling and storage practices have changed markedly since the early days of the pecan industry in the United States. The shift from an "inshell" industry to one where the major portion of the crop is sold with the shell removed, has speeded progress in handling and storage practices. Removal of the shell, which controls, to a significant extent, the kernel's micro-climate, hastens the rate at which detrimental processes occur. In addition, quality is much more easy to estimate visibly.

Due to the low respiratory rate of the kernels (i.e. 0.2 to 0.8 mg CO_2/kg·hr), the pecan has been considered a relatively stable commodity. This is not, in fact, the case. From the optimum time of harvest the kernels begin to undergo a series of irreversible biochemical changes which are detrimental to quality. Management of these changes, which are mainly a function of the kernel's unique composition, presents a series of challenges which vary considerably from the postharvest problems encountered with many other horticultural crops.

In the following paper, the major quality components of pecans will be outlined together with our current understanding of the factors affecting them. In addition, the quality criteria and handling and storage practices presently used by the industry will be discussed as well as projections for the future.

Components of Quality and Factors Affecting Them

Color: The color of the testa of pecan kernels has long been used as a measure of the quality of the kernel. As the kernel ages, there is a characteristic darkening of the surface color (15), an increase in rancidity of oils in the kernel (21), and a loss of the nuts aromatic, flavor and textural properties. This surface discoloration is generally from a light golden color to a very dark red-brown and in many cases coincides with, although not necessarily parallels, the general decline in the biochemical quality of the kernel. Color is not, however, always a good measure of quality. Color, at optimum harvest, varies substantially with cultivar (17). With many of the darker colored cultivars other quality attributes are not necessarily low in proportion to color.

While kernel color is definitely not a precise means of ascertaining nut quality, it is used extensively since it is easy to measure. Initially kernels were almost exclusively sorted visually into the various USDA color grades by trained personnel in the shelling plants. The advent of electronic sorting, however, has made color grading automated and relatively easy. This ease and degree of accuracy has, in turn, tended to increase the emphasis placed on the use of kernel color in determining quality.

The majority of color development of the kernel during maturation on the tree occurs after the onset of the dehiscence or shuck split period (5). While many factors can affect the coloration of pecan kernels in the field and during harvest (14), only postharvest color changes will be discussed. These can be classed into two general groups: I) those which represent normal color changes during aging, and II) abnormal or atypical color changes.

I. Factors affecting normal changes in kernel color: Many of the detrimental processes occurring in the pecan after harvest are progressive and cumulative. Lowering the kernel moisture level to 3.5 to 4.0% as rapidly as possible after harvest is essential for minimizing the rate at which these detrimental processes occur (13, 34). To accomplish this, forced air drying is commonly used; however, drying should not be attempted at temperatures above 38°C since exposure to excessive heat also increases kernel darkening. Storage of nuts in high moisture atmospheres can result in the re-equilibration of the kernel moisture content leading to an acceleration of kernel darkening. Storage of dried pecans at 58 to 64% relative humidity results in an equilibrium moisture level in the kernels of 3.5 to 4%.

Storage of pecans at low temperatures greatly minimizes undesirable color and flavor changes (4, 5, 6, 11, 30). In general, the lower the storage temperature, the greater the storage life of the kernel. Temperatures above 4.4°C result in relatively rapid discoloration, while temperatures at or below -17.8°C will maintain quality for several years.

Oxygen has been shown to be a major contributor to the discoloration of pecan kernels during storage (16). The detrimental effect of oxygen is concentration and temperature dependent. Partial replacement of oxygen with nitrogen appears to be superior to replacement with carbon dioxide (25) in retail packages of pecans. A final storage level of approximately 2% oxygen gives excellent results for extending the optimum color life of pecan kernels (16). Use of low oxygen storage is much more critical during the retail sales period when the nuts are subjected to nonrefrigerated temperatures.

Vacuum or subatmospheric pressure storage of pecans has been known for some time to retard the loss of nut quality (5). The rate of change in kernel color is decreased by storage in sealed glass containers at 406 mm Hg. Recent work has shown that the beneficial effect of subatmospheric pressure storage on detrimental changes in kernel color is due to the alteration of the internal oxygen partial pressure rather than the vacuum *per se* (16).

Light, especially in the red wavelengths of the visible spectrum, is detrimental to the color of shelled pecans (31). Storage of shelled pecans in red cellophane containers decreased the rate of kernel darkening when compared to clear, amber or green containers. Sunlight, in addition, is more detrimental than fluorescent light (10). The rate of unfavorable color change increases with increased light transmission of the package, product moisture content and length of storage.

Cultivar of pecan is one of the more obvious factors that affect the color of pecan kernels (2, 12, 17). While the wide variation in kernel color among the many commercial cultivars has long been recognized by both growers and shellers, postharvest color stability also varies widely (17). As a consequence, it may be advantageous to process and store under refrigated conditions those cultivars with a lower color stability before those which are more resistant to darkening.

II. Abnormal kernel color changes: Accidental exposure to ammonia gas can transform kernels from a state of optimum color to a black unmarketable product within minutes (19, 20, 22). The extent to which the kernels are discolored varies with the moisture content of the kernels, concentration of the gas, length of exposure (14), temperature and age of the nut (36). This adverse color transformation is localized in the testa or skin of the kernel while the interior portion of the kernel remains normal. After adequate aeration, the flavor of the exposed kernels is generally not adversely affected; however, the loss of visual aesthetic appeal renders the product unmarketable. Fortunately the incidence of ammonia damage has declined substantially in recent years with the general trend away from ammonia as a cold storage refrigerant. There are, however, a sufficient number of cases of accidental exposure each year to still consider it a significant problem for the pecan industry.

Other abnormal color changes occur when the kernels are exposed to alkali or strong acids. Alkaline conditions tend to produce black kernels while strong acids result in a deep red-brown coloration, similar in color to kernels which have undergone considerable oxidation.

Rancidity: A plant species can be called "oleaginous" when its fruit, seeds or other plant parts contain at least 15-20% oil. Pecan kernels are 55 to 75% lipid. While being a major factor in the unique flavor and nutritional qualities of the nut, this present a series of postharvest problems in that the oils are susceptible to rancidity.

An analysis of the lipids of six cultivars of pecan showed that of the mean lipid content (73%), triglycerides comprised 71.25%, with complex lipids, monoglycerides, α,β-diglyceride, α,α'-diglyceride and sterol representing only 0.38, 0.53, 0.64, 0.22 and 0.16%, respectively (26). Of the ten major fatty acids (Table 1), unsaturated acids were predominant, representing 63.19 gm/100 gm of nutmeat versus 6.31 gm/100 gm for saturated fatty acids (26, 27). Oleic and linoleic comprised $\sim$84% of the total fatty acid composition of the kernels indicating the relative degree of unstability of the oils to oxidation.

TABLE 1. FATTY ACID COMPOSITION OF PECANS[b]

Abbreviation	Systematic name	Common name	Concentration gm/100 gm nut mean[a]
1. Saturated fatty acids			
10:0	decanoic	capric	<0.20
12:0	dedecanoic	lauric	<0.20
14:0	tetracecanoic	myristic	1.20
15:0	pentadecanoic		<0.2
16:0	hexadecanoic	palmitic	4.09
17:0	heptadecanoic	margaric	0.27
18:0	octadecanoic	stearic	1.51
20:0	eicosanoic	arachidic	0.44
21:0	heneicosanoic		<0.20
2. Unsaturated fatty acids			
12:1	dedecenoic		<0.20
14:1	tetradecenoic		<0.20
14:2	tetradecadienoic		<0.20
15:1	pentadecenoic		<0.20
15:2	pentadecadienoic		<0.20
16:1	hexadecenoic	palmitolic	0.42
16:2	hexadecadienoic		<0.20
17:1	heptadecenoic		0.26
17:2	heptadecadienoic		<0.20
18:1	octadecenoic	oleic	37.90
18:2	octadecadienoic	linoleic	22.53
18:3	octadecatrienoic	α-linolenic	1.53
20:1	eicosenoic		0.54
20:2	eicosodienoic		<0.20

[a] Data for the major fatty acids represent means of six cultivars.

[b] Data from Senter and Horvat (26, 27).

The fatty acid composition is known to vary considerably with cultivar, year and during the maturation process of the nut (21, 23). Rudolph (23) suggests that the final ratio between oleate and linoleate at maturity may represent differences in the turnover number between their respective synthesis enzymes. "...saturated fats are present in the pecan at a steady state level and are being continually desaturated as rapidly as they are formed. The enzyme systems for monodesaturation to oleic acid are very active, but the synthesis is stimulated only during periods of rapid oil synthesis. The enzyme responsible for desaturation of oleate to linoleate is always present but has a much lower turnover number. Hence, there is always some conversion of oleate to linoleate in maturing pecans, but this net synthesis is nullified by the very rapid synthesis of oleic acid which occurs during the periods of oil synthesis." This hypothesis is supported by the parallel in concentrations of linoleate and oleate during the early stages of maturation.

The three major processes for the development of rancidity in the pecan appear to be a) the oxidation of double bonds of unsaturated glycerides to form peroxides, b) hydrolysis of component glycerides into free fatty acids and c) β-oxidation of free saturated fatty acids. In most pathways of lipid deterioration, the initial step is the attack of an unsaturated fatty acyl group by oxygen to form a peroxide. Farmer et al. (7) described the sequence which begins with an induction period after the exposure of the lipids to air. During this period natural antioxidants, such as tocopherols in the pecan, are consumed and free radicals or their precursors begin to accumulate. The radicals eventually reach a high enough concentration to end the induction period and a period of rapid oxygen absorption follows. Several factors can shorten the length of the induction period and increase the subsequent rate of oxidation. These include:

a) increases in temperature of the lipids,
b) irradiation,
c) increases in the surface to volume ratio,
d) increases in the number of double bonds in the fatty acid chain, i.e. high levels of linoleic in relation to oleic,
e) elevated oxygen partial pressures,
f) the presence of catalysts or initiators such as preformed hydroperoxides or transition metals,
g) low levels of natural antioxidants, i.e. γ-tocopherol.

Research on pecan storage has concentrated on several techniques for retarding the development of rancidity. One of the first used was cold storage of the product, whether shelled or in the shell. Storage of pecans at 0°C and 70-75% RH dramatically prolongs the useful life of the raw product (4, 6, 33, 35). Decreasing the ambient oxygen partial pressure whether by subatmospheric or controlled atmosphere storage, also decreases the rate at which quality is lost (5, 16). Both endogenous and exogenous antioxidants appear to affect the rate at which rancidification occurs. Pecans contain tocopherols which are at their highest (600 μg/gm oil) in the kernel around 6 weeks prior to maturity (23). This level declines to 100-200 μg/gm oil at maturation. This may represent the utilization of tocopherol in the process of triglyceride synthesis or merely the early synthesis of tocopherol with subsequent dilution during the period of rapid oil synthesis. In contrast to many nuts (e.g. filberts, walnuts, brazil nuts, almonds, chestnuts and peanuts), in pecans where the predominant fatty acid is oleate, the γ-tocopherol isomer was found almost exclusively (i.e. >95%) in lieu of the α-tocopherol isomer (18).

While the precise role and degree of beneficial effects of tocopherols on preventing oxidation of pecan oils are not known, some evidence points toward their positive value. For pecan oils with a constant % linoleate, keeping time increased in a linear fashion with tocopherol concentration up to 800 μg/gm of oil at which point a plateau was reached (24).

The storage life of pecan nuts and oils can be extended if exogenous antioxidants are added. Butylated hydroxyanisole (BHA), butylated hydroxytoluene (BHT) and propyl gallate are several that have been used with varying levels of success (3, 9).

Light, especially in the red wavelengths of the visible spectrum, is also detrimental to pecan quality (31) and sunlight is more detrimental than fluorescent light (10). The rate of unfavorable changes in kernel quality increases as the light transmission properties of the package increase.

Excessive drying of the kernels is also thought to accelerate rancidity (34). The optimum kernel moisture level for retarding the development of rancidity is 4.5%.

Kernel breakage: Shelled pecans are graded into half-kernels, pieces and particles and dust based on size and the degree of intactness of the kernel (28). Half-kernels are the separate halves of the nut which have no more than one-eighth of its original volume missing. Pieces represent a portion of the original kernel which is less than seven-eighths of a kernel, but will not pass through a round opening two-sixteenths of an inch in diameter. Particles and dust are fragments of kernels which will pass through a round opening two-sixteenths in diameter. These classes are subsequently graded into one of six U.S. grades (U.S. No. 1 halves, U.S. No. 1 halves and pieces, U.S. Commercial halves, U.S. No. 1 pieces, U.S. Commercial pieces and Unclassified) based on additional criteria such as color, freedom from damage, absence of foreign material, etc.

While kernel or piece size has a significant effect on wholesale market value, the points in the wholesale and retail marketing chain where breakage occurs have not been well studied. It is evident that significant breakage does occur at some point in the handling chain. In one study (32), 31% of the U.S. No. 1 halves were broken by the time they reached the retail level. The major point of breakage initially is during the shelling operation where cultivar, size grading before shelling and preconditioning are important factors. Steam conditioning has been shown to increase the precentage of halves obtained during shelling by approximately 15% (8). Retail quality problems, however, cannot be attributed to shelling breakage in that grading occurs after that point. Inferior samples obtained at retail outlets reflect either inaccurate grading or breakage during handling and storage operations.

Two factors that appear to be associated with the breakage problem are kernel moisture content and the use of loose packs for retail sales. As the shelled kernels dehydrate they become brittle and break more readily. Optimum kernel moisture level for storage is 3.5 to 4.0%. The actual percent moisture in the kernel, however, will increase or decrease according to the relative humidity of the ambient environment. Equilibrium moisture content for 3.5% kernel moisture is reached at 58% RH (35). physical interaction between shelled halves in the currently used loose Packages, whether from vertical drops or vibrational effects, appears to also be a contributing factor to breakage. This area, unfortunately, has yet to be studied.

Flavor: Pecans in the raw state have a characteristic flavor and are often consumed without roasting. The oil content of the kernels varies from 50 to 75% and is closely associated with flavor (35). Nuts with the highest oil content typically are rated highest in flavor. Volatile constituents are also of critical importance as flavor components of the pecan. The major volatiles of both raw and roasted pecans have been identified (Table 2). Four low molecular weight alcohols, five low molecular weight aldehydes and a lactone are the major constituents of the natural pecan aroma (8). It is thought that their origin may occur via the oxidation of unsaturated fatty acids which are found in significant levels in the kernels. In roasted kernels nineteen carbonyls, pyridine, eight pyrazines, seven acids, five alcohols, and one lactone have been identified by GC-MS (29). The alkyl-pyrazines appeared to provide the backbone of the roasted nut character, however, the carbonyl and basic compounds were also important. No single specific compound could be identified as possessing the roasted nut aroma, which appears to be comprised of a complex cross-section of volatiles.

The characteristic volatile flavor components are the first quality attributes of the kernels to be lost during the general decline in quality after harvest (34). Little is known about the factors governing the synthesis of these plant products or their loss from the tissue. While the use of vacuum storage has significant advantages for the prevention of rancidity and color loss (16), its use has not been tested with regard to the volatile components of the nut.

Kernel texture: Texture profile analysis of pecans has not been attempted. Both kernel moisture level and oil leakage are known to affect the kernel's textural properties, however, these changes have not been assessed relative to consumer preferences.

Insects and diseases in stored pecans: With improper storage conditions both insects and diseases can pose serious quality problems. Larvae that feed on pecan kernels in storage are produced by the almond moth (*Ephestia cautella* Wlkr.) and the Indian-meal moth (*Plodia interpunctella* Hbr.). Beetles which feed on pecans are the saw-toothed grain beetle (*Oryzaephilus surinamensis* L.), cadelle beetles (*Tenebroides mauritanicus* L.), the confused flour beetle (*Tribolium confusum* Duv.), and the red flour beetle (*Tribolium castaneum* Hbst.). Control measures include storage of the nuts at temperatures below 9°C and when necessary, fumigation with carbon dioxide and ethylene oxide (9:1) or ethylene dichloride and carbon tetrachloride (7:3) or methyl bromide (34).

TABLE 2. VOLATILE COMPOUNDS IDENTIFIED FROM RAW AND ROASTED PECANS.

Raw[a]	Roasted[b]
Alcohols	Carbonyls
1-heptanol	Ethanal
1-hexanol	Propanal
1-octanol	Butanal
1-pentanol	Pentanal
Aldehydes	Hexanal
amylfurane	Heptanal
hexanol	Octanal
heptanol	2-Hexenal
nonanal	2-Heptenal
octanal	2-Decenal
Lactones	2-Undecenal
α-caprolactone	Acrolein
	2,4-Heptadienal
	2,4-Decadienal
	Furfural
	Glyoxal
	Pyruvaldehyde
	Diacetyl
	2,3-Pentanedione
	Basic compounds
	Pyridine
	2-Methylpyrazine
	2,5-Dimethylpyrazine
	2,6-Dimethylpyrazine
	2,3-Dimethylpyrazine
	2-Ethyl-6-methylpyrazine
	2-Ethyl-5-methylpyrazine
	2,3,5-Trimethylpyrazine
	2,5-Dimethyl-3-ethylpyrazine
	Acids
	Acetic acid
	Propionic acid
	Pentanoic acid
	4-Methyl-pentanoic acid
	Hexanoic acid
	Heptanoic acid
	Octanoic acid
	Alcohols
	Ethanol
	1-Pentanol
	1-Hexanol
	1-Heptanol
	1-Octanol
	Lactone
	γ-Octalactone

[a] Data from Rudolph (23).

[b] Data from Wang and Odell (29).

Pecans are subject to molds and bacterial and fungal contamination. While there are numerous molds that can invade the nut, Aspergillus flavus and A. parasiticus are of considerable interest since both can produce carcinogenic aflatoxins. Escherichia coli and Salmonella are the most common bacterial contaminations encountered. Both bacteria and molds on pecans are controlled by the use of chlorine in the water the pecans are soaked in before cracking, fumigation when necessary with propylene oxide, and with temperature and relative humidity control during storage (11). Pink rot (Cephalothecium roseum Corde.), while contracted in the orchard, can cause serious storage problems. The kernels obtain a translucent, waterlogged appearance with discoloration and softening. Maintaining the kernels at a moisture level below 4% is the best postharvest means of control.

Quality Criteria Used by the Pecan Industry

Of the many components of quality in the pecan, size and color represent the primary parameters utilized by the industry. Nut meats are separated initially based on the degree of intactness of the kernel. Kernels with one-eighth or more of their volume missing are considered pieces. The pieces are subsequently separated into the following size classifications.

Size classification	Maximum diameter (will pass through round opening of following diameter)	Minimum diameter (will not pass through round opening of following diameter)
		Inch
Mammoth pieces	No limitation	8/16
Extra large pieces	9/16 inch	7/16
Halves and pieces	No limitation	5/16
Large pieces	8/16 inch	5/16
Medium pieces	6/16 inch	3/16
Small pieces	4/16 inch	2/16
Midget pieces	3/16 inch	1/16
Granules	2/16 inch	1/16

Intact halves are also sold based on size classifications.

Size classifications for halves	Number of halves per pound
Mammoth	250 or less
Junior mammoth	251-300
Jumbo	301-350
Extra large	351-450
Large	451-550
Medium	551-650
Small (topper)	651-750
Midget	751 or more

Color sorting utilizes automated electronic equipment that senses and removes discolored kernels. Typically pecans are separated into the following four USDA color classes:

(A) "Light" means that the kernel is mostly golden color or lighter, with not more than 25% of the surface darker than golden, and none of the surface darker than light brown.

(B) "Light amber" means that the kernel has more than 25% of its surface light brown, but not more than 25% of the surface darker than light brown and none of the surface darker than medium brown.

(C) "Amber" means that the kernel has more than 25% of the surface medium brown, but not more than 25% of the surface darker than medium brown, and none of the surface darker than dark brown (very dark-brown or blackish-brown discoloration).

(D) "Dark amber" means that the kernel has more than 25% of the surface dark brown, but not more than 25% of the surface darker than dark brown (very dark-brown or blackish-brown discoloration)(28).

Pecan pieces that are considered excessively dark are occasionally chopped to a smaller size to enhance their overall lightness. While the actual color of the pigmented surface remains the same, its percentage of the surface area exposed declines in relation to the cream colored interior of the nut. Hence, the smaller pieces appear lighter.

Secondary quality factors which influence sales are the degree of filling of the kernel, freedom from damage, freedom from foreign material and cultivar.

Current Handling Practices

Wholesale handling of pecans: Due in part to the value of the crop and level of competition in the industry, wholesale storage practices are on the whole quite good. Shelled pecans are normally stored at 0°C, 3.5-4.0% kernel moisture. A storage temperature of -17.8°C is recommended, however, if the nuts are to be held longer than one year. Due to the low storage temperature, alteration of the storage gas oxygen concentration is not needed for maintenance of high quality.

Packaging in the final retail pack can occur at either the wholesale or retail distributor level. Since significant quality problems are found at the retail level, an analysis of the point of origin of these problems would be advantageous.

Retail handling of pecans: The handling of pecans at the retail level has not been adequately studied. It has been shown, however, that the quality of nuts found at the retail level is significantly substandard (32), the degree of which varied with the area of the United States where the pecans were marketed. Of all the pecans sampled, 50.6% did not meet the standards for U.S. No. 1 grade. Poor kernel fill, rancid flavor, dark color, excessive particles and dust and excessive pieces were listed as the major defects in quality.

We purchased samples of the three major brands of pecans that were available in Athens' grocery stores and monitored the conditions the pecans were exposed to in the respective stores to gain some insight into retail marketing problems. It should be noted from the outset that this inquiry represents a small sampling of the total volume of pecans sold annually in the United States. Although extrapolation of these results to other brands, geographic locations or times of the year is tenuous, the results do illustrate several interesting facts that may be reasonably widespread.

The pecans were packaged loose in cellophane pouches of 64 to 80 gm (Table 3). Grading the pecan halves to measure the extent of breakage indicated that 28 to 37% of the pecans by weight were not of the grade claimed on the package. This level of breakage is comparable to that reported by Williams et al. (32). In that, loose packs may contribute to breakage, the excess air volume with the packages was estimated by measuring the package volume before and after a partial vacuum (360 mm Hg) was applied to the interior. These volumes ranged from 50 cc in the packages containing 80 gm to 100 cc in packages containing 64 gm of pecans. The additional space, along with allowing unrestricted physical interaction between kernels during transport and handling, significantly increases the amount of oxygen available to chemically react with the pecans.

With the exception of an occasional package that had been perforated or improperly sealed, the internal oxygen concentrations within the packages were within reason, although with two brands (A and C) not entirely adequate. Oxygen ranged from 1.65% in brand B to 4.6% in brand C. As indicated earlier, oxygen concentration is a critical factor in the rate of development of rancidity and in the loss of color. The internal carbon dioxide concentration was inversely related with 1.91% in brand B and 0.44% in brand C. Moisture levels in the nuts were low ranging from 2.6 to 3.4% although comparable to those found previously (32). The respiratory rate of the kernels reflected, in part, these differences in kernel moisture level. Rates ranged from 0.77 mg CO_2/kg·hr with brand B (3.4% H_2O) to 0.3 mg CO_2/kg·hr for brand C (2.6% H_2O). It should be noted that the respiratory rates were measured under ambient laboratory conditions ($\sim$21% O_2, 21°C) and would be significantly lower at the oxygen concentration in which the kernels were packaged.

The environmental conditions to which the pecans were exposed on the retail level varied considerably (Table 4). Ambient temperature at the location of the pecans in the store ranged from 22°C to 25°C. Storage at these temperatures is known to significantly enhances the rate at which quality diminishes. Light, which is detrimental to nut quality, varied in intensity and quality. Pecans placed near windows had substantially higher (3X) light levels than those placed further back in the store. In addition, sunlight is known to be substantially more detrimental to nut quality than fluorescent.

In summarizing the retail conditions we measured, temperature was excessively high for prolonged storage, light intensity and quality were in some cases distinctly undesirable, and package oxygen levels ranged from good to fair. In addition, none of the pecans sampled were equal to the USDA grade advertised on the package. These results indicate that there is substantial room for improvement in the marketing of pecans at the retail level. Hopefully future research will address these problems.

TABLE 3. NUT AND PACKAGING CONDITIONS OF 3 MAJOR BRANDS OF SHELLED PECAN HALVES COLLECTED FROM RETAIL STORES IN ATHENS, GEORGIA ON 7-6-81.[1/]

Brand	Type of kernel	Wt package (gm)	% grade #1[2/] (% by wt)	Excess gas[3/] volume in package (cc)	Package O_2 conc. (%)	Package CO_2 conc. (%)	Kernel[4/] respiratory rate (mg CO_2/Kg·hr)	Kernel moisture level (%)	Packaging material
A	Grade 1 halves	80	73.4	50	4.48	0.28	0.34	3.0	4 mil laminated polyethylene and polyolefin
B	Grade 1 halves	64	63.1	100	1.65	1.91	0.77	3.4	4 mil laminated polyethylene and polyolefin
C	Grade 1 halves	64	67.5	100	4.60	0.44	0.30	2.6	4 mil laminated polyethylene and polyolefin

1/ All measurements represent means of 3 packages per brand.

2/ Based on broken pieces only.

3/ Based on the internal gas volume before and after applying a partial vacuum (360 mm Hg).

4/ Respiratory rate at 21% O_2, $21^{\circ}C$.

TABLE 4. MARKETING CONDITIONS OF PECANS IN FOUR RETAIL SUPERMARKETS IN ATHENS, GA (7-6-81).

Store	Temperature (°C)	Light intensity (F.C.)	Type of light
1	25°	54	fluorescent
2	22°	175	sunlight + fluorescent
3	24°	77	fluorescent + sunlight
4	25°	58	fluorescent

Trends for the future: While predicting future trends for any commodity is not without considerable risk, there are several changes that I feel confident will occur in the pecan industry. I believe we will see significant changes in how we measure the quality of pecans and the level of precision at which quality is measured. Greater and more precise use of modified atmosphere packaging will be used where the internal respiratory rate and weight of nuts will be balanced against the oxygen diffusion rate and surface area of the packaging material. I would also predict that we will begin to see the use of flexible vacuum packs which will stabilize the kernels within the package as well as altering the oxygen partial pressure.

References

1. Agricultural Statistics. USDA. Statistics of fruit, tree nuts, and horticultural specialties. 1975, Chapter V.

2. Amling, H. J., K. A. Marcus, J. E. Barrett, and N. R. McDaniel. 1975. Nut quality of selected pecan varieties grown in south Alabama. Auburn Univ. Agric. Expt. Sta. Circ. 225, 19 pp.

3. Anon. 1958. Flavor preservation of pecans can be a factor in bigger market. Peanut J. and Nat. World 37:7, 34-35.

4. Blackmon, G. H. 1932. Cold storage of pecans. Fla. Expt. Sta. Ann. Rept., pp. 102-106.

5. Blackmon, G. H. 1933. Pecan cold storage experiments. Ga. Fla. Pecan Grs. Assoc. Proc., pp. 6-11.

6. Brison, F. R. 1945. The storage of shelled pecans. Texas Agri. Expt. Sta. Bull. 667:1-16.

7. Farmer, E. H., G. F. Bloomfield, A. Sundralingam, and D. A. Sutton. 1942. The course and mechanism of autoxidation reactions in olefinic and polyolefinic substances, including rubber. Trans. Faraday Soc. 38:348-345.

8. Forbus, W. R., Jr. and S. D. Senter. 1976. Conditioning pecans to improve shelling efficiency and storage stability. J. Food Sci. 41: 794-798.

9. Godkin, W. J., H. G. Beattie, and W. H. Cathcart. 1951. Retardation of rancidity in pecans. Food Tech. 5(10):442-447.

10. Heaton, E. K. and A. L. Shewfelt. 1976. Pecan quality: Effect of light exposure on kernel color and flavor. Lebensm.-Wiss. u.-Technol. 9:201-206.

11. Heaton, E. K., A. L. Shewfelt, A. E. Badenhop and L. R. Beuchat. 1977. Pecans: Handling, storage, processing and utilization. Ga. Agric. Res. Bull. 197, 79 pp.

12. Heaton, E. K. and J. G. Woodroof. 1955. Effects of varieties and other factors on extending the shelf-life of pecans. Southeastern Pecan Grs. Assoc. Proc. 48:47-49.

13. Heaton, E. K. and J. G. Woodroof. 1970. Humidity and weight loss in cold stored pecans. A.S.H.R.A.E.J. 12:49.

14. Kays, S. J. 1979. Pecan kernel color changes during maturation, harvest, storage and distribution. Pecan Quarterly 13(3):4-12, 34.

15. Kays, S. J. and D. M. Wilson. 1977. Chronological sequence of pigment development in the kernels of pecan cultivar 'Stuart'. Scientia Hort. 6:213-222.

16. Kays, S. J. and D. M. Wilson. 1977. Effect of subatmospheric pressure and oxygen partial pressure on the color stability of stored pecan kernels. Lebensm.-Wiss. u.-Technol. 10:109-110.

17. Kays, S. J. and D. M. Wilson. 1978. Genotype variation in pecan kernel color and color stability during storage. J. Amer. Soc. Hort. Sci. 103:137-141.

18. Lambertsen, G., H. Myklestad, and O. R. Braekkan. 1962. Tocopherols in nuts. J. Sci. Food and Agric. 13:617-620.

19. Medlock, O. C. 1931. Pecan storage. Ala. Expt. Sta. Ann. Rept. 42:50-51.

20. Medlock, O. C. 1933. Effect of different temperatures, humidities, and free ammonia on pecans in storage. Nat. Pecan Grs. Assoc. Proc. 32:21-28.

21. Odell, G. V., H. A. Hinrichs, P. S. Wang, C. J. Rudolph, P. Koehler, P. S. Juneja, D. Hopfer, and L. Chu. 1971. Chemical and biochemical study of factors responsible for rancidity and off-flavor in pecans. Rept. to USDA Crops Res. Div., Project SS318J, 107 pp.

22. Rose, D. H. 1939. Effect of ammonia on nuts in storage. Ice and Regrig. 96:147-148.

23. Rudolph, C. J., Jr. 1971. Factors responsible for flavor and off-flavor development in pecans. Ph.D. Thesis, Dept. of Biochemistry, Okla. State Univ., 233 pp.

24. Rudolph, C. J., G. V. Odell, and H. A. Hinrichs. 1971. Chemical studies on pecan composition and factors responsible for the typical flavor of pecans. Proc. of the 41st Okla. Pecan Grs. Assoc., pp. 63-70.

25. Sacharow, S. 1971. Packaging confections and nutmeats. Food Prod. Dev. 5:78, 80, 82.

26. Senter, S. D. and R. J. Horvat. 1976. Lipids of pecan nutmeats. J. Food Sci. 41:1201-1203.

27. Senter, S. D. and R. J. Horvat. 1978. Minor fatty acids from pecan kernel lipids. J. Food Sci. 43:1614-1615.

28. U.S.D.A. 1969. U.S. standards for grades of shelled pecans. U.S.D.A. Com. Mkt. Serv., Washington, D. C.

29. Wang, P. S. and G. V. Odell. 1972. Characterization of some volatile constituents of roasted pecans. J. Agric. and Food Chem. 20:206-210.

30. Wells, A. W. 1951. The storage of edible nuts. U.S.D.A.-A.R.S. Bur. Plt. Ind., Soils and Agric. Eng. H. T. and S. Rept. 240.

31. Wells, A. W. and H. R. Barker. 1959. Extending the market life of packaged shelled nuts. U.S.D.A. Mktg. Res. Rept. 329:1-14.

32. Williams, F. W., M. G. LaPlante, and E. K. Heaton. 1973. Evaluation of quality of pecans in retail markets. J. Amer. Soc. Hort. Sci. 98: 460-462.

33. Wright, R. C. 1941. Investigations on the storage of nuts. U.S.D.A. Tech. Bull. 770.

34. Woodroof, J. G. 1979. Tree Nuts: Production, Processing, Products. AVI, Westport, Conn., 371 pp.

35. Woodroof, J. G. and E. K. Heaton. 1961. Pecans for processing. Ga. Agric. Expt. Sta. Bull. 80, 91 pp.

36. Woodroof, J. G. and E. K. Heaton. 1966. Ammonia damage to pecans. Proc. SE Pecan Grs. Assoc. 59:128-135, 146-147.

CONTROL OF TREE FRUIT INSECTS WITH MODIFIED ATMOSPHERES

A. P. Gaunce, C. V. G. Morgan and M. Meheriuk
Agriculture Canada, Research Station
Summerland, B.C. V0H 1Z0

The export of fruit to other countries is often restricted by phytosanitary regulations. The restriction is required because the importing country is free or nearly free of a serious pest that may be present on the fruit from the exporting country. Postharvest fumigation of export fruit is mandatory to guarantee complete mortality of all pests present on the fruit.

Typical fumigation procedures expose the fruit at ambient or near ambient temperatures for several hours to methyl bromide or ethylene dibromide at concentrations of 1/2 to several pounds per 1000 cubic feet of air space in air tight chambers. The fumigants are highly toxic and special handling precautions must be exercised to prevent exposure of people to these gaseous products. Residues can pose an additional problem in the use of the chemical fumigants. Detectable residues of ethylene dibromide have been found in cold stored fruit treated two months earlier. Many fumigants are known to be or are suspected of being carcinogenic and thus, it is difficult to obtain permission to evaluate them. The future of fumigants currently in use is perhaps tenuous as any indication of carcinogenicity will initiate their withdrawal. Another problem with fumigants is that the dosage required for 100% mortality may be detrimental to fruit quality.

Work at the Research Station in Summerland, B.C. has shown that cold storage can kill some fruit insects. Cold storage periods from one to several months, depending upon cultivar, were required for elimination of San José scale on apples(1). Control of scale was better with CA storage than with regular storage on late harvested apple cultivars (table 1). Apples infested with scale were placed in the two types of storage and a sample of fruit containing about 1000 scale in the black-cap stage were examined every four months for scale mortality. The increased mortality with CA storage was thought to be a consequence of either the higher CO_2 or lower O_2 in the atmosphere.

We found similar biocidal effects with mites on apples under CA storage (table 2). The diapause or over-wintering stage of the European Red Mite is eggs which are deposited mainly in the calyx and stem ends of the apples. Fruit samples with 1000 to 3000 eggs were examined at monthly intervals for ability of the eggs to hatch. Eggs reach their peak hatchability after 4 to 5 months of regular cold storage. Although CA storage reduced survival substantially it did not provide 100% mortality (table 2). McDaniel spider mites which over-winter as adults in the calyx and stem ends were not affected by CA storage.

A requirement for export of B.C. apples to South Africa was that they be free of both European red and McDaniel spider mites. A procedure for fumigation of apples with ethylene dibromide which killed all mites with minimal reduction in fruit quality was available. However, registration for its use on export fruit could not be obtained because of residual traces of the fumigant on the fruit. It was decided that high CO_2 or low

O_2 fumigations be tried as an acceptable alternative. What had not been attempted were shorter exposure times but at higher temperatures.

Treatment of European red mite on Delicious and Spartan apples with close to 100% CO_2 or N_2 at 21-24°C provided a 100% kill in 2 days (table 3). There was only slight mortality with the 1 day exposure. A mixture of 60% CO_2 with N_2 was equally effective but 60% CO_2 with air provided no mortality even after 7 days exposure. This indicated that the biocidal effect resulted from low oxygen. McDaniel spider mite was quite resistant and a 7 day exposure to 100% CO_2 failed to achieve complete mortality. It is obvious that this procedure is not practical for elimination of both species of mites but it may be useful in regions where McDaniel spider mite is absent.

The effects of CO_2 were also tested on a number of other insects of economic importance. Apples infested with San José scale were exposed to 100% CO_2 in plastic bags in growth chambers at three temperatures(2). Over 1000 scales per sample were examined for mortality at various exposure times. In the first experiment the O_2 concentrations (measured by the Orstat analyzer) were not as low as expected and a second experiment was conducted in which the flow of CO_2 was increased to reduce the level of oxygen. The highest temperature and low O_2 levels killed scale in a 24 hr. exposure (table 4). Cold storage temperatures (2°C) allowed numerous survivors even after a 5 day exposure. The intermediate temperature killed scale in 2 to 3 days.

Codling moth is also sensitive to high CO_2 and low O_2. Samples of about 100 larvae in artificial media rather than on fruit were used in the experiment. The diapause or over-wintering larvae are mature (5th instar) larvae which leave the apples to spin-up and survive the winter in a cocoon. The non-diapause larvae were mostly mature larvae destined to pupate and emerge as adults. It can be seen that survival decreased with exposure time (table 5). Moribund larvae moved slugglishly when touched and probably would not survive beyond another day.

The above results are encouraging from the standpoint of pest mortality but cannot be practical if fruit quality suffers. McIntosh, Golden Delicious, Delicious and Spartan apples were harvested at commercial maturity, brought to room temperature (22°C), placed in sealed boxes and flushed with 100% CO_2. Control fruit was held at the same temperature. The apples were treated for 3 days, placed in cold storage and examined 2 to 3 months later for quality. Firmness retention was significantly better with the CO_2 treated apples compared to control fruit (table 6). No differences were found in soluble solids but titratible acidity was higher in Delicious and Spartan apples treated with CO_2. Significant internal CO_2 injury was found in McIntosh but not in Golden Delicious apples. Considerable flesh browning occurred in the CO_2 treated Spartan apples.

In another experiment Golden Delicious, Delicious and Spartan apples were removed from cold storage in December and exposed to 100% CO_2, 60% CO_2 with N_2, or 60% CO_2 with air for 4 and 7 days at room temperature. Oxygen was monitored by gas chromatography. Control fruit was left in cold storage. Treated fruit was placed in cold storage and assessed for quality one month later. Results of the experiment are summarized in table 7. Firmness in CO_2 treated Golden Delicious did not differ much from that of the control fruit but substantial injury was apparent in the 4 day CO_2 treatment. Internal CO_2 injury was similar to that described by Meheriuk(3). Skin browning which only appeared on Golden Delicious was dark, leather-like in texture and was somewhat depressed. Internal breakdown, seen only in Golden Delicious, seemed to be associated with skin browning in that it was usually present below the injured skin. Diffuse internal flesh browning was a stellar pattern of slightly brown to deep brown marks radiating out from the core. Injury was always more severe with the 7 day CO_2 treatment.

Spartans and Delicious seemed to be more resistant to injury but did not suffer from diffuse internal flesh browning. Reduction in firmness from CO_2 treatments (probably due to the period of storage at elevated temperature) was more pronounced than with Golden Delicious. The diffuse flesh browning appeared to be associated with the lowest O_2 concentrations whereas the other forms of injury were associated with high CO_2 concentrations.

All the apples, particularly Golden Delicious, had a peculiar taste, possibly from a high level of acetaldehyde.

In summary, several insects on harvested fruit can be killed by low O_2 treatments. Conditions necessary to achieve complete mortality may also affect fruit quality. Perhaps the current low oxygen storage (less than 1.5%) offers a means to retain fruit quality and act as a biocide for insects on the fruit. Not investigated is the use of CO as a biocidal agent. Work along these avenues is in progress at the Research Station in Summerland, B.C.

References

(1) Morgan, C.V.G. and B.J. Angle 1967. Mortality of the San José Scale (Homoptera: Diaspididae) on Stored Apples of Different Varieties and Harvest Dates. Can. Ent. 99 : 971-974.

(2) Morgan, C.V.G. and A.P. Gaunce 1975. Carbon Dioxide as a Fumigant Against the San José Scale (Homoptera: Diaspididae) on Harvested Apples. Can. Ent. 107 : 935-936.

(3) Meheriuk, M. 1977. Treatment of Golden Delicious Apples with CO_2 Prior to CA Storage. Can. J. Plant Sci. 57 : 467-471.

TABLE 1

NUMBER OF DAYS REQUIRED FOR COMPLETE MORTALITY
OF THE BLACK-CAP STAGE OF SAN JOSÉ SCALE HELD UNDER REGULAR AND
CONTROLLED ATMOSPHERE CONDITIONS

Cultivar	Regular Air Storage	Controlled Atmosphere Storage	
		Commercial	Laboratory
Rome Beauty	85	98	57
Newtown	155	126	113
Winesap	140	122	95

Adapted from Morgan, C.V.G. and Angle, B.G. Can. Ent. 99 : 971-974(1967)

TABLE 2

SURVIVAL OF OVER-WINTERING MITES ON APPLES HELD UNDER
REGULAR AND CONTROLLED ATMOSPHERE STORAGE CONDITIONS

Mite Species	Months	% Survival in Storage	
		Regular	C.A.
European Red	2	9	0.4
	3	45	0.3
	4	85	1.3
	5	87	1.2
McDaniel Spider	2	85	91
	3	92	95
	4	94	86
	5	56	85

TABLE 3

SURVIVAL OF OVER-WINTERING MITES ON APPLES EXPOSED TO HIGH CO_2 AND LOW O_2 ATMOSPHERES

Treatment	Days	% Live Mites	
		Spartan	Delicious
European Red Mite			
100% N_2	1	20	12
	2	0	0
100% CO_2	1	22	36
	2	0	0
60% CO_2 + 40% N_2	1	25	62
	2	0	0
Control		47	65
European Red Mite			
60% CO_2 + 40% Air	1	-	26
	2	-	26
	7	-	29
60% CO_2 + 40% N_2	1	25	3.1
	2	0	0
Control		66	23
McDaniel Spider Mite			
100% CO_2	2	-	14
	3	-	17
	4	-	5.1
	7	-	0.9
Control			99.5

TABLE 4

SURVIVAL OF SAN JOSE SCALE EXPOSED TO HIGH CO_2

Temp (°C)	% O_2 in Atmosphere Containing CO_2	% Live Scale					
		Days					
		1	2	3	4	5	Control
2	1.1			33	34	11	41
12	1.5		3.7	0	0		44
22	2.3	11	0				45
2	<1			32	29	22	55
12	<1		0	0	0		39
22	<1	0	0				41

Adapted from Morgan, C.V.G. and Gaunce, A.P. <u>Can.Ent.</u> <u>107</u> : 935-936(1975)

TABLE 5

SURVIVAL OF CODLING MOTH LARVAE
EXPOSED TO 95% CO_2 AT 27°C

Phase	Hours of Exposure	% Live	% Moribund
Diapause	0	87	7
	9	70	20
	24	10	53
	33	0	38
	48	0	1
Non-diapause	0	95	0
	9	91	0
	33	0	54
	48	0	1

TABLE 6

EFFECT OF CO_2 ON APPLES HARVESTED AT COMMERCIAL MATURITY (97% CO_2 FOR 3 DAYS AT 21°C)

Cultivar	Treatment	Firmness (lbs)	Acidity (%)	Injury %	Injury Type
McIntosh	CO_2	13.1 *[1]	N.S.[1]	34.8 *	Internal CO_2
	Control	12.1		0.0	
Golden Delicious	CO_2	12.4 *	N.S.	6.5	Internal CO_2
	Control	10.5		0.0	
Delicious	CO_2	15.8 *	.242 *	0	Scald
	Control	13.6	.208	6.3	
Spartan	CO_2	17.9 *	.437 *	48.0	Flesh Browning
	Control	15.1	.400	0.0	

1. *, Significant difference from control at 95% level
   NS, No significant difference between treatment and control

TABLE 7

EFFECT OF HIGH CO_2 ON APPLES REMOVED FROM COLD STORAGE IN DECEMBER

Treatment and Duration	% O_2 Present	Firmness (lbs)	% Injury: Internal CO_2	External CO_2	Breakdown	Internal Browning
Golden Delicious						
4 days 60% CO_2/air	6	11.7	0	0	4	0
60% CO_2/N_2	0.5	10.8	0	16*	0	0
100% CO_2	1	9.9	0	25	0	0
7 days 60% CO_2/air	6	10.8	5	58	26	0
60% CO_2/N_2	1	11.0	0	39	4	22
100% CO_2	1.5	10.8	71	100*	100	0
Control	20	11.2	0	0	0	0
Delicious						
4 days 60% CO_2/air	6	18.8	0	0	0	0
60% CO_2/N_2	0.5	16.1	0	0	0	32*
100% CO_2	1	18.0	8*	0	0	0
7 days 60% CO_2/air	6	14.8	0	0	0	13*
60% CO_2/N_2	1	15.6	0	0	0	89**
100% CO_2	1.5	14.8	5	0	0	9
Control	20	20.0	0	0	0	0

* denotes slight category
** denotes severe category